STE

Test Prep

# ACS
# Physical Chemistry

## Practice Questions with Detailed Explanations

3rd edition

**Customer Satisfaction Guarantee**

Your feedback is important because we strive to provide the highest quality educational materials. Email us comments or suggestions.

info@sterling-prep.com

We reply to emails – check your spam folder



3 2 1

ISBN-13: 979-8-8855732-6-9

Sterling Test Prep materials are available at quantity discounts.

Contact info@sterling–prep.com

Sterling Test Prep
6 Liberty Square #11
Boston, MA 02109



Published by Sterling Education

Printed in the U.S.A.

# STERLING
Test Prep

Thousands of students use our study aids to achieve high test scores!

Scoring well on the ACS Physical Chemistry exam is a challenging task. Solving targeted practice questions improves your ability to analyze information and distinguish between similar answer choices, which is a more effective strategy than mere memorization. This book helps you master physical chemistry content, so you can build your knowledge and quickly choose the correct answer.

These 1,235 high-yield practice questions cover physical chemistry topics tested on the ACS Physical Chemistry exam for a targeted preparation. The detailed explanations provide step-by-step solutions to quantitative questions and teach the essential physical chemistry principles needed to answer conceptual test questions. Read the explanations carefully to understand how they apply to the question and learn important concepts and the relationships between them. From the foundations of physical and chemical properties of matter to complex mechanisms of atomic particles, you will develop a better understanding of chemical phenomena on microscopic and macroscopic levels.

Experienced chemistry instructors analyzed the test content and developed this practice material that builds knowledge and skills crucial for success on the ACS. Our test preparation experts structured the content to match the current exam for targeted and effective learning.

With this practice material, you will significantly improve your test score.

250504akp

*ACS Physical Chemistry Review* provides comprehensive and targeted coverage of physical chemistry topics tested on the ACS exam. The content covers foundational principles and theories necessary to understand the material and answer test questions.

*ACS Organic Chemistry Practice Questions* provides high-yield practice questions covering topics tested on the ACS Organic Chemistry exam. Develop the ability to apply your knowledge and quickly choose the correct answer to increase your test score.

**Visit our Amazon**

If you benefited from this book, please leave a review on Amazon so others can learn from your input. Reviews help us understand our customers' needs and experiences while keeping our commitment to quality.

# Table of Contents

## Table of Contents (*continued*)

# PRACTICE QUESTIONS

# 1 – Phases and Phase Equilibria

## Practice Set 1: Questions 1–20

**1.** Which statement regarding vapor pressure is NOT true?

I. Solids do not have a vapor pressure
II. Vapor pressure of a pure liquid does not depend on the amount of vapor present
III. Vapor pressure of a pure liquid does not depend on the amount of liquid present

**A.** I only
**B.** II only
**C.** III only
**D.** I and II only

**2.** How does the volume of a fixed sample of gas change if the temperature is doubled at constant pressure?

**A.** Decreases by a factor of 2
**B.** Increases by a factor of 4
**C.** Doubles
**D.** Remains the same

**3.** When a solute is added to a pure solvent, the boiling point [ ] and freezing point [ ]?

**A.** decreases … decreases
**B.** decreases … increases
**C.** increases … decreases
**D.** increases … increases

**4.** Consider the phase diagram for $H_2O$. The termination of the gas-liquid transition at which distinct gas or liquid phases do NOT exist is the:

**A.** critical point
**B.** endpoint
**C.** triple point
**D.** condensation point

**5.** The van der Waals equation of state for a real gas is expressed as $[P + n^2a / V^2]\cdot(V - nb) = nRT$. The van der Waals constant, $a$, represents a correction for:

**A.** negative deviation in the measured value of P from that of an ideal gas due to the attractive forces between the molecules of a real gas
**B.** positive deviation in the measured value of P from that of an ideal gas due to the attractive forces between the molecules of a real gas
**C.** negative deviation in the measured value of P from that of an ideal gas due to the finite volume of space occupied by molecules of a real gas
**D.** positive deviation in the measured value of P from that of an ideal gas due to the finite volume of space occupied by molecules of a real gas

**6.** What is the value of the ideal gas constant, expressed in units (torr × mL) / mole × K?

**A.** 62.4
**B.** 62,400
**C.** 0.0821
**D.** 1 / 0.0821

**7.** If the pressure and the temperature of a gas are halved, the volume is:

**A.** halved
**B.** the same
**C.** doubled
**D.** quadrupled

**8.** If container X is occupied by 2.0 moles of $O_2$ gas, while container Y is occupied by 10.0 grams of $N_2$ gas, and containers are maintained at 5.0 °C and 760 torrs, then:

**A.** container X must have a volume of 22.4 L
**B.** the average kinetic energy of the molecules in X is equal to the average kinetic energy of the molecules in Y
**C.** container Y must be larger than container X
**D.** the average speed of the molecules in container X is greater than that of the molecules in container Y

**9.** Among the following choices, how tall should a properly designed Torricelli mercury barometer be?

**A.** 100 in
**B.** 380 mm
**C.** 76 mm
**D.** 800 mm

**10.** Heat is released to the surroundings when volatile solvents X and Y are mixed in equal proportions. If pure X has a higher boiling point than pure Y, which of the following statements is NOT true?

**A.** The vapor pressure of the mixture is lower than that of pure Y
**B.** The vapor pressure of the mixture is lower than that of pure X
**C.** The boiling point of the mixture is lower than that of pure X
**D.** The boiling point of the mixture is lower than that of pure Y

**11.** For the balanced reaction 2 Na + $Cl_2 \rightarrow$ 2 NaCl, which of the following is a gas?

I. Na II. $Cl_2$ III. NaCl

**A.** I only
**B.** II only
**C.** III only
**D.** I and II only

**12.** How does a real gas deviate from an ideal gas?

I. Molecules occupy a significant amount of space
II. Intermolecular forces may exist
III. Pressure is created from molecular collisions with the walls of the container

**A.** I only
**B.** II only
**C.** I and II only
**D.** II and III only

**13.** A 2.75 L sample of He gas has a pressure of 0.950 atm. What is the pressure of the gas if the volume is reduced to 0.450 L?

**A.** 5.80 atm
**B.** 0.520 atm
**C.** 0.230 atm
**D.** 0.960 atm

**14.** Avogadro's law, in its alternate form, is similar in form to:

I. Boyle's law
II. Charles' law
III. Gay-Lussac's law

**A.** I only
**B.** II only
**C.** III only
**D.** II and III only

 

**15.** What is the relationship between the pressure and volume of a fixed amount of gas at constant temperature?

**A.** directly proportional
**B.** equal
**C.** inversely proportional
**D.** decreased by a factor of 2

**16.** What is the term for a change of state from a liquid to a gas?

**A.** vaporization
**B.** melting
**C.** deposition
**D.** condensing

**17.** The combined gas law can NOT be written as:

**A.** $V_2 = V_1 \times P_1 / P_2 \times T_2 / T_1$
**B.** $P_1 = P_2 \times V_2 / V_1 \times T_1 / T_2$
**C.** $T_2 = T_1 \times P_1 / P_2 \times V_2 / V_1$
**D.** $V_1 = V_2 \times P_2 / P_1 \times T_1 / T_2$

**18.** Which singular molecule is most likely to show a dipole-dipole interaction?

**A.** $CH_4$
**B.** H–C≡C–H
**C.** $SO_2$
**D.** $CO_2$

**19.** What happens if the pressure of a gas above a liquid increases, such as by pressing a piston above a liquid?

**A.** Pressure goes down, and the gas moves out of the solvent
**B.** Pressure goes down, and the gas goes into the solvent
**C.** The gas is forced into solution, and the solubility increases
**D.** The solution is compressed, and the gas is forced out of the solvent

**20.** Which of the following two variables are present in mathematical statements of Avogadro's law?

**A.** P and V
**B.** n and V
**C.** n and T
**D.** V and T

## Practice Set 2: Questions 21–40

**21.** Which of the following acids has the lowest boiling point elevation?

| Monoprotic Acids | *K*a |
|---|---|
| Acid I | $1.4 \times 10^{-8}$ |
| Acid II | $1.6 \times 10^{-9}$ |
| Acid III | $3.9 \times 10^{-10}$ |

**A.** I
**B.** II
**C.** III
**D.** Requires more information

**22.** An ideal gas differs from a real gas because the molecules of an ideal gas have:

**A.** no attraction to each other
**B.** no kinetic energy
**C.** molecular weight equal to zero
**D.** appreciable volumes

**23.** Which of the following is the definition of standard temperature and pressure?

**A.** 0 K and 1 atm
**B.** 298.15 K and 750 mmHg
**C.** 273.15 K and $10^5$ Pa
**D.** 273 °C and 750 torr

**24.** At constant volume, as the temperature of a gas sample is decreased, the gas deviates from ideal behavior. Compared to the pressure predicted by the ideal gas law, actual pressure would be:

**A.** higher because of the volume of the gas molecules
**B.** higher because of intermolecular attractions between gas molecules
**C.** lower because of the volume of the gas molecules
**D.** lower because of intermolecular attractions among gas molecules

**25.** A flask contains a mixture of $O_2$, $N_2$ and $CO_2$. The pressure exerted by $N_2$ is 320 torr and by $CO_2$ is 240 torr. If the total pressure of the gas mixture is 740 torr, what is the percent pressure of $O_2$?

**A.** 14% **B.** 18% **C.** 21% **D.** 24%

**26.** A balloon contains 40 grams of He with a pressure of 1,000 torrs. When He is released from the balloon, the new pressure is 900 torr, and the volume is half the original. If the temperature is the same, how many grams of He remain in the balloon?

**A.** 18 grams
**B.** 22 grams
**C.** 10 grams
**D.** 40 grams

**27.** Which of the following is a true statement regarding evaporation?

**A.** Increasing the surface area of the liquid decreases the rate of evaporation
**B.** The temperature of the liquid changes during evaporation
**C.** Decreasing the surface area of the liquid increases the rate of evaporation
**D.** Molecules with greater kinetic energy escape from the liquid

**28.** Under which conditions does a real gas behave most nearly like an ideal gas?

**A.** High temperature and high-pressure
**B.** High temperature and low-pressure
**C.** Low temperature and low-pressure
**D.** Low temperature and high-pressure

**29.** Which of the following compounds has the highest boiling point?

**A.** $CH_3OH$
**B.** $CH_3CH_2CH_2CH_2CH_2OH$
**C.** $CH_3CH_2CH_2C(OH)HOH$
**D.** $CH_3CH_2OCH_2CH_2CH_3$

**30.** According to the kinetic theory of gases, which of the following is the average kinetic energy of the gas particles directly proportional to?

**A.** temperature
**B.** molar mass
**C.** volume
**D.** pressure

**31.** Under ideal conditions, which of the following gases is least likely to behave as an ideal gas?

**A.** $CF_4$
**B.** $CH_3OH$
**C.** $N_2$
**D.** $O_3$

 

**32.** Which statement is true regarding gases when compared to liquids?

**A.** Gases have lower compressibility and higher density
**B.** Gases have lower compressibility and lower density
**C.** Gases have higher compressibility and higher density
**D.** Gases have higher compressibility and lower density

**33.** When nonvolatile solute molecules are added to a solution, vapor pressure of the solution:

**A.** stays the same
**B.** increases
**C.** decreases
**D.** is directly proportional to the second power of the amount added

**34.** Which of the following laws states that the pressure exerted by a mixture of gases equals the sum of the individual gas pressures?

**A.** Gay-Lussac's law
**B.** Dalton's law
**C.** Charles's law
**D.** Boyle's law

**35.** Which characteristics best describe a solid?

**A.** Definite volume; the shape of the container; no intermolecular attractions
**B.** Volume and shape of the container; no intermolecular attractions
**C.** Definite shape and volume; strong intermolecular attractions
**D.** Definite volume; the shape of the container; moderate intermolecular attractions

**36.** Which of the following statements is true if three 2.0 L flasks are filled with $H_2$, $O_2$ and He at STP?

**A.** There are twice as many He atoms as $H_2$ or $O_2$ molecules
**B.** There are four times as many $H_2$ or $O_2$ molecules as He atoms
**C.** Each flask contains the same number of atoms
**D.** The number of $H_2$ or $O_2$ molecules is the same as the number of He atoms

**37.** Which of the following atoms could interact through a hydrogen bond?

**A.** The hydrogen of an amine and the oxygen of an alcohol
**B.** The hydrogen on an aromatic ring and the oxygen of carbon dioxide
**C.** The oxygen of a ketone and the hydrogen of an aldehyde
**D.** The oxygen of methanol and hydrogen on the methyl carbon of methanol

**38.** Which of the following laws states that the pressure and volume are inversely proportional for gas at a constant temperature?

**A.** Dalton's law
**B.** Gay-Lussac's law
**C.** Boyle's law
**D.** Charles's law

**39.** As an automobile travels the highway, the temperature of the air inside the tires [ ] and the pressure [ ]?

**A.** increases … decreases
**B.** increases … increases
**C.** decreases … decreases
**D.** decreases … increases

**40.** Using the following unbalanced chemical reaction, what volume of $H_2$ gas at 780 mmHg and 23 °C is required to produce 12.5 L of $NH_3$ gas at the same temperature and pressure?

$$N_2\ (g) + H_2\ (g) \rightarrow NH_3\ (g)$$

**A.** 21.4 L
**B.** 15.0 L
**C.** 18.8 L
**D.** 13.0 L

---

**Practice Set 3: Questions 41–60**

---

**41.** A mixture of gases containing 16 g of $O_2$, 14 g of $N_2$ and 88 g of $CO_2$ is collected above water at a temperature of 23 °C. The total pressure is 1 atm, and the vapor pressure of water is 38 torr. What is the partial pressure exerted by $CO_2$?

**A.** 283 torr **B.** 367 torr **C.** 481 torr **D.** 549 torr

**42.** Which of the following demonstrates colligative properties?

I. Freezing point II. Boiling point III. Vapor pressure

**A.** I only
**B.** II only
**C.** III only
**D.** I, II and III

**43.** What is the proportionality relationship between the pressure of a gas and its volume?

**A.** directly
**B.** inversely
**C.** pressure is raised to the $2^{nd}$ power
**D.** pressure raised to the $\sqrt{2}$ power

**44.** Which of the following statements about gases is correct?

**A.** Formation of homogeneous mixtures, regardless of the nature of non-reacting gas components
**B.** Relatively long distances between molecules
**C.** High compressibility
**D.** All the above

**45.** Which of the following describes a substance in the solid physical state?

I. It compresses negligibly
II. It has a fixed volume
III. It has a fixed shape

**A.** I only
**B.** II only
**C.** I and III only
**D.** I, II and III

**46.** Which of the following is NOT a unit used in measuring pressure?

**A.** kilometers Hg
**B.** millimeters Hg
**C.** atmosphere
**D.** Pascal

**47.** Which of the following compounds has the highest boiling point?

**A.** $CH_4$
**B.** $CHCl_3$
**C.** $CH_3COOH$
**D.** $NH_3$

**48.** Identify the decreasing ordering of attractions among particles in the three states of matter.

**A.** gas > liquid > solid
**B.** gas > solid > liquid
**C.** solid > liquid > gas
**D.** liquid > solid > gas

**49.** A sample of $N_2$ gas occupies a volume of 190 mL at STP. What volume will it occupy at 660 mmHg and 295 K?

**A.** 1.15 L
**B.** 0.760 L
**C.** 0.214 L
**D.** 1.84 L

**50.** A nonvolatile liquid would have:

**A.** a highly explosive propensity
**B.** strong attractive forces between molecules
**C.** weak attractive forces between molecules
**D.** a high vapor pressure at room temperature

**51.** A closed-end manometer was constructed from a U-shaped glass tube. It was loaded with mercury so that the closed side was filled to the top, 820 mm above the neck, while the open end was 160 mm above the neck. The manometer was taken into a chamber used for training astronauts. What is the highest pressure that can be read with assurance on this manometer?

**A.** 66.0 torr
**B.** 660 torr
**C.** 220 torr
**D.** 760 torr

**52.** Vessels X and Y each contain 1.00 L of a gas at STP, but vessel X contains oxygen, while vessel Y contains nitrogen. Assuming the gases behave as ideal, they have the same:

I. number of molecules II. density III. kinetic energy

**A.** II only
**B.** III only
**C.** I and III only
**D.** I, II and III

**53.** What condition must be satisfied for the noble gas Xe to exist in the liquid phase at 180 K, a temperature significantly greater than its normal boiling point?

**A.** external pressure > vapor pressure of xenon
**B.** external pressure < vapor pressure of xenon
**C.** external pressure = partial pressure of water
**D.** temperature is increased quickly

**54.** Matter is nearly incompressible in which of these states?

I. solid II. Liquid III. gas

**A.** I only
**B.** II only
**C.** III only
**D.** I and II only

**55.** Which of the following laws states that volume and temperature are directly proportional for a gas at constant pressure?

**A.** Gay-Lussac's law
**B.** Dalton's law
**C.** Charles' law
**D.** Boyle's law

**56.** Which transformation describes sublimation?

**A.** solid → liquid
**B.** solid → gas
**C.** liquid → solid
**D.** liquid → gas

**57.** What is the ratio of the diffusion rate of $O_2$ molecules to the diffusion rate of $H_2$ molecules if six moles of $O_2$ gas and six moles of $H_2$ gas are placed in a large vessel, and the gases and vessels are at the same temperature?

**A.** 4:1 **B.** 1:4 **C.** 12:1 **D.** 1:1

**58.** How many neon gas molecules are present in 6 liters at 10 °C and 320 mmHg? (Use the ideal gas constant R = 0.0821 L·atm $K^{-1}$ $mol^{-1}$)

**A.** (320 mmHg / 760 atm)·(0.821)·(6 L) / (6 × $10^{23}$)·(283 K)
**B.** (320 mmHg / 760 atm)·(6 L)·(6 × $10^{23}$) / (0.0821)·(283 K)
**C.** (320 mmHg)·(6 L)·(283 K)·(6 × $10^{23}$)
**D.** (320 mmHg / 760 atm)·(6 L)·(283 K)·(6 × $10^{23}$) / (0.821)

**59.** Which gas has the greatest density at STP:

**A.** $CO_2$ **B.** $O_2$ **C.** $N_2$ **D.** NO

**60.** A hydrogen bond is a special type of:

**A.** dipole–dipole attraction involving hydrogen-bonded to another hydrogen atom
**B.** attraction involving molecules that contain hydrogens
**C.** dipole-dipole attraction involving hydrogen-bonded to a highly electronegative atom
**D.** dipole–dipole attraction involving hydrogen-bonded to other atoms

 

---

**Practice Set 4: Questions 61–80**

---

**61.** What is the mole fraction of $H_2$ in a gaseous mixture of 9.50 g of $H_2$ and 14.0 g of Ne in a 4.50-liter container maintained at 37.5 °C?

**A.** 0.13 **B.** 0.43 **C.** 0.67 **D.** 0.87

**62.** Which of the following is true about Liquid A if the vapor pressure of Liquid A is greater than that of Liquid B?

**A.** Liquid A boils at a lower temperature
**B.** Liquid A has a higher heat of vaporization
**C.** Liquid A forms stronger bonds
**D.** Liquid A boils at a higher temperature

**63.** When 25 g of non-ionizable compound X is dissolved in 1 kg of camphor, the freezing point of the camphor falls 2.0 K. What is the approximate molecular weight of compound X? (Use $K_{camphor} = 40$)

**A.** 50 g/mol
**B.** 500 g/mol
**C.** 5,000 g/mol
**D.** 5,500 g/mol

**64.** The boiling point of a liquid is the temperature:

**A.** where sublimation occurs
**B.** where the vapor pressure of the liquid is less than the atmospheric pressure over the liquid
**C.** where the vapor pressure of the liquid equals the atmospheric pressure over the liquid
**D.** where the rate of sublimation equals evaporation

**65.** The van der Waals equation $[(P + n^2a / v^2)\cdot(V - nb) = nRT]$ is used to describe nonideal gases. The terms $n^2a / v^2$ and $nb$ stand for, respectively:

**A.** volume of gas molecules and intermolecular forces
**B.** nonrandom movement and intermolecular forces between gas molecules
**C.** nonelastic collisions and volume of gas molecules
**D.** intermolecular forces and volume of gas molecules

**66.** What are the units of the gas constant R?

**A.** atm·K/L·mol
**B.** atm·L/mol·K
**C.** mol·L/atm·K
**D.** mol·K/L·atm

**67.** A sample gas occupying a volume of 120.0 mL at STP was placed in a different vessel with a volume of 155.0 mL, in which the pressure was measured at 0.80 atm. What was its temperature?

**A.** 9.1 °C
**B.** 43.1 °C
**C.** 4.1 °C
**D.** 93.6 °C

**68.** Why does a beaker of water boil at 22 °C when placed in a closed chamber, and a vacuum pump is used to evacuate the air from the chamber?

**A.** The vapor pressure decreases
**B.** Air is released from the water
**C.** The atmospheric pressure decreases
**D.** The vapor pressure increases

**69.** According to the kinetic theory, what happens to the kinetic energy of gaseous molecules when the temperature of a gas decreases?

**A.** Increase as does velocity
**B.** Remains constant, as does velocity
**C.** Increases and velocity decreases
**D.** Decreases, as does velocity

**70.** A sample of $SO_3$ gas is decomposed to $SO_2$ and $O_2$: $2\ SO_3\ (g) \rightarrow 2\ SO_2\ (g) + O_2\ (g)$

If the pressure of $SO_2$ and $O_2$ is 1,250 torrs, what is the partial pressure of $O_2$ in torrs?

**A.** 417 torrs
**B.** 1,040 torrs
**C.** 1,250 torrs
**D.** 884 torrs

**71.** For the balanced reaction $2\ Na + Cl_2 \rightarrow 2\ NaCl$, which of the following is a solid?

I. Na  II. Cl  III. NaCl

**A.** I only
**B.** II only
**C.** III only
**D.** I and III only

**72.** According to Charles's law, what happens to gas as temperature increases?

**A.** volume decreases
**B.** volume increases
**C.** pressure decreases
**D.** pressure increases

**73.** How does the pressure of a sample of gas change if the moles of gas remain constant while the volume is halved and the temperature is quadrupled?

**A.** increase by a factor of 8
**B.** quadruple
**C.** decrease by a factor of 4
**D.** decrease by a factor of 2

**74.** For a fixed quantity of gas, gas laws describe the relationships between pressure and which two variables?

**A.** chemical identity; mass
**B.** volume; chemical identity
**C.** temperature; volume
**D.** temperature; size

**75.** What is the term for a direct change of state from a solid to a gas?

**A.** sublimation
**B.** vaporization
**C.** condensation
**D.** deposition

**76.** The average speed at which a methane molecule effuses at 28.5 °C is 631 m/s. The average speed at which a krypton molecule effuses at the same temperature is:

**A.** 123 m/s
**B.** 276 m/s
**C.** 312 m/s
**D.** 421 m/s

**77.** A chemical reaction A (*s*) → B (*s*) + C (*g*) occurs when substance A is vigorously heated. The molecular mass of the gaseous product was determined from the following experimental data:

Mass of A before reaction: 5.2 g

Mass of A after reaction: 0 g

Mass of residue B after cooling and weighing when no more gas evolved: 3.8 g

When all the gas C evolved, it was collected and stored in a 668.5 mL glass vessel at 32.0 °C, and the gas exerted a pressure of 745.5 torr

Use the ideal gas constant R equals 0.0821 L·atm $K^{-1}$ $mol^{-1}$

From this data, determine the apparent molecular mass of *C*, assuming it behaves as an ideal gas:

**A.** 6.46 g/mol
**B.** 46.3 g/mol
**C.** 53.9 g/mol
**D.** 72.2 g/mol

**78.** Which of the following describes a substance in the liquid physical state?

I. It has a variable shape
II. It compresses negligibly
III. It has a fixed volume

**A.** I only
**B.** II only
**C.** I and III only
**D.** I, II and III

**79.** Which is the strongest form of intermolecular attraction between water molecules?

**A.** ion-dipole
**B.** covalent bonding
**C.** induced dipole-induced dipole
**D.** hydrogen bonding

**80.** When a helium balloon is placed in a freezer, the temperature in the balloon [ ] and the volume [ ]?

**A.** increases … increases
**B.** increases … decreases
**C.** decreases … increases
**D.** decreases … decreases

---

**Practice Set 5: Questions 81–100**

---

**81.** How does the volume of a fixed sample of gas change if the pressure is doubled?

**A.** Decreases by a factor of 2
**B.** Increases by a factor of 4
**C.** Doubles
**D.** Remains the same

**82.** What is the term for the frequency and energy of gas molecules colliding with the walls of the container?

I. partial pressure II. vapor pressure III. gas pressure

**A.** I only
**B.** II only
**C.** III only
**D.** I and II only

**83.** The freezing point changes by 10 K when an unknown amount of toluene is added to 100 g of benzene. Find the number of moles of toluene added from the given data. (Use $K_{benzene}$ = 5.0 and $K_{toluene}$ = 8.4)

**A.** 0.14 **B.** 0.20 **C.** 0.23 **D.** 0.27

**84.** A container is labeled "Ne, 5.0 moles" but has no pressure gauge. By measuring the temperature and determining the volume of the container, a chemist uses the ideal gas law to estimate the pressure inside the container. If the container was mislabeled and contained 5.0 moles of He, not Ne, how would this affect the scientist's estimate?

**A.** The estimate is correct because the identity of the gas is irrelevant
**B.** The estimate is too high
**C.** The estimate is slightly too low
**D.** The estimate is significantly lower because of the large difference in molecular mass

**85.** According to the kinetic theory, which is NOT true of ideal gases?

**A.** For a sample of gas molecules, average kinetic energy is directly proportional to temperature
**B.** There are no attractive or repulsive forces between gas molecules
**C.** Collisions among gas molecules are perfectly elastic
**D.** There is no transfer of kinetic energy during collisions between gas molecules

**86.** At what temperature is degrees Celsius equivalent to degrees Fahrenheit?

**A.** 0 **B.** –10 **C.** –25 **D.** –40

**87.** Which statement about the boiling point of water is NOT correct?

**A.** At sea level and a pressure of 760 mmHg, the boiling point is 100 °C
**B.** In a pressure cooker, shorter cooking times are achieved due to the change in boiling point
**C.** The boiling point is greater than 100 °C in a pressure cooker
**D.** The boiling point is less than 100 °C for locations at low elevations

**88.** Which of the following compounds has the lowest boiling point?

**A.** $CH_4$
**B.** $CHCl_3$
**C.** $CH_3CH_2OH$
**D.** $NH_3$

**89.** As the pressure is increased on solid $CO_2$, the melting point is:

**A.** decreased
**B.** unchanged
**C.** increased
**D.** inversely proportional to the square root of the change

**90.** A vessel contains 32 g of $CH_4$ gas and 12.75 g of $NH_3$ gas at a combined pressure of 2.4 atm. What is the partial pressure of $NH_3$ gas?

**A.** 0.30 atm
**B.** 0.66 atm
**C.** 0.44 atm
**D.** 1.88 atm

**91.** Which of the following statements best describes a liquid?

**A.** Definite shape, but indefinite volume
**B.** Indefinite shape, but definite volume
**C.** Indefinite shape and volume
**D.** Definite shape and volume

**92.** Assuming constant pressure, if a volume of nitrogen gas at 420 K decreases from 100 mL to 50 mL, what is the final temperature in Kelvin?

**A.** 630 K **B.** 420 K **C.** 210 K **D.** 150 K

**93.** Which of the following laws states that pressure and Kelvin temperature are directly proportional for a gas at constant volume?

**A.** Gay-Lussac's law
**B.** Dalton's law
**C.** Charles' law
**D.** Boyle's law

**94.** The Gay–Lussac's law of increased pressure due to increased temperature is explained using kinetic molecular theory, stating that the pressure must increase because the molecules:

**A.** increase in size
**B.** move slower
**C.** decrease in size
**D.** strike the container walls more often

**95.** Which of the following terms does NOT involve the solid state?

**A.** solidification
**B.** sublimation
**C.** evaporation
**D.** melting

**96.** Which of the following compounds exhibit dipole-dipole intermolecular forces primarily?

**A.** $CO_2$
**B.** $F_2$
**C.** $CH_3–O–CH_3$
**D.** $CH_3CH_3$

**97.** Which of the following increases the pressure of a gas?

**A.** Decreasing the volume
**B.** Increasing the number of molecules
**C.** Increasing temperature
**D.** All the above

**98.** According to Avogadro's law, the volume of gas [ ] as the [ ] increases while [ ] is held constant.

**A.** increases… temperature… pressure, and number of moles
**B.** decreases… pressure… temperature, and number of moles
**C.** increases… pressure… temperature, and number of moles
**D.** increases… number of moles… pressure and temperature

**99.** A gas initially filled a 3.0 L container. Heat was added to the gas, raising its temperature from 100 K to 150 K and increasing its pressure from 3.0 to 4.5 atm. What is the new volume of the gas?

**A.** 1.4 L
**B.** 3.0 L
**C.** 2.0 L
**D.** 4.5 L

**100.** Consider a 10.0 liter sample of helium and a 10.0 liters sample of neon, both at 23 °C, 2.0 atm. Which statement regarding these samples is NOT true?

**A.** The density of the neon sample is greater than the density of the helium sample
**B.** Each sample contains the same number of moles of gas
**C.** Each sample weighs the same amount
**D.** Each sample contains the same number of atoms of gas

 

*Notes for active learning*

*Notes for active learning*

## 2 – Thermochemistry

---

**Practice Set 1: Questions 1–20**

---

**1.** Which statement regarding the symbol ΔG is NOT true?

**A.** Specifies the enthalpy of the reaction
**B.** Refers to the free energy of the reaction
**C.** Predicts the spontaneity of a reaction
**D.** Describes the effect of enthalpy and entropy on a reaction

**2.** What happens to the kinetic energy of a gas molecule when the gas is heated?

**A.** Depends on the gas
**B.** Kinetic energy increases
**C.** Kinetic energy decreases
**D.** Kinetic energy remains constant

**3.** How much heat energy (in Joules) is required to heat 21.0 g of copper from 21.0 °C to 68.5 °C? (Use specific heat *c* of Cu = 0.382 J/g·°C)

**A.** 462 J  **B.** 188 J  **C.** 522 J  **D.** 381 J

**4.** For *n* moles of gas, which term expresses the kinetic energy?

**A.** *n*PA, where *n* = number of moles of gas, P = total pressure, and A = surface area of container walls
**B.** ½*n*PA, where *n* = number of moles of gas, P = total pressure, and A = surface area of container walls
**C.** 3/2 *n*RT, where *n* = number of moles of gas, R = ideal gas constant, and T = absolute temperature
**D.** $MV^2$, where M = molar mass of the gas and V = volume of the container

**5.** Which of the following is NOT an endothermic process?

**A.** Condensation of water vapor
**B.** Boiling liquid
**C.** Water evaporating
**D.** Ice melting

**6.** What is true of an endothermic reaction if it causes a decrease in the entropy (S) of the system?

**A.** Only occurs at low temperatures when ΔS is insignificant
**B.** Occurs if coupled to an endergonic reaction
**C.** Never occurs because it decreases ΔS of the system
**D.** Never occurs because ΔG is positive

**7.** Which terms describe energy contained in an object or transferred to an object?

I. chemical    II. electrical    III. heat

**A.** I only
**B.** II only
**C.** I and II only
**D.** I, II and III

**8.** The greatest entropy is observed for which 10 g sample of $CO_2$?

**A.** $CO_2$ (*g*)
**B.** $CO_2$ (*aq*)
**C.** $CO_2$ (*s*)
**D.** $CO_2$ (*l*)

**9.** What is the term for a reaction that proceeds by absorbing heat energy?

**A.** Isothermal reaction
**B.** Exothermic reaction
**C.** Endothermic reaction
**D.** Spontaneous

**10.** A chemical reaction has ΔH = X, ΔS = Y and ΔG = X – RY and occurs at R K. The reaction is:

**A.** spontaneous
**B.** at equilibrium
**C.** nonspontaneous
**D.** cannot be determined

**11.** The thermodynamic systems that have high stability tend to demonstrate:

**A.** maximum ΔH and maximum ΔS
**B.** maximum ΔH and minimum ΔS
**C.** minimum ΔH and maximum ΔS
**D.** minimum ΔH and minimum ΔS

**12.** Whether a reaction is endothermic or exothermic is determined by:

**A.** energy balance between bond breaking and bond forming, resulting in net loss or gain of energy
**B.** the presence of a catalyst
**C.** the activation energy
**D.** the physical state of the reaction system

**13.** What role does entropy play in chemical reactions?

**A.** The entropy change determines whether the reaction occurs spontaneously
**B.** The entropy change determines whether the chemical reaction is favorable
**C.** The entropy determines how much product is actually produced
**D.** The entropy change determines whether the reaction is exothermic or endothermic

**14.** Calculate the value of $\Delta H°$ of the reaction using the provided bond energies.

$H_2C{=}CH_2\ (g) + H_2\ (g) \rightarrow H_3C{-}CH_3\ (g)$

C–C: 348 kJ C≡C: 960 kJ

C=C: 612 kJ C–H: 412 kJ H–H: 436 kJ

**A.** –348 kJ
**B.** +134 kJ
**C.** –546 kJ
**D.** –124 kJ

**15.** The bond dissociation energy is:

I. useful in estimating the enthalpy change in a reaction
II. the energy required to break a bond between two gaseous atoms
III. the energy released when a bond between two gaseous atoms is broken

**A.** I only
**B.** II only
**C.** I and II only
**D.** I and III only

**16.** Based on the following reaction, which statement is true?

$N_2 + O_2 \rightarrow 2\ NO$ (Use the value for enthalpy, $\Delta H = 43.3$ kcal)

**A.** 43.3 kcal are consumed when 2.0 mole of $O_2$ reacts
**B.** 43.3 kcal are consumed when 2.0 moles of NO are produced
**C.** 43.3 kcal are produced when 1.0 g of $N_2$ reacts
**D.** 43.3 kcal are consumed when 2.0 g of $O_2$ reacts

 

**17.** Which of the following properties of gas is/are a state function?

I. temperature　　II. heat　　III. work

**A.** I only
**B.** I and II only
**C.** II and III only
**D.** I, II and III

**18.** The following statements concerning temperature change as a substance is heated are correct, EXCEPT:

**A.** As a liquid is heated, its temperature rises until its boiling point is reached
**B.** During the time a liquid is changing to the gaseous state, the temperature gradually increases until all the liquid is changed
**C.** As a solid is heated, its temperature rises until its melting point is reached
**D.** During the time for a solid to melt into a liquid, the temperature remains constant

**19.** Calculate the value of $\Delta H°$ of reaction for:

$$O{=}C{=}O\ (g) + 3\ H_2\ (g) \rightarrow CH_3{-}O{-}H\ (g) + H{-}O{-}H\ (g)$$

Use the following bond energies, $\Delta H°$:

C–C: 348 kJ　　C=C: 612 kJ　　C≡C: 960 kJ　　C–H: 412 kJ

C–O: 360 kJ　　C=O: 743 kJ　　H–H: 436 kJ　　H–O: 463 kJ

**A.** –348 kJ
**B.** +612 kJ
**C.** –191 kJ
**D.** –769 kJ

**20.** Which statement(s) is/are true for $\Delta S$?

I. $\Delta S$ of the universe is conserved
II. $\Delta S$ of a system is conserved
III. $\Delta S$ of the universe increases with each reaction

**A.** I only
**B.** II only
**C.** III only
**D.** I and II only

---

**Practice Set 2: Questions 21–40**

---

**21.** Which of the following reaction energies is the most endothermic?

**A.** 360 kJ/mole
**B.** –360 kJ/mole
**C.** 88 kJ/mole
**D.** –88 kJ/mole

**22.** A fuel cell contains hydrogen and oxygen gas that react explosively, and the energy converts water to steam, which drives a turbine to turn a generator that produces electricity. The fuel cell and the steam represent which forms of energy, respectively?

**A.** Electrical and heat energy
**B.** Electrical and chemical energy
**C.** Chemical and heat energy
**D.** Chemical and mechanical energy

**23.** Which of the following is true for the ΔG of formation for $N_2$ (*g*) at 25 °C?

**A.** 0 kJ/mol
**B.** positive
**C.** negative
**D.** 1 kJ/mol

**24.** Which of the statements best describes the following reaction?

$$HC_2H_3O_2\ (aq) + NaOH\ (aq) \rightarrow NaC_2H_3O_2\ (aq) + H_2O\ (l)$$

**A.** Acetic acid and NaOH solutions produce sodium acetate and $H_2O$
**B.** Aqueous solutions of acetic acid and NaOH produce aqueous sodium acetate and $H_2O$
**C.** Acetic acid and NaOH solutions produce sodium acetate solution and $H_2O$
**D.** Acetic acid and NaOH produce sodium acetate and $H_2O$

**25.** If a chemical reaction is spontaneous, which value must be negative?

**A.** $C_p$
**B.** $\Delta S$
**C.** $\Delta G$
**D.** $\Delta H$

**26.** What purpose is the hollow walls in a closed hollow-walled container that effectively maintains the temperature inside?

**A.** To trap air trying to escape from the container, which minimizes convection
**B.** To act as an effective insulator, which minimizes convection
**C.** To act as an effective insulator, which minimizes conduction
**D.** To provide an additional source of heat for the container

**27.** If the heat of reaction is exothermic, which of the following is always true?

**A.** The energy of the reactants is greater than the products
**B.** The energy of the reactants is less than the products
**C.** The reaction rate is fast
**D.** The reaction rate is slow

**28.** How much heat must be absorbed to evaporate 16 g of $NH_3$ to its condensation point at −33 °C? (Use heat of condensation for $NH_3$ = 1,380 J/g)

**A.** 86.5 J
**B.** 2,846 J
**C.** 118 J
**D.** 22,080 J

**29.** Which of the reactions is the most exothermic, if the following energy profiles have the same scale? (Use the notation of R = reactants and P = products)

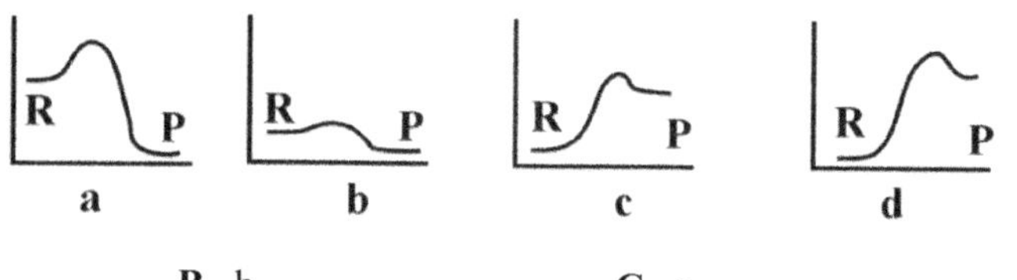


**A.** a
**B.** b
**C.** c
**D.** d

**30.** What is the heat of formation of $NH_3$ (*g*) of the following reaction:

$$2\,NH_3(g) \rightarrow N_2(g) + 3\,H_2(g)$$

(Use $\Delta H° = 92.4$ kJ/mol)

**A.** −92.4 kJ/mol
**B.** −184.4 kJ/mol
**C.** 46.2 kJ/mol
**D.** −46.2 kJ/mol

**31.** If a chemical reaction has a positive ΔH and a negative ΔS, the reaction tends to be:

**A.** at equilibrium
**B.** nonspontaneous
**C.** spontaneous
**D.** irreversible

**32.** What happens to the entropy of a system as the components of the system are introduced to a larger number of possible arrangements, such as when liquid water transforms into water vapor?

**A.** Entropy of a system is solely dependent upon the amount of material undergoing reaction
**B.** Entropy of a system is independent of introducing the components of the system to a larger number of possible arrangements
**C.** Entropy increases because there are more ways for the energy to disperse
**D.** Entropy decreases because there are fewer ways in which the energy can disperse

**33.** Which quantities are needed to calculate heat energy released as water turns to ice at 0 °C?

**A.** The heat of condensation for water and the mass
**B.** The heat of vaporization for water and the mass
**C.** The heat of fusion for water and the mass
**D.** The heat of solidification for water and the mass

**34.** A nuclear power plant uses $^{235}U$ to convert water to steam, which drives a turbine that turns a generator to produce electricity. What are the initial and final forms of energy, respectively?

**A.** Heat energy and electrical energy
**B.** Nuclear energy and electrical energy
**C.** Chemical energy and mechanical energy
**D.** Chemical energy and heat energy

**35.** Which statement(s) is/are correct for the entropy?

I. Higher for a sample of gas than for the same sample of liquid
II. A measure of the disorder in a system
III. Available energy for conversion into mechanical work

**A.** I only
**B.** II only
**C.** III only
**D.** I and II only

**36.** Given the following data, what is the heat of formation for ethanol?

$C_2H_5OH + 3\ O_2 \rightarrow 2\ CO_2 + 3\ H_2O$ : $\Delta H = 327.0$ kcal/mole

$H_2O \rightarrow H_2 + \frac{1}{2}\ O_2$ : $\Delta H = +68.3$ kcal/mole

$C + O_2 \rightarrow CO_2$ : $\Delta H = -94.1$ kcal/mole

**A.** –720.1 kcal
**B.** –327.0 kcal
**C.** +62.6 kcal
**D.** +720.1 kcal

**37.** Based on the reaction shown, which statement is true?

$S + O_2 \rightarrow SO_2 + 69.8$ kcal

**A.** 69.8 kcal are consumed when 32.1 g of sulfur reacts
**B.** 69.8 kcal are produced when 32.1 g of sulfur reacts
**C.** 69.8 kcal are consumed when 1 g of sulfur reacts
**D.** 69.8 kcal are produced when 1 g of sulfur reacts

**38.** In the reaction, $2\ H_2\ (g) + O_2\ (g) \rightarrow 2\ H_2O\ (g)$, entropy is:

**A.** increasing
**B.** the same
**C.** inversely proportional
**D.** decreasing

**39.** For an isolated system, which of the following is NOT exchanged between the system and its surroundings?

I. Temperature II. Matter III. Energy

**A.** I only
**B.** II only
**C.** II and III only
**D.** I, II and III

**40.** Which is a state function?

I. $\Delta G$ II. $\Delta H$ III. $\Delta S$

**A.** I only
**B.** II only
**C.** III only
**D.** I, II and III

---

**Practice Set 3: Questions 41–60**

---

**41.** To simplify comparisons, the energy value of fuels is expressed in units of:

**A.** kcal/g
**B.** kcal/L
**C.** J/kcal
**D.** kcal/mol

**42.** Which process is slowed when an office worker places a lid on a hot cup of coffee?

I. Radiation II. Conduction III. Convection

**A.** I only
**B.** II only
**C.** III only
**D.** I and III only

**43.** Which statement below is true for a spontaneous chemical reaction?

**A.** $\Delta S_{sys} - \Delta S_{surr} = 0$
**B.** $\Delta S_{sys} + \Delta S_{surr} > 0$
**C.** $\Delta S_{sys} + \Delta S_{surr} < 0$
**D.** $\Delta S_{sys} + \Delta S_{surr} = 0$

**44.** A fuel cell contains hydrogen and oxygen gas that react explosively, and the energy converts water to steam, which drives a turbine to turn a generator that produces electricity. What energy changes are employed in the process?

I. Mechanical → electrical energy
II. Heat → mechanical energy
III. Chemical → heat energy

**A.** I only
**B.** II only
**C.** I and II only
**D.** I, II and III

**45.** Determine the value of $\Delta E°_{rxn}$ for this reaction, whereby the standard enthalpy of reaction ($\Delta H°_{rxn}$) is –311.5 kJ mol$^{-1}$:

$C_2H_2\ (g) + 2\ H_2\ (g) \rightarrow C_2H_6\ (g)$

**A.** –306.5 kJ mol$^{-1}$
**B.** –318.0 kJ mol$^{-1}$
**C.** +346.0 kJ mol$^{-1}$
**D.** +306.5 kJ mol$^{-1}$

**46.** For a closed system, what can be exchanged between the system and its surroundings?

I. Heat　　II. Matter　　III. Energy

**A.** I only
**B.** II only
**C.** III only
**D.** I and III only

**47.** Which reaction is accompanied by an increase in entropy?

**A.** $Na_2CO_3$ (*s*) + $CO_2$ (*g*) + $H_2O$ (*g*) → 2 $NaHCO_3$ (*s*)
**B.** BaO (*s*) + $CO_2$ (*g*) → $BaCO_3$ (*s*)
**C.** $CH_4$ (*g*) + $H_2O$ (*g*) → CO (*g*) + 3 $H_2$ (*g*)
**D.** ZnS (*s*) + 3/2 $O_2$ (*g*) → ZnO (*s*) + $SO_2$ (*g*)

**48.** Under standard conditions, which reaction has the largest difference between the energy of the reaction and enthalpy?

**A.** C (*graphite*) → C (*diamond*)
**B.** C (*graphite*) + $O_2$ (*g*) → C (*diamond*)
**C.** 2 C (*graphite*) + $O_2$ (*g*) → 2 CO (*g*)
**D.** C (*graphite*) + $O_2$ (*g*) → $CO_2$ (*g*)

**49.** Which type of reaction tends to be the most stable?

I. Isothermic　　II. Exergonic　　III. Endergonic

**A.** I only
**B.** II only
**C.** III only
**D.** I and II only

**50.** Which is true for the thermodynamic functions G, H, and S in $\Delta G = \Delta H - T\Delta S$?

**A.** G refers to the universe, H to the surroundings, and S to the system
**B.** G, H, and S refer to the system
**C.** G and H refer to the surroundings, and S to the system
**D.** G and H refer to the system, and S to the surroundings

**51.** Which of the following statements is true for the following reaction? (Use the change in enthalpy, $\Delta H° = -113.4$ kJ/mol and the change in entropy, $\Delta S° = -145.7$ J/K mol)

$2\ NO\ (g) + O_2\ (g) \rightarrow 2\ NO_2\ (g)$

**A.** The reaction is at equilibrium at 25 °C under standard conditions
**B.** The reaction is spontaneous at only high temperatures
**C.** The reaction is spontaneous only at low temperatures
**D.** The reaction is spontaneous at all temperatures

**52.** Which of the following expressions defines enthalpy? (Use the conventions: $q$ = heat, U = internal energy, P = pressure and V = volume)

**A.** $q - \Delta U$ **B.** $U + q$ **C.** U + PV **D.** $\Delta U$

**53.** Which statement is true regarding entropy?

I. It is a state function
II. It is an extensive property
III. It has an absolute zero value

**A.** I only
**B.** III only
**C.** I and II only
**D.** I and III only

**54.** Where does the energy released during an exothermic reaction originate from?

**A.** The kinetic energy of the surrounding
**B.** The kinetic energy of the reacting molecules
**C.** The potential energy of the reacting molecules
**D.** The thermal energy of the reactants

**55.** The species in the reaction $KClO_3\ (s) \rightarrow KCl\ (s) + 3/2\ O_2\ (g)$ have the values for standard enthalpies of formation at 25 °C. At constant physical states, assume that the values of $\Delta H°$ and $\Delta S°$ are constant throughout a broad temperature range. Which of the following conditions may apply to the reaction? (Use $KClO_3\ (s)$ with $\Delta H_f° = -391.2\ kJ\ mol^{-1}$ and KCl ($s$) with $\Delta H_f° = -436.8\ kJ\ mol^{-1}$)

**A.** Nonspontaneous at low temperatures but spontaneous at high temperatures
**B.** Spontaneous at low temperatures but nonspontaneous at high temperatures
**C.** Nonspontaneous at temperatures over a broad temperature range
**D.** Spontaneous at temperatures over a broad temperature range

 

**56.** What is the standard enthalpy change for the reaction?

$P_4\,(s) + 6\,Cl_2\,(g) \rightarrow 4\,PCl_3\,(l)$  $\Delta H^\circ = -1{,}289$ kJ

$3\,P_4\,(s) + 18\,Cl_2\,(g) \rightarrow 12\,PCl_3\,(l)$

**A.** –3,867 kJ
**B.** –1,345 kJ
**C.** –366 kJ
**D.** 1,289 kJ

**57.** Which of the following reactions is endothermic?

**A.** $PCl_3 + Cl_2 \rightarrow PCl_5 +$ *heat*
**B.** $2\,NO_2 \rightarrow N_2 + 2\,O_2 +$ *heat*
**C.** $CH_4 + NH_3 +$ *heat* $\rightarrow HCN + 3\,H_2$
**D.** $NH_3 + HBr \rightarrow NH_4Br$

**58.** When the system undergoes a spontaneous reaction, is it possible for the entropy of a system to decrease?

**A.** No, because this violates the second law of thermodynamics
**B.** No, because this violates the first law of thermodynamics
**C.** Yes, but only if the reaction is endothermic
**D.** Yes, but only if the entropy gain of the environment is greater than the entropy loss in the system

**59.** A 500 ml beaker of distilled water is placed under a bell jar and then covered by a layer of opaque insulation. After several days, some of the water evaporated. What kind of system is the contents of the bell jar?

**A.** endothermic
**B.** isolated
**C.** closed
**D.** open

**60.** Which law explains the observation that the amount of heat transfer accompanying a change in one direction is equal in magnitude but opposite in sign to the amount of heat transfer in the opposite direction?

**A.** Law of Conservation of Energy
**B.** Law of Definite Proportions
**C.** Avogadro's law
**D.** Boyle's law

 

---

**Practice Set 4: Questions 61–80**

---

**61.** What is the term for a reaction that proceeds by releasing heat energy?

**A.** Endothermic reaction
**B.** Isothermal reaction
**C.** Exothermic reaction
**D.** Nonspontaneous

**62.** If a stationary gas has a kinetic energy of 500 J at 25 °C, what is its kinetic energy at 50 °C?

**A.** 125 J
**B.** 450 J
**C.** 540 J
**D.** 1,120 J

**63.** Which is NOT true for entropy in a closed system according to the equation $\Delta S = Q / T$?

**A.** Entropy is a measure of energy dispersal of the system
**B.** The equation is only valid for a reversible process
**C.** Changes due to heat transfer are greater at low temperatures
**D.** Disorder of the system decreases as heat is transferred out of the system

**64.** In which of the following physical changes are both processes exothermic?

**A.** Melting and condensation
**B.** Freezing and condensation
**C.** Sublimation and evaporation
**D.** Freezing and sublimation

**65.** A fuel cell contains hydrogen and oxygen gas that react explosively, and the energy converts water to steam, which drives a turbine to turn a generator that produces electricity. What are the initial and final forms of energy, respectively?

**A.** Chemical and electrical energy
**B.** Nuclear and electrical energy
**C.** Chemical and mechanical energy
**D.** Chemical and heat energy

**66.** Which must be true concerning a solution that reaches an equilibrium where chemicals are mixed in a redox reaction?

**A.** $\Delta G° = \Delta G$
**B.** $E = 0$
**C.** $\Delta G° < 1$
**D.** $K = 1$

 

**67.** A solid sample at room temperature spontaneously sublimes, forming a gas. Which of the state changes is accompanied by this change in state in the sample?

**A.** Entropy decreases and energy increases
**B.** Entropy increases and energy decreases
**C.** Entropy and energy decrease
**D.** Entropy and energy increase

**68.** At constant temperature and pressure, a negative $\Delta G$ indicates that the:

**A.** reaction is nonspontaneous
**B.** reaction is fast
**C.** reaction is spontaneous
**D.** reaction is endothermic

**69.** The process of $H_2O\ (g) \rightarrow H_2O\ (l)$ is nonspontaneous under the pressure of 760 torr and temperatures of 378 K because:

**A.** $\Delta H > T\Delta S$
**B.** $\Delta G < 0$
**C.** $\Delta H > 0$
**D.** $\Delta H < T\Delta S$

**70.** Consider the contribution of entropy to the spontaneity of the reaction. As written, the reaction is [ ], and the entropy of the system [ ].

$$2\ Al_2O_3\ (s) \rightarrow 4\ Al\ (s) + 3\ O_2\ (g),\ \Delta G = +138\ \text{kcal}$$

**A.** non-spontaneous … decreases
**B.** non-spontaneous … increases
**C.** spontaneous … decreases
**D.** spontaneous … increases

**71.** The $\Delta G$ of a reaction is the maximum energy that the reaction releases to do:

**A.** P–V work only
**B.** work and release heat
**C.** any type of work
**D.** non P–V work only

**72.** Which of the following represent forms of internal energy?

I. bond energy II. thermal energy III. gravitational energy

**A.** I only
**B.** II only
**C.** I and II only
**D.** I and III only

 

**73.** If it takes energy to break bonds and energy is gained when forming bonds, how can some reactions be exothermic while others are endothermic?

**A.** Some products have more energy than others and require energy to be formed
**B.** Some reactants have more energetic bonds than others and release energy
**C.** It is the number of bonds that is determinative. Since all bonds have the same amount of energy, the net gain or net loss of energy depends on the number of bonds
**D.** It is the amount of energy that is determinative. Some bonds are stronger than others, so there is a net gain or net loss of energy when formed

**74.** Which constant is represented by A in the following calculation to determine how much heat is required to convert 60 g of ice at −25 °C to steam at 320 °C?

$$\text{Total heat} = [(\text{A})\cdot(60\text{ g})\cdot(25\text{ °C})] + [(\text{heat of fusion})\cdot(60\text{ g}]) + $$
$$+ [(4.18\text{ J/g}\cdot\text{°C})\cdot(60\text{ g})\cdot(100\text{ °C})] + [(\text{B})\cdot(60\text{ g})] + [(\text{C})\cdot(60\text{ g})\cdot(220\text{ °C})]$$

**A.** Specific heat of ice
**B.** Heat of vaporization of water
**C.** Heat of condensation
**D.** Heat capacity of steam

**75.** The heat of formation of water vapor is:

**A.** positive, but greater than the heat of formation for $H_2O$ (*l*)
**B.** positive and smaller than the heat of formation for $H_2O$ (*l*)
**C.** negative, but greater than the heat of formation for $H_2O$ (*l*)
**D.** negative and smaller than the heat of formation for $H_2O$ (*l*)

**76.** Entropy can be defined as the amount of:

**A.** equilibrium in a system
**B.** chemical bonds changed during a reaction
**C.** energy required to initiate a reaction
**D.** disorder in a system

**77.** Which is true of an atomic fission bomb according to the conservation of mass and energy law?

**A.** The mass of the bomb and the fission products are identical
**B.** A small amount of mass is converted into energy
**C.** The energy of the bomb and the fission products are identical
**D.** The mass of the fission bomb is greater than the mass of the products

**78.** Once an object enters a black hole, astronomers consider it to have left the universe, which means the universe is:

**A.** entropic **B.** isolated **C.** closed **D.** open

**79.** Which ranking from lowest to highest entropy per gram of NaCl is correct?

**A.** NaCl (*s*) < NaCl (*l*) < NaCl (*aq*) < NaCl (*g*)
**B.** NaCl (*s*) < NaCl (*l*) < NaCl (*g*) < NaCl (*aq*)
**C.** NaCl (*g*) < NaCl (*aq*) < NaCl (*l*) < NaCl (*s*)
**D.** NaCl (*s*), NaCl (*aq*), NaCl (*l*), NaCl (*g*)

**80.** From the given bond energies, how many kJ of energy are released or absorbed from the reaction of one mole of $N_2$ with three moles of $H_2$ to form two moles of $NH_3$?

$$N{\equiv}N + H{-}H + H{-}H + H{-}H \rightarrow NH_3 + NH_3$$

H–N: 389 kJ/mol H–H: 436 kJ/mol N≡N: 946 kJ/mol

**A.** –80 kJ/mol released
**B.** +89.5 kJ/mol absorbed
**C.** –946 kJ/mol released
**D.** +895 kJ/mol absorbed

*Notes for active learning*

*Notes for active learning*

# 3 – Thermodynamics

Practice Set 1: Questions 1–20

1. Compared to the initial value, what is the resulting pressure for an ideal gas compressed isothermally to one-third of its initial volume?

A. Equal
B. Three times larger
C. Larger, but less than three times larger
D. More than three times larger

2. A uniform hole in a brass plate has a diameter of 1.2 cm at 25 °C. What is the diameter of the hole when the plate is heated to 225 °C? (Use the coefficient of linear thermal expansion for brass = $19 \times 10^{-6}$ $K^{-1}$)

A. 2.2 cm
B. 2.8 cm
C. 1.2 cm
D. 1.6 cm

3. A student heats 90 g of water using 50 W of power, with 100% efficiency. How long does it take to raise the temperature of the water from 10 °C to 30 °C? (Use the specific heat of water $c$ = 4.186 J/g·°C)

A. 232 s
B. 81 s
C. 59 s
D. 151 s

4. A runner generates 1,260 W of thermal energy. If her heat is to be dissipated only by evaporation, how much water does she shed in 15 minutes of running? (Use latent heat of vaporization of water $L_v = 22.6 \times 10^5$ J/kg)

A. 500 g
B. 35 g
C. 350 g
D. 50 g

**5.** A Carnot-efficiency engine is operated as a heat pump to heat a room in the winter. The heat pump delivers heat to the room at the rate of 32 kJ per second. It maintains the room at a temperature of 293 K when the outside temperature is 237 K. The power requirement for the heat pump under these operating conditions is:

**A.** 6,100 W
**B.** 3,400 W
**C.** 7,300 W
**D.** 14,300 W

**6.** How much heat is needed to melt a 55 kg sample of ice at 0 °C? (Use latent heat of fusion for water $L_f = 334$ kJ/kg and latent heat of vaporization $L_v = 2{,}257$ kJ/kg)

**A.** 0 kJ
**B.** $1.8 \times 10^4$ kJ
**C.** $3 \times 10^5$ kJ
**D.** $4.6 \times 10^6$ kJ

**7.** Metals are good heat conductors and good electrical conductors because of the:

**A.** relatively high densities of metals
**B.** high elasticity of metals
**C.** ductility of metals
**D.** looseness of outer electrons in metal atoms

**8.** Solar houses are designed to retain the heat absorbed during the day so that the stored heat can be released during the night. A botanist produces steam at 100 °C during the day and then allows the steam to cool to 0 °C and freeze during the night. How many kilograms of water are needed to store 200 kJ of energy for this process? (Use the latent heat of vaporization of water $L_v = 22.6 \times 10^5$ J/kg, the latent heat of fusion of water $L_f = 33.5 \times 10^4$ J/kg, and the specific heat capacity of water $c = 4{,}186$ J/kg·K)

**A.** 0.066 kg
**B.** 0.103 kg
**C.** 0.482 kg
**D.** 1.18 kg

**9.** The heat required to change a substance from the solid to the liquid state is the heat of:

**A.** condensation
**B.** freezing
**C.** fusion
**D.** vaporization

 

**10.** A rigid container holds 0.2 kg of hydrogen gas. How much heat is needed to change the temperature of the gas from 250 K to 280 K? (Use specific heat of hydrogen gas = 14.3 J/g·K)

**A.** 46 kJ
**B.** 72 kJ
**C.** 56 kJ
**D.** 86 kJ

**11.** An aluminum electric tea kettle with a mass of 500 g is heated with a 500 W heating coil. How many minutes are required to heat 1 kg of water from 18 °C to 98 °C in the tea kettle? (Use the specific heat of aluminum = 900 J/kg·K and specific heat of water = 4,186 J/kg·K)

**A.** 16 min
**B.** 12 min
**C.** 8 min
**D.** 4 min

**12.** Heat is added at a constant rate to a pure substance in a closed container. The temperature of the substance as a function of time is shown in the graph. If $L_f$ = latent heat of fusion and $L_v$ = latent heat of vaporization, what is the value of the ratio $L_v / L_f$ for this substance?

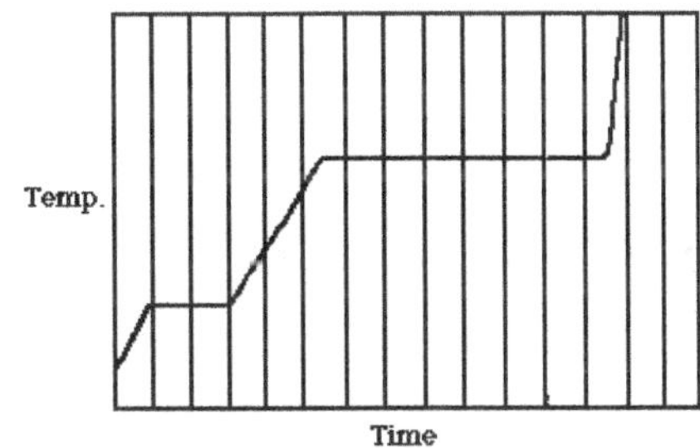


**A.** 3.5
**B.** 7.2
**C.** 4.5
**D.** 5.0

**13.** The moderate temperatures of islands throughout the world have much to do with water's:

**A.** high evaporation rate
**B.** high specific heat capacity
**C.** vast supply of thermal energy
**D.** poor conductivity

**14.** A 4.5 g lead BB moving at 46 m/s penetrates a woodblock and comes to rest inside the block. If the BB absorbs half of the kinetic energy, what is the change in the temperature of the BB? (Use the specific heat of lead = 128 J/kg·K)

**A.** 2.8 K
**B.** 3.6 K
**C.** 1.6 K
**D.** 4.1 K

**15.** The heat required to change a substance from the liquid to the vapor state is the heat of:

**A.** melting
**B.** condensation
**C.** vaporization
**D.** fusion

**16.** A Carnot engine operating between a reservoir of liquid mercury at its melting point and a colder reservoir extracts 18 J of heat from the mercury. It does 5 J of work during each cycle. What is the temperature of the colder reservoir? (Use the melting temperature of mercury = 233 K)

**A.** 168 K
**B.** 66 K
**C.** 57 K
**D.** 82 K

**17.** A 920 g empty iron pan is put on a stove. How much heat must the iron pan absorb in joules to raise its temperature from 18 °C to 96 °C? (Use the specific heat for iron = 113 cal/kg·°C and 1 cal = 4.186 J)

**A.** 50,180 J
**B.** 81,010 J
**C.** 63,420 J
**D.** 33,940 J

**18.** When a solid melts, what change occurs in the substance?

**A.** Heat energy dissipates
**B.** Heat energy enters
**C.** Temperature increases
**D.** Temperature decreases

**19.** Which of the following is an accurate statement about the work done for a cyclic process conducted in a gas? (Use P for pressure and V for volume on the graph)

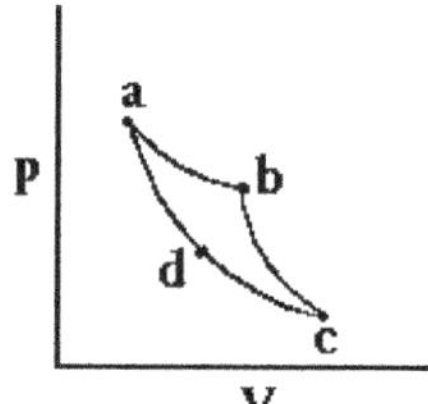


**A.** It is equal to the area under *ab* minus the area under *dc*
**B.** It is equal to the area under the curve *adc*
**C.** It is equal to the area enclosed by the cyclic process
**D.** It equals zero

**20.** Substance A has higher specific heat than substance B. With other factors equal, which substance requires more energy to be heated to the same temperature?

**A.** Substance A
**B.** Substance B
**C.** Both require the same amount of heat
**D.** Depends on the density of each substance

 

---

**Practice Set 2: Questions 21–40**

---

**21.** A 6.5 g meteor hits the Earth at a speed of 300 m/s. If the meteor's kinetic energy is entirely converted to heat, how much does its temperature rise? (Use the specific heat of the meteor = 120 cal/kg·°C and the conversion of 1 cal = 4.186 J)

**A.** 134 °C
**B.** 68 °C
**C.** 120 °C
**D.** 90 °C

**22.** When a liquid freezes, what change occurs in the substance?

**A.** Heat energy dissipates
**B.** Heat energy enters
**C.** Temperature increases
**D.** Temperature decreases

**23.** A monatomic ideal gas ($C_V = 3/2$ R) undergoes an isothermal expansion at 300 K as the volume increases from 0.05 $m^3$ to 0.2 $m^3$. The final pressure is 130 kPa. What is the heat transfer of the gas? (Use the ideal gas constant R = 8.314 J/mol·K)

**A.** –14 kJ
**B.** 36 kJ
**C.** 14 kJ
**D.** –21 kJ

**24.** What is the maximum temperature rise expected for 1 kg of water falling from a waterfall with a vertical drop of 30 m? (Use the acceleration due to gravity $g = 9.8$ m/s$^2$ and the specific heat of water = 4,186 J/kg·K)

**A.** 0.1 °C
**B.** 0.06 °C
**C.** 0.15 °C
**D.** 0.07 °C

**25.** When 0.75 kg of water at 0 °C freezes, what is the change in entropy of the water? (Use the latent heat of fusion of water $L_f = 33{,}400$ J/kg)

**A.** –92 J/K
**B.** –18 J/K
**C.** 44 J/K
**D.** 80 J/K

 

**26.** When a bimetallic bar made of a copper and iron strip is heated, the copper part of the bar bends toward the iron strip. The reason for this is:

**A.** copper expands more than iron
**B.** iron expands more than copper
**C.** iron gets hotter before copper
**D.** copper gets hotter before iron

**27.** In a flask, 110 g of water is heated using 60 W of power, with perfect efficiency. How long does it take to raise the temperature of the water from 20 °C to 30 °C? (Use the specific heat of water $c$ = 4,186 J/kg·K)

**A.** 132 s
**B.** 57 s
**C.** 9.6 s
**D.** 77 s

**28.** When a liquid evaporates, what change occurs in the substance?

**A.** Heat energy dissipates
**B.** Heat energy enters
**C.** Temperature increases
**D.** Temperature decreases

**29.** A flask of liquid nitrogen is at a temperature of –243 °C. If the nitrogen is heated until the average energy of the particles is doubled, what is the new temperature?

**A.** 356 °C
**B.** –356 °C
**C.** –213 °C
**D.** 134 °C

**30.** Which of the following relationships is true for all types of Carnot heat engines?

I. $\eta = 1 - T_C / T_H$

II. $\eta = 1 - | Q_C / Q_H |$

III. $T_C / T_H = Q_C / Q_H$

**A.** I only
**B.** II only
**C.** III only
**D.** I, II and III

**31.** A substance has a density of 1,800 kg/m$^3$ in the liquid state. At atmospheric pressure, the substance has a boiling point of 170 °C. The vapor has a density of 6 kg/m$^3$ at the boiling point at atmospheric pressure. What is the change in the internal energy of 1 kg of the substance as it vaporizes at atmospheric pressure? (Use the heat of vaporization $L_v = 1.7 \times 10^5$ J/kg)

**A.** 180 kJ
**B.** 170 kJ
**C.** 6 kJ
**D.** 12 kJ

 

**32.** If an aluminum rod that is at 5 °C is heated until it has twice the thermal energy, its temperature is:

**A.** 10 °C
**B.** 56 °C
**C.** 278 °C
**D.** 283 °C

**33.** A thermally isolated system comprises a hot piece of aluminum and a cold piece of copper, with the aluminum and the copper in thermal contact. The specific heat capacity of aluminum is more than double that of copper. Which object experiences the greater temperature change during the time the system takes to reach thermal equilibrium?

**A.** Both experience the same magnitude of temperature change
**B.** The mass of each is required
**C.** The copper
**D.** The aluminum

**34.** In liquid water of a given temperature, the water molecules move randomly at different speeds. Electrostatic forces of cohesion tend to hold them together. However, occasionally one molecule gains enough energy through multiple collisions to pull away and escape the liquid. Which of the following is an illustration of this phenomenon?

**A.** When a large steel suspension bridge is built, gaps are left between the girders
**B.** When a body gets too warm, it produces sweat to cool itself down
**C.** Increasing the atmospheric pressure over a liquid causes the boiling temperature to decrease
**D.** If the snow begins to fall when Mary is skiing, she feels colder than before it started to snow

**35.** A 2,200 kg water sample at 0 °C is cooled to –30 °C and freezes in the process. How much heat is liberated during this process? (Use heat of fusion for water $L_f$ = 334 kJ/kg, heat of vaporization $L_v$ = 2,257 kJ/kg and specific heat for ice = 2,050 J/kg·K)

**A.** 328,600 kJ
**B.** 190,040 kJ
**C.** 637,200 kJ
**D.** 870,100 kJ

**36.** Object 1 has three times the specific heat capacity and four times the mass of Object 2. The same amount of heat is transferred to the two objects. If Object 1 changes by an amount of ΔT, what is the change in temperature of Object 2?

**A.** 12ΔT
**B.** 3ΔT
**C.** ΔT
**D.** (3/4)ΔT

**37.** What is the change of entropy associated with 8 kg of water freezing to ice at 0 °C? (Use the latent heat of fusion $L_f = 80$ kcal/kg)

**A.** 1.4 kcal/K
**B.** 0 kcal/K
**C.** –2.3 kcal/K
**D.** –1.4 kcal/K

**38.** The graph shows a PV diagram for 5.1 g of oxygen gas in a sealed container. The temperature of $T_1$ is 20 °C. What are the values for temperatures of $T_3$ and $T_4$, respectively? (Use the gas constant R= 8.314 J/mol·K)

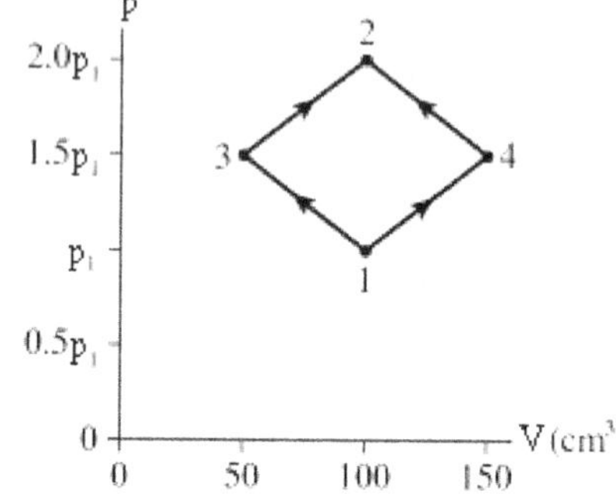


**A.** –53 °C and 387 °C
**B.** –14 °C and 34 °C
**C.** 210 °C and 640 °C
**D.** 12 °C and 58 °C

**39.** On a cold day, a piece of steel feels much colder to the touch than a piece of plastic. This is due to the difference in which one of the following physical properties of these materials?

**A.** Emissivity
**B.** Thermal conductivity
**C.** Density
**D.** Specific heat

**40.** What is the term for a process when a gas is allowed to expand as heat is added to it at constant pressure?

**A.** Isochoric
**B.** Isobaric
**C.** Adiabatic
**D.** Isothermal

---

**Practice Set 3: Questions 41–60**

---

**41.** A Carnot engine is used as an air conditioner to cool a house in the summer. The air conditioner removes 20 kJ of heat per second from the house and maintains the inside temperature at 293 K, while the outside temperature is 307 K. What is the power required for the air conditioner?

**A.** 2.3 kW
**B.** 3.22 kW
**C.** 1.6 kW
**D.** 0.96 kW

**42.** Heat energy is measured in units of:

I. Joules    II. calories    III. work

**A.** I only
**B.** II only
**C.** I and II only
**D.** III only

**43.** The process in which heat flows by the mass movement of molecules from one place to another is:

I. conduction    II. convection    III. radiation

**A.** I only
**B.** II only
**C.** III only
**D.** I and II only

**44.** A Carnot-efficiency engine extracts 515 J of heat from a high-temperature reservoir during each cycle and ejects 340 J of heat to a low-temperature reservoir during the same cycle. What is the efficiency of the engine?

**A.** 34%
**B.** 21%
**C.** 53%
**D.** 17%

**45.** The figure shows 0.008 mol of gas that undergoes the process $1 \rightarrow 2 \rightarrow 3$. What is the volume of $V_3$?

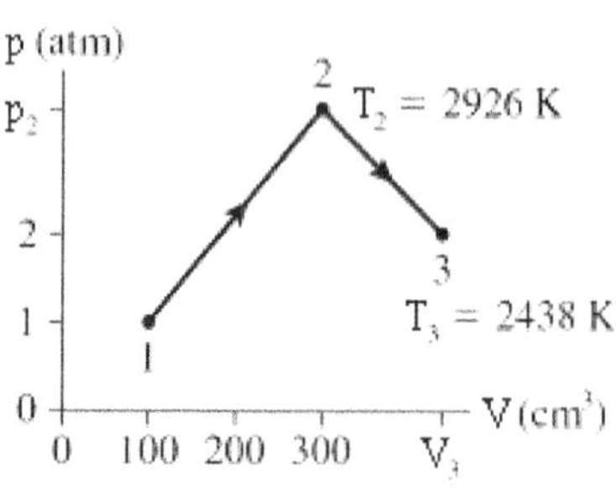


(Use ideal gas constant R = 8.314 J/mol·K and 1 atm = 101,325 Pa)

**A.** 435 $cm^3$
**B.** 568 $cm^3$
**C.** 656 $cm^3$
**D.** 800 $cm^3$

**46.** When a gas expands adiabatically:

**A.** it does no work
**B.** work is done on the gas
**C.** the internal (thermal) energy of the gas decreases
**D.** the internal (thermal) energy of the gas increases

**47.** Why is it that when a swimmer gets out of a swimming pool and stands in a breeze dripping wet, he feels much colder than when he dries off?

**A.** This is a physiological effect resulting from the skin's sensory nerves
**B.** To evaporate a gram of water from his skin requires heat, and most of this heat flows out of his body
**C.** The moisture on his skin has good thermal conductivity
**D.** Water has a relatively small specific heat

**48.** Which method of heat flow requires the movement of energy through solid matter to a new location?

I. Conduction     II. Convection     III. Radiation

**A.** I only
**B.** II only
**C.** III only
**D.** I and II only

**49.** An ideal gas is compressed via an isobaric process to one-third of its initial volume. Compared to the initial pressure, the resulting pressure is:

**A.** more than three times greater
**B.** nine times greater
**C.** three times greater
**D.** the same

**50.** Which of the following would be the best radiator of thermal energy?

**A.** A metallic surface
**B.** A black surface
**C.** A white surface
**D.** A shiny surface

**51.** A brass rod is 59.1 cm long, and an aluminum rod is 39.3 cm long when both rods are at an initial temperature of 0 °C. The rods are placed with 1.1 cm between them, and the distance between the far ends of the rods is maintained at 99.5 cm. The temperature is raised until the two rods are barely in contact.

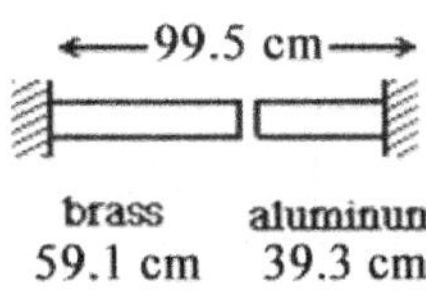


In the figure, what is the temperature at which contact of the rods barely occurs? (Use the coefficient of linear expansion of brass = $2 \times 10^{-5}$ K$^{-1}$ and the coefficient of linear expansion of aluminum = $2.4 \times 10^{-5}$ K$^{-1}$)

**A.** 424 °C
**B.** 588 °C
**C.** 518 °C
**D.** 363 °C

**52.** At room temperature, a person loses energy to the surroundings at the rate of 60 W. If an equivalent food intake compensates for this energy loss, how many kilocalories does he need to consume every 24 hours? (Use the conversion of 1 cal = 4.186 J)

**A.** 1,240 kcal
**B.** 1,660 kcal
**C.** 600 kcal
**D.** 880 kcal

**53.** What primary heat transfer mechanism is when one end of an iron bar become hot when the other end is placed in a flame?

**A.** Convection
**B.** Forced convection
**C.** Radiation
**D.** Conduction

 

Questions **54-55** are based on the following:

Two experiments are performed to determine the calorimetric properties of alcohol with a melting point of –10 °C. In the first trial, a 220 g cube of frozen alcohol, at the melting point, is added to 350 g of water at 26 °C in a Styrofoam container. When thermal equilibrium is reached, the alcohol-water solution is at a temperature of 5 °C. In the second trial, an identical cube of alcohol is added to 400 g of water at 30 °C, and the temperature at thermal equilibrium is 10 °C. (Use the specific heat of water = 4,190 J/kg·K and assume no heat exchange between the Styrofoam container and the surroundings).

**54.** What is the specific heat capacity of the alcohol?

**A.** 2,150 J/kg·K
**B.** 2,475 J/kg·K
**C.** 1,175 J/kg·K
**D.** 1,820 J/kg·K

**55.** What is the heat of fusion of the alcohol?

**A.** $7.2 \times 10^3$ J/kg
**B.** $1.9 \times 10^5$ J/kg
**C.** $5.2 \times 10^4$ J/kg
**D.** $10.3 \times 10^4$ J/kg

**56.** The silver coating on the glass surface of a Thermos bottle reduces the energy that is transferred by:

I. conduction II. convection III. radiation

**A.** I only
**B.** II only
**C.** III only
**D.** I and II only

**57.** A person consumes a snack containing 16 kcal. What is the power this food produces if it is to be expended during exercise in 5 hours? (Use the conversion 1 cal = 4.186 J)

**A.** 0.6 W
**B.** 3.7 W
**C.** 9.7 W
**D.** 96.3 W

**58.** If 50 kcal of heat is added to 5 kg of water, what is the resulting temperature change? (Use the specific heat of water = 1 kcal/kg·°C)

**A.** 10 °C
**B.** 20 °C
**C.** 5 °C
**D.** 40 °C

**59.** How much heat is needed to melt a 30 kg sample of ice at 0 °C? (Use latent heat of fusion for water $L_f$ = 334 kJ/kg and latent heat of vaporization for water $L_v$ = 2,257 kJ/kg)

**A.** 0 kJ
**B.** $5.6 \times 10^4$ kJ
**C.** $1 \times 10^4$ kJ
**D.** $2.4 \times 10^6$ kJ

**60.** A 6 kg aluminum rod is originally at 12 °C. If 160 kJ of heat is added to the rod, what is its final temperature? (Use the specific heat capacity of aluminum = 910 J/kg·K)

**A.** 32 °C
**B.** 41 °C
**C.** 54 °C
**D.** 23 °C

---

**Practice Set 4: Questions 61–80**

---

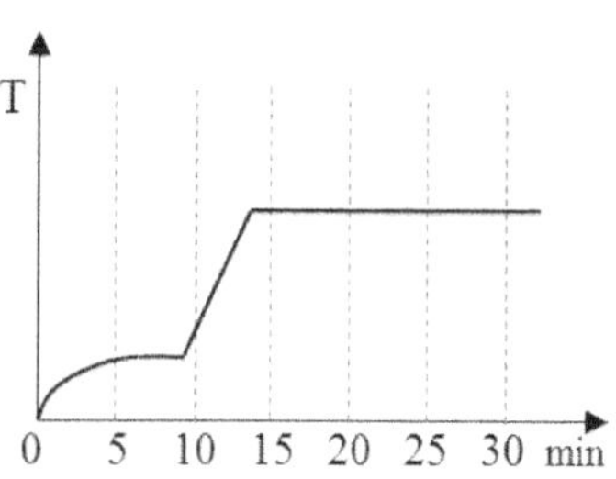


**61.** A solid sample of a pure compound is contained in a closed, well-insulated container. Heat is added at a constant rate, and the sample temperature is recorded. The resulting data is shown in the graph. Which of the following statements is true?

**A.** After 5 minutes, the sample was a mixture of solid and liquid
**B.** The heat capacity of the solid phase is greater than that of the liquid phase
**C.** The sample never boiled
**D.** The heat of fusion is greater than the heat of vaporization

**62.** Warm air rises because it tends to move to regions of less:

I. density II. pressure III. friction

**A.** I only
**B.** II only
**C.** III only
**D.** I and II only

**63.** A machine part consists of 0.15 kg of iron and 0.20 kg of copper. How much heat must be added to the material to raise its temperature from 23 °C to 58 °C? (Use the specific heat for iron = 470 J/kg·K and the specific heat for copper = 390 J/kg·K)

**A.** 2,800 J
**B.** 5,198 J
**C.** 910 J
**D.** 5,200 J

**64.** How much heat is required to raise the temperature of 300 g of lead by 20 °C? (Use the specific heat $c$ of lead = 0.11 kcal/kg·°C)

**A.** 6 kcal
**B.** 0.33 cal
**C.** 660 cal
**D.** 66 cal

**65.** A large and a small container are filled with water of the same temperature. In comparison, they have:

**A.** the same amounts of internal energy less external energy
**B.** different amounts of internal energy
**C.** the same amounts of internal energy and external energy
**D.** the same amounts of internal energy

**66.** A 200 L electric water heater uses 4 kW. Assuming no heat loss, how many hours would it take to heat the water in this tank from 28 °C to 80 °C? (Use the specific heat of water = 4,186 J/kg·K, density of water = 1,000 kg/m$^3$ and the conversion of 1 L = 0.001 m$^3$)

**A.** 12 h
**B.** 9 h
**C.** 1.5 h
**D.** 3 h

**67.** The First Law of Thermodynamics is equivalent to:

**A.** the law of conservation of energy
**B.** Newton's First Law of motion
**C.** the law of conservation of momentum
**D.** Newton's Third Law of motion

**68.** A glass windowpane is 3.1 m high, 1.8 m wide, and 9 mm thick. The temperature at the inner surface of the glass is 22 °C and at the outer surface 7 °C. Given the properties of glass listed below, how much heat is lost through the window every hour?

| | |
|---|---|
| Density | 2,150 kg/m$^3$ |
| Specific Heat | 790 J/kg·°C |
| Coefficient of Linear Thermal Expansion | $8.1 \times 10^{-6}$ °C$^{-1}$ |
| Thermal Conductivity | 0.8 W/m·°C |

**A.** $8.7 \times 10^5$ J
**B.** $7.4 \times 10^3$ J
**C.** $2.7 \times 10^7$ J
**D.** $4.3 \times 10^4$ J

**69.** What happens because of a temperature difference?

**A.** No energy moves unless it is warm enough, at least above the freezing temperature
**B.** Energy moves slowly from colder to warmer regions
**C.** Energy moves from colder to warmer if the difference is less than 3%
**D.** Energy moves from warmer to colder regions

**70.** How much heat must be removed from 435 g of water at 30 °C to change it into ice at –8 °C? (Use specific heat of ice = 2,090 J/kg·K, latent heat of fusion of water $L_f = 33.5 \times 10^4$ J/kg and the specific heat of water = 4,186 J/kg·K)

**A.** 208 kJ
**B.** 420 kJ
**C.** 105 kJ
**D.** 145 kJ

**71.** How much heat is required to melt 400 g of ice? (Use the latent heat of fusion of ice $L_f = 80$ kcal/kg)

**A.** 32 kcal
**B.** 16 kcal
**C.** 320 kcal
**D.** 7.7 kcal

**72.** A Carnot engine operates between a high-temperature reservoir at 420 K and a river with water at 270 K. If it absorbs 3,650 J of heat each cycle, how much work per cycle does the Carnot engine perform?

**A.** 851 J
**B.** 1,846 J
**C.** 2,822 J
**D.** 1,303 J

**73.** If 800 kJ of heat is added to 800 g of water originally at 70 °C, how much water is left in the container? (Use a latent heat of vaporization of water $L_v = 22.6 \times 10^5$ J/kg and the specific heat of water $c = 4,186$ J/kg·K)

**A.** 321 g
**B.** 490 g
**C.** 433 g
**D.** 253 g

**74.** When the first law of thermodynamics, $Q = \Delta U - \text{W}$, is applied to an ideal gas that is taken through an adiabatic process:

**A.** $Q = 0$
**B.** $\text{W} > \Delta U$
**C.** $\Delta U = 0$
**D.** $\text{W} = 0$

**75.** A heat pump pumps heat into a greenhouse at a rate of 50 kW. If work is being done to run this heat pump at a rate of 7.5 kW, what is the coefficient of performance of the heat pump?

**A.** 0.9
**B.** 1.7
**C.** 3.1
**D.** 6.7

**76.** Energy transfer by convection is primarily restricted to:

I. liquids II. gases III. solids

**A.** I only
**B.** II only
**C.** III only
**D.** I and II only

**77.** A 740 kg copper bar is put into a furnace for melting. The initial temperature of the copper is 250 K. How much heat must the furnace produce to melt the copper bar completely? (Use the specific heat of copper $c$ = 386 J/kg·K, the heat of fusion of copper $L_f$ = 205,000 J/kg, and the melting point of copper = 1,357 K)

**A.** $3.5 \times 10^7$ kJ
**B.** $5.6 \times 10^5$ kJ
**C.** $4.7 \times 10^5$ kJ
**D.** $3.3 \times 10^{11}$ kJ

**78.** An investigator uses 8 grams of water which is initially at 100 °C. The water is poured into a cavity in a very large block of ice initially at 0 °C. How many grams of ice melt before thermal equilibrium is attained? (Use the latent heat of fusion for ice $L_f$ = 80 kcal/kg)

**A.** 80 g
**B.** 10 g
**C.** 100 g
**D.** 800 g

**79.** A Carnot heat engine operating between a warmer unknown temperature and a reservoir of boiling helium at 1.9 K has an efficiency of 13%. What is the warmer temperature?

**A.** 3.2 K
**B.** 0.2 K
**C.** 1.4 K
**D.** 2.2 K

**80.** A specific Carnot engine extracts 1,200 J of heat from a high-temperature reservoir and discharges 800 J of heat to a low-temperature reservoir. What is the efficiency of this engine?

**A.** 18%
**B.** 74%
**C.** 33%
**D.** 44%

==========================================================================

**Practice Set 5: Questions 81–106**

==========================================================================

**81.** What is the rate of heat flow through a 35 × 55 cm glass pane that is 6 mm thick when the outside temperature is –10 °C, and the inside temperature is 30 °C? (Use the specific heat of glass $c$ = 0.180 cal/g·°C and the thermal conductivity of glass $k$= 0.105 W/m·K)

**A.** 56 W
**B.** 135 W
**C.** 84 W
**D.** 96 W

**82.** A 0.3 kg ice cube at 0 °C has sufficient heat added to result in total melting, and the resulting water is heated to 60 °C. How much total heat is added? (Use the latent heat of fusion for water $L_f$ = 334 kJ/kg, the latent heat of vaporization for water $L_v$ = 2,257 kJ/kg and the specific heat of water = 4.186 kJ/kg·K)

**A.** 73 kJ
**B.** 48 kJ
**C.** 176 kJ
**D.** 144 kJ

**83.** 50 g of lead of specific heat 0.11 kcal/kg·°C at 100 °C, is put into 80 g of water at 0 °C. What is the final temperature of the mixture? (Use the specific heat of water = 1 kcal/kg·°C at 0 °C)

**A.** 22.8 °C
**B.** 32.2 °C
**C.** 4.2 °C
**D.** 6.4 °C

**84.** A gas follows the PV trajectory shown in the figure. How much work is done per cycle by the gas if $P_0$ = 4.8 atm?

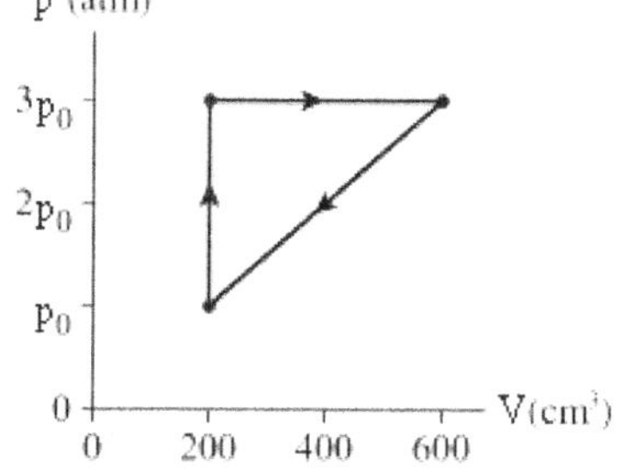


**A.** 840 J
**B.** 84 J
**C.** 192 J
**D.** 344 J

**85.** The water flowing over a large dam drops 60 m. If all the gravitational potential energy is converted to thermal energy, by what temperature does the water rise? (Use acceleration due to gravity $g$ = 10 m/s$^2$ and specific heat of water = 4,186 J/kg·K)

**A.** 0.34 °C
**B.** 0.44 °C
**C.** 0.09 °C
**D.** 0.14 °C

**86.** A lamp radiates 80 J/s when the room is set at 25 °C, but it radiates 95 J/s when the room temperature drops to 20 °C. What is the temperature of the lamp?

**A.** 37 K
**B.** 48 °C
**C.** 101 °C
**D.** 84 °C

**87.** A Carnot heat engine receives 6,000 J of heat and loses 4,000 J in each cycle. What is the efficiency of this engine?

**A.** 33%
**B.** 46%
**C.** 54%
**D.** 18%

**88.** A 40 m long steel pipe, installed when the temperature was 15 °C, is used to transport superheated steam at 160 °C. The pipe is allowed to expand freely when the steam is transported. What is the increase in the length of the pipe? (Use a coefficient of linear expansion for steel $\alpha = 1.2 \times 10^{-5}\ K^{-1}$)

**A.** 52 mm
**B.** 44 mm
**C.** 70 mm
**D.** 37 mm

**89.** When 2 kg of steam at 100 °C condenses to water at 100 °C, what is the change in entropy of the steam? (Use the latent heat of vaporization of water $L_v = 22.6 \times 10^5$ J/kg)

**A.** $-6.1 \times 10^3$ J/K
**B.** $12.1 \times 10^3$ J/K
**C.** 0 J/K
**D.** $-12.1 \times 10^3$ J/K

**90.** Change in internal energy is:

**A.** $Q + W$
**B.** $Q_{out} / W_{in}$
**C.** $Q / m$
**D.** $1 - Q_{cold} / Q_{hot}$

**91.** On a cold winter day, the outside temperature is –20 °C, and the inside temperature is maintained at 20 °C. There is a net heat flow to the outside through the home of 25 kW. What is the change of entropy of the air outside the house due to this heat flow?

**A.** 85 W/K
**B.** 99 W/K
**C.** 13 W/K
**D.** 25 W/K

**92.** A substance has a melting point of 20 °C and a heat of fusion of $3.6 \times 10^4$ J/kg. The boiling point is 150 °C, and the heat of vaporization is $7.2 \times 10^4$ J/kg at a pressure of one atmosphere. What is the quantity of heat released by 3.4 kg of the substance when it is cooled from 160 °C to 75 °C at a pressure of one atmosphere? (Use the specific heat for the solid phase = 600 J/kg·K, the specific heat for the liquid phase = 1,000 J/kg·K and the specific heat for the gaseous phase = 400 J/kg·K)

**A.** 448 kJ
**B.** 325 kJ
**C.** 249 kJ
**D.** 513 kJ

**93.** A mass of 0.2 kg ethanol in the liquid state at its melting point of −114.4 °C, is frozen at atmospheric pressure. What is the change in the entropy of the ethanol as it freezes? (Use the heat of fusion of ethanol $L_f = 1.04 \times 10^5$ J/kg)

**A.** −360 J/K
**B.** 54 J/K
**C.** −131 J/K
**D.** −220 J/K

**94.** Isobaric work is

**A.** $Q - W$
**B.** $P\Delta V$
**C.** $P\Delta T$
**D.** $V\Delta P$

**95.** An adiabatic process is performed on 9 moles of an ideal gas. The initial temperature is 315 K, and the initial volume is 0.70 $m^3$. The final volume is 0.30 $m^3$. What is the amount of heat absorbed by the gas? (Use the adiabatic constant for the gas = 1.44)

**A.** −18 kJ
**B.** 32 kJ
**C.** 0 kJ
**D.** −32 kJ

**96.** An 80-gram aluminum calorimeter contains 360 g of water at an equilibrium temperature of 20 °C. A 180 g piece of metal, initially at 305 °C, is added to the calorimeter. The final temperature at equilibrium is 35 °C. Assume there is no external heat exchange. What is the specific heat capacity of the metal? (Use the specific heat capacity of aluminum = 910 J/kg·K and the specific heat of water = 4,190 J/kg·K)

**A.** 260 J/kg·K
**B.** 324 J/kg·K
**C.** 488 J/kg·K
**D.** 410 J/kg·K

**97.** A chemist uses 120 g of water heated using 65 W of power with 100% efficiency. How much time is required to raise the temperature of the water from 20 °C to 50 °C? (Use specific heat of water = 4.186 J/g·°C)

**A.** 136 s **B.** 182 s **C.** 93 s **D.** 232 s

**98.** The second law of thermodynamics leads to the following conclusion:

**A.** the average temperature of the universe is increasing over time
**B.** it is theoretically possible to convert heat into work with 100% efficiency
**C.** disorder in the universe is increasing over time
**D.** total energy of the universe remains constant

**99.** A glass beaker of unknown mass contains 65 ml of water. The system absorbs 1,800 cal of heat, and the temperature rises 20 °C. What is the mass of the beaker? (Use a specific heat of glass = 0.18 cal/g·°C and the specific heat of water = 1 cal/g·°C)

**A.** 342 g
**B.** 139 g
**C.** 546 g
**D.** 268 g

**100.** What is the change in entropy when 20 g of water at 100 °C is turned into steam at 100 °C? (Use the latent heat of vaporization of water $L_v = 22.6 \times 10^5$ J/kg)

**A.** –346 J/K **B.** 346 J/K **C.** –80.8 J/K **D.** 121 J/K

*Notes for active learning*

# 4 – Fluid Statics and Dynamics

---

**Practice Set 1: Questions 1–20**

---

Questions **1-3** are based on the following:

A container has a vertical tube with an inner radius of 20 mm connected to the container at its side. An unknown liquid reaches level A in the container and level B in the tube. Level A is 5 cm higher than level B. The liquid supports a 20 cm high column of oil between levels B and C that has a density of 850 kg/m$^3$. (Use acceleration due to gravity $g = 9.8$ m/s$^2$)

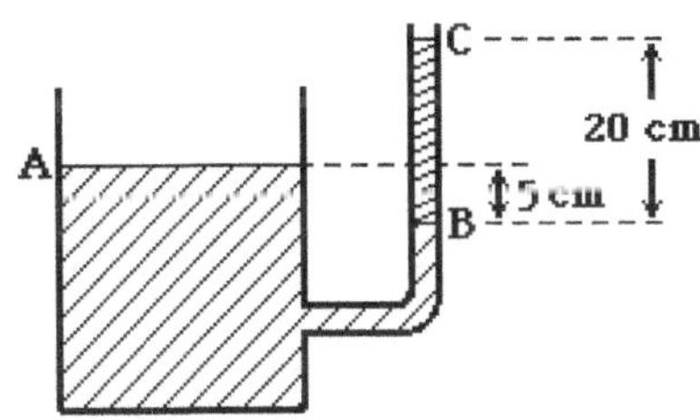


**1.** What is the density of the unknown liquid?

**A.** 2,800 kg/m$^3$
**B.** 2,100 kg/m$^3$
**C.** 3,400 kg/m$^3$
**D.** 3,850 kg/m$^3$

**2.** The gauge pressure at level B is closest to:

**A.** 1,250 Pa
**B.** 1,830 Pa
**C.** 340 Pa
**D.** 1,666 Pa

**3.** What is the mass of the oil?

**A.** 210 g
**B.** 453 g
**C.** 620 g
**D.** 847 g

**4.** A cubical block of stone is lowered at a steady rate into the ocean by a crane, always keeping the top and bottom faces horizontal. Which of the following graphs best describes the gauge pressure P on the bottom of this block as a function of time $t$ if the block just enters the water at time $t = 0$ s?

**A.** 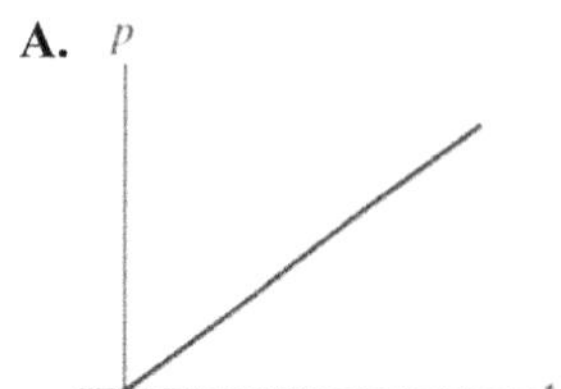


**C.** 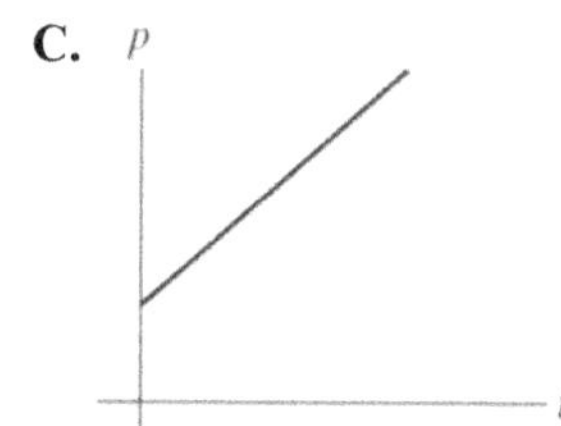


**B.** 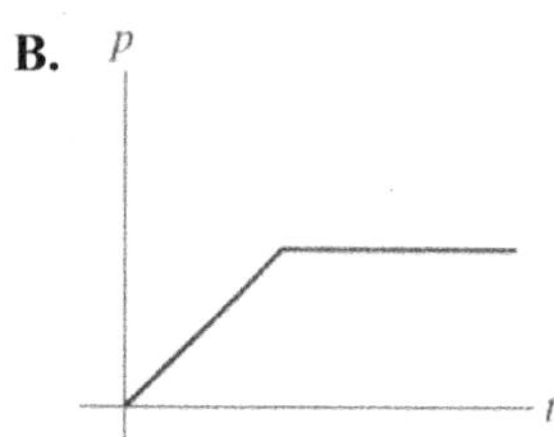


**D.** 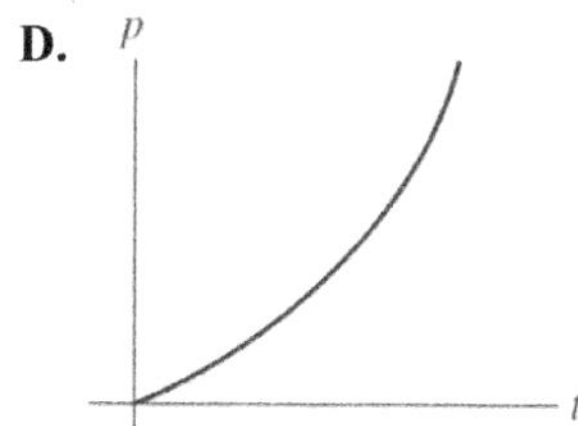


**5.** Consider a small hole in the bottom of a tank that is 19 cm in diameter and is filled with water to a height of 80 cm. What is the speed that the water exits the tank through the hole? (Use acceleration due to gravity $g = 9.8$ m/s$^2$)

**A.** 8.6 m/s
**B.** 12 m/s
**C.** 14.8 m/s
**D.** 4 m/s

**6.** An ideal gas at standard temperature and pressure is compressed until its volume is half the initial volume, and then it is allowed to expand until its pressure is half the initial pressure. This is achieved while holding the temperature constant. If the initial internal energy of the gas is U, the final internal energy of the gas is:

**A.** U/2
**B.** U/3
**C.** U
**D.** 2U

**7.** A viscous oil flows through a narrow pipe at a constant velocity. By what factor does the flow rate increase if the diameter of the pipe is doubled?

**A.** 4
**B.** $\sqrt{2}$
**C.** 6
**D.** 8

**8.** An object is sinking in a fluid. What is the weight of the fluid displaced by the sinking object when the object is completely submerged?

**A.** Dependent on the viscosity of the liquid
**B.** Equal to the weight of the object
**C.** Less than the weight of the object
**D.** Zero

**9.** A wire circle and solid circle of the same diameter are resting on the surface of the water. Which one can have the larger maximum mass without sinking?

**A.** The solid circle, by a factor of 2
**B.** The solid circle, by a factor of 4
**C.** The wire circle, by a factor of 2
**D.** They have the same maximum mass

**10.** When atmospheric pressure increases, what happens to the absolute pressure at the bottom of a pool?

**A.** It does not change
**B.** It increases by double the amount
**C.** It increases by the same amount
**D.** It increases by half the amount

**11.** When soup gets cold, it often tastes greasy because oil spreads out on the surface of the soup instead of staying in small globules. This is explained in terms of the:

**A.** increase in the surface tension of water with a decreasing temperature
**B.** Archimedes' principle
**C.** decrease in the surface tension of water with a decreasing temperature
**D.** Joule-Thomson effect

**12.** A particular grade of motor oil, which has a viscosity of 0.3 $N{\cdot}s/m^2$, flows through a 1 m long tube with a radius of 3.2 mm. What is the average speed of the oil if the drop in pressure over the length of the tube is 225 kPa?

**A.** 0.82 m/s
**B.** 0.96 m/s
**C.** 1.2 m/s
**D.** 1.4 m/s

**13.** An object whose weight is 60 N is floating at the surface of a container of water. How much of the object's volume is submerged? (Use acceleration due to gravity $g = 10\ \text{m/s}^2$)

**A.** 0.006 $m^3$
**B.** 0.06 $m^3$
**C.** 0.6 $m^3$
**D.** 6%

**14.** What is the volume flow rate of a fluid if it flows at 2.5 m/s through a pipe of diameter 3 cm?

**A.** 0.9 $m^3/s$
**B.** $1.8 \times 10^{-3}\ m^3/s$
**C.** $5.7 \times 10^{-4}\ m^3/s$
**D.** $5.7 \times 10^{-3}\ m^3/s$

**15.** What is the specific gravity of a cork that floats with three-quarters of its volume in and one-quarter of its volume out of the water?

**A.** 0.25
**B.** 0.5
**C.** 0.75
**D.** 2

**16.** What volume does 600 g of cottonseed oil occupy if the density of cottonseed oil is 0.93 $g/cm^3$?

**A.** 255 $cm^3$
**B.** 360 $cm^3$
**C.** 470 $cm^3$
**D.** 645 $cm^3$

**17.** The kinetic theory of a monatomic gas suggests the average kinetic energy per molecule is:

**A.** 1/3 $k_B$T
**B.** 2 $k_B$T
**C.** 3/2 $k_B$T
**D.** 2/3 $k_B$T

**18.** An object is weighed in air, and it is also weighed while totally submerged in water. If it weighs 150 N less when submerged, find the volume of the object. (Use acceleration due to gravity $g = 10\ \text{m/s}^2$ and density of water $\rho = 1{,}000\ \text{kg/m}^3$)

**A.** 0.0015 $m^3$
**B.** 0.015 $m^3$
**C.** 0.15 $m^3$
**D.** 1 $m^3$

**19.** When a water container is placed on a laboratory scale, the scale reads 140 g. Now a 30 g piece of copper is suspended from a thread and lowered into the water without contacting the bottom of the container. What does the scale read? (Use acceleration due to gravity $g = 9.8$ m/s$^2$, density of water $\rho = 1$ g/cm$^3$ and density of copper $\rho = 8.9$ g/cm$^3$)

**A.** 122 g
**B.** 168 g
**C.** 143 g
**D.** 110 g

**20.** An ideal, incompressible fluid flows through a 6 cm diameter pipe at 1 m/s. There is a 3 cm decrease in diameter within the pipe. What is the speed of the fluid in this constriction?

**A.** 3 m/s
**B.** 1.5 m/s
**C.** 8 m/s
**D.** 4 m/s

===

**Practice Set 2: Questions 21–40**

===

**21.** Two blocks are submerged in a fluid. Block A has dimensions 2 cm high × 3 cm wide × 4 cm long, and Block B is 2 cm × 3 cm × 8 cm. Both blocks are submerged with their large faces pointing up and down (i.e., the blocks are horizontal), and they are submerged to the same depth. Compared to the fluid pressure on the bottom of Block A, the bottom of Block B experiences:

**A.** equal fluid pressure
**B.** greater fluid pressure
**C.** exactly double the fluid pressure
**D.** less fluid pressure

**22.** A piece of thread of diameter $d$ is in the shape of a rectangle (length $l$, width $w$) and is lying on the surface of the water in a beaker. If A is the surface tension of the water, what is the maximum weight that the thread can have without sinking?

**A.** $Ad(l + w) / \pi$
**B.** $A(l + w)$
**C.** $4A(l + w)$
**D.** $A(l + w) / 2$

**23.** What is the difference between the pressure inside and outside a tire called?

**A.** Absolute pressure
**B.** Fluid pressure
**C.** Atmospheric pressure
**D.** Gauge pressure

**24.** Which of the following is NOT a unit of pressure?

**A.** atm
**B.** $N \cdot m^2$
**C.** inches of mercury
**D.** Pascal

**25.** Which of the following is a dimensionless number?

I. Reynolds number
II. specific gravity
III. shear stress

**A.** I only
**B.** II only
**C.** III only
**D.** I and II only

 

**26.** Water flows out of a large reservoir through a 5 cm diameter pipe. The pipe connects to a 3 cm diameter pipe that is open to the atmosphere, as shown.

4.0 m

What is the speed of the water in the 5 cm pipe? Treat the water as an ideal incompressible fluid. (Use the acceleration due to gravity $g = 9.8$ m/s$^2$)

**A.** 2.6 m/s
**B.** 3.2 m/s
**C.** 4.8 m/s
**D.** 8.9 m/s

**27.** If the pressure acting on an ideal gas at constant temperature is tripled, what is the resulting volume of the ideal gas?

**A.** Increased by a factor of two
**B.** Remains the same
**C.** Reduced to one-third
**D.** Increased by a factor of three

**28.** A circular plate with an area of 1 m$^2$ covers a drain hole at the bottom of a tank of water 1 m deep. How much force is required to lift the cover if it weighs 1,500 N? (Use the acceleration due to gravity $g = 10$ m/s$^2$)

**A.** 4,250 N
**B.** 9,550 N
**C.** 16,000 N
**D.** 11,500 N

**29.** A bowling ball that weighs 80 N is dropped into a swimming pool filled with water. If the buoyant force on the bowling ball is 20 N when the ball is 1 m below the surface (and sinking), what is the normal force exerted by the bottom of the pool on the ball when it comes to rest 4 m below the surface?

**A.** 0 N
**B.** 60 N
**C.** 50 N
**D.** 70 N

**30.** A block of unknown material is floating in a fluid, half-submerged. If the specific gravity of the fluid is 1.6, what is the block's density? (Use the specific gravity = $\rho_{fluid} / \rho_{water}$ and density of water $\rho = 1{,}000$ kg/m$^3$)

**A.** 350 kg/m$^3$
**B.** 800 kg/m$^3$
**C.** 900 kg/m$^3$
**D.** 1,250 kg/m$^3$

**31.** A solid sphere of mass 9.2 kg, made of metal whose density is 3,650 kg/m$^3$, hangs by a cord. When the sphere is immersed in a liquid of unknown density, the tension in the cord is 42 N. What is the density of the liquid? (Use acceleration due to gravity $g$ = 9.8 m/s$^2$)

**A.** 1,612 kg/m$^3$
**B.** 1,468 kg/m$^3$
**C.** 1,950 kg/m$^3$
**D.** 1,742 kg/m$^3$

**32.** In a closed container of fluid, object A is submerged at 6 m from the bottom, and object B is submerged at 12 m from the bottom. Compared to object A, object B experiences:

**A.** less fluid pressure
**B.** double the fluid pressure
**C.** equal fluid pressure
**D.** triple the fluid pressure

**33.** Ideal, incompressible water flows at 14 m/s in a horizontal pipe with a 3.5 × 104 Pa pressure. If the pipe widens to twice its original radius, what is the pressure in the wider section? (Use density of water $\rho$ = 1,000 kg/m$^3$)

**A.** $7.6 \times 10^4$ Pa
**B.** $12.7 \times 10^4$ Pa
**C.** $2 \times 10^5$ Pa
**D.** $11.1 \times 10^3$ Pa

**34.** Determine the speed at which water exits a tank through a tiny hole in the bottom of the tank that is 20 cm in diameter and filled with water to a height of 50 cm. (Use the acceleration due to gravity $g$ = 9.8 m/s$^2$)

**A.** 3.1 m/s
**B.** 17.8 m/s
**C.** 31.2 m/s
**D.** 21.6 m/s

**35.** An 80 kg man would weigh 784 N if there were no atmosphere. By how much does the buoyancy due to air reduces the man's weight? (Use the density of the man = 1 g/cm$^3$, the density of the air = $1.2 \times 10^{-3}$ g/cm$^3$, $m$ = 80 kg and the acceleration due to gravity $g$ = 9.8 m/s$^2$)

**A.** 0.58 N
**B.** 0.32 N
**C.** 0.94 N
**D.** 2.8 N

**36.** Diffusion is described by which law?

**A.** Dulong's
**B.** Faraday's
**C.** Kepler's
**D.** Graham's

**37.** An object has a volume of 4.2 $m^3$ and weighs 41,800 N. What is its apparent weight in water? (Use acceleration due to gravity $g = 9.8$ m/s$^2$ and density of water $\rho = 1{,}000$ kg/m$^3$)

**A.** 1,140 N
**B.** 230 N
**C.** 800 N
**D.** 640 N

**38.** A pump uses a piston 12 cm in diameter that moves 3 cm/s. What is the fluid velocity in a tube that is 2 mm in diameter?

**A.** 218 cm/s
**B.** 88 cm/s
**C.** 136 cm/s
**D.** 108 m/s

**39.** What is the specific gravity of an object floating with one-tenth of its volume out of the water?

**A.** 0.3
**B.** 0.9
**C.** 1.3
**D.** 2.1

**40.** If each of the factors listed below was changed by 15%, which would have the greatest effect on the flow rate?

**A.** Fluid density
**B.** Pressure difference
**C.** Radius of the pipe
**D.** Fluid viscosity

## Practice Set 3: Questions 41–60

**41.** A 680 g steel hammer ($m_h$) is tied to a string that is hung from a force meter. A 5 kg container of water ($m_w$) sits on a scale. The hammer is lowered completely into the water but above the bottom. What does the force meter read? (Use the density of steel $\rho = 7.9$ g/cm$^3$, density of water $\rho = 1$ g/cm$^3$ and acceleration due to gravity $g = 10$ m/s$^2$)

**A.** 5.9 N
**B.** 8.4 N
**C.** 10.7 N
**D.** 5.2 N

**42.** An external pressure applied to an enclosed fluid that is transmitted unchanged to every point within the fluid is:

**A.** Torricelli's law
**B.** Bernoulli's principle
**C.** Archimedes' principle
**D.** Pascal's principle

**43.** A pipe with a circular cross-section has water flowing within it from point I to point II. The pipe radius at point I is 12 cm, while the radius at point II is 6 cm. If at the end of point I the flow rate is 0.09 m$^3$/s, what is the flow rate at the end of point II?

**A.** 0.6 m$^3$/s
**B.** 0.09 m$^3$/s
**C.** 0.15 m$^3$/s
**D.** 1.2 m$^3$/s

**44.** Consider a brick totally immersed in water, with the long edge of the brick vertical. Which statement describes the pressure on the brick?

**A.** Greatest on the sides of the brick
**B.** Greatest on the top of the brick
**C.** Smallest on the sides with largest area
**D.** Greatest on the bottom of the brick

**45.** Water is flowing in a drainage channel of a rectangular cross-section. The width of the channel is 14 m, the depth of water is 7 m, and the flow speed is 3 m/s. What is the mass flow rate of the water? (Use the density of water $\rho = 1{,}000$ kg/m$^3$)

**A.** $2.9 \times 10^5$ kg/s
**B.** $4.8 \times 10^4$ kg/s
**C.** $6.2 \times 10^5$ kg/s
**D.** $9.3 \times 10^4$ kg/s

 

**46.** Which answer best describes what happens to a spherical lead ball with a density of 11.3 $g/cm^3$ when placed in a tub of mercury with a density of 13.6 $g/cm^3$?

**A.** It sinks slowly to the bottom of the mercury
**B.** It floats with about 17% of its volume above the surface of the mercury
**C.** It floats with its top exactly even with the surface of the mercury
**D.** It floats with about 83% of its volume above the surface of the mercury

**47.** What is the pressure 6 m below the surface of the ocean? (Use density of water $\rho = 10^3$ $kg/m^3$, atmospheric pressure $P_{atm} = 1.01 \times 10^5$ Pa and acceleration due to gravity $g = 10$ $m/s^2$)

**A.** $1.6 \times 10^5$ Pa
**B.** $0.8 \times 10^5$ Pa
**C.** $2.7 \times 10^4$ Pa
**D.** $3.3 \times 10^4$ Pa

**48.** Density is:

**A.** inversely proportional to mass and volume
**B.** proportional to mass and inversely proportional to the volume
**C.** inversely proportional to mass and proportional to the volume
**D.** proportional to mass and volume

**49.** Which of the following would be expected to have the smallest bulk modulus?

**A.** Solid plutonium
**B.** Liquid water
**C.** Helium vapor
**D.** Liquid mercury

**50.** A 14,000 N car is raised using a hydraulic lift. The lift consists of a U-tube with arms of unequal areas, initially at the same level. The lift is filled with oil with a density of 750 $kg/m^3$ with tight-fitting pistons at each end. The narrower arm has a radius of 6 cm, while the wider arm of the U-tube has a radius of 16 cm. The car rests on the piston on the wider arm of the U-tube. What is the force that must be applied to the smaller piston to lift the car after it has been raised 1.5 m? (Ignore the weight of the pistons and use the acceleration due to gravity $g = 9.8$ $m/s^2$)

**A.** 4,568 N
**B.** 3,832 N
**C.** 2,094 N
**D.** 1,379 N

**51.** A polar bear of mass 240 kg stands on a floating ice 100 cm thick. What is the minimum area of the ice that will just support the bear? (Use the specific gravity of ice = 0.98 and the specific gravity of saltwater = 1.03)

**A.** 2.6 $m^2$
**B.** 4.9 $m^2$
**C.** 4.8 $m^2$
**D.** 11.2 $m^2$

**52.** If atmospheric pressure increases by an amount ΔP, which of the following statements about the pressure in a large pond is true?

**A.** The gauge pressure increases by ΔP
**B.** The absolute pressure increases by ΔP
**C.** The absolute pressure increases, but by an amount less than ΔP
**D.** The absolute pressure does not change

**53.** A cubical box with 25 cm sides is immersed in a fluid. The pressure at the top surface of the box is 108 kPa, and the pressure on the bottom surface is 114 kPa. What is the density of the fluid? (Use the acceleration due to gravity $g = 9.8\ m/s^2$)

**A.** 980 $kg/m^3$
**B.** 1,736 $kg/m^3$
**C.** 2,452 $kg/m^3$
**D.** 2,794 $kg/m^3$

**54.** Fluid flows through a 19 cm long tube with a radius of 2.1 mm at an average speed of 1.8 m/s. What is the viscosity of the fluid if the pressure drop is 970 Pa?

**A.** 0.036 $N{\cdot}s/m^2$
**B.** 0.013 $N{\cdot}s/m^2$
**C.** 0.0044 $N{\cdot}s/m^2$
**D.** 0.0016 $N{\cdot}s/m^2$

**55.** The pressure differential across the cross-section of a condor's wing due to the difference in airflow is explained by:

**A.** Torricelli's law
**B.** Poiseuille's law
**C.** Bernoulli's equation
**D.** Newton's First Law

**56.** An air bubble underwater has the same pressure as the water. As the air bubble rises toward the surface (with its temperature remaining constant), the volume of the air bubble:

**A.** increases
**B.** decreases
**C.** remains constant
**D.** depends on the rate it rises

 

**57.** At a depth of about 1,060 m in the ocean, the pressure has increased by 110 atmospheres (about $10^7$ N/m$^2$). By how much has 1 m$^3$ of water been compressed by this pressure? (Use the bulk modulus B of water = $2.3 \times 10^9$ N/m$^2$)

**A.** $2.7 \times 10^{-3}$ m$^3$
**B.** $4.3 \times 10^{-3}$ m$^3$
**C.** $5.2 \times 10^{-3}$ m$^3$
**D.** $7.6 \times 10^{-2}$ m$^3$

**58.** The hydraulic lift is a practical application of:

**A.** Huygens' principle
**B.** Pascal's principle
**C.** Fermat's principle
**D.** Kepler's law

**59.** What is the gauge pressure in the water at the deepest point of the Pacific Ocean, which is 11,030 m? (Use the density of seawater $\rho$ = 1,025 kg/m$^3$ and the acceleration due to gravity $g$ = 9.8 m/s$^2$)

**A.** $1.1 \times 10^8$ Pa
**B.** $3.1 \times 10^8$ Pa
**C.** $4.2 \times 10^7$ Pa
**D.** $7.6 \times 10^7$ Pa

**60.** A man is breathing through a snorkel while swimming in the ocean. When his chest is about 1 meter underwater, he has a difficult time breathing. What is the net pressure that his lungs must expand against for him to breathe? (Use the atmospheric pressure $P_{atm} = 1.01 \times 10^5$ Pa, the density of water $\rho = 10^3$ kg/m$^3$, the density of air $\rho$ = 1.2 kg/m$^3$ and the acceleration due to gravity g = 9.8 m/s$^2$)

**A.** $3.2 \times 10^5$ Pa
**B.** $1.1 \times 10^5$ Pa
**C.** $4.1 \times 10^5$ Pa
**D.** $1 \times 10^4$ Pa

---

**Practice Set 4: Questions 61–80**

---

**61.** What is the wall thickness of a hollow steel ball of diameter 3 m that barely floats in water? (Use the density of steel $\rho_{steel} = 7.87$ g/cm$^3$ and the density of water $\rho_{water} = 10^3$ kg/m$^3$)

**A.** 3 cm
**B.** 18 cm
**C.** 7 cm
**D.** 26 cm

**62.** The Bernoulli effect describes the lift force on an airplane wing. Wings must be designed to ensure that air molecules:

**A.** move more rapidly past the upper surface of the wing than past the lower surface
**B.** flow around wings that are smooth enough for an easy flow of the air
**C.** are deflected upward when they hit the wing
**D.** are deflected downward when they hit the wing

**63.** Two horizontal pipes (A and B) are the same length, but pipe B has twice the diameter of pipe A. Water undergoes viscous flow in both pipes, subject to the same pressure difference across the lengths of the pipes. If the flow rate in pipe A is Q, what is the flow rate in pipe B?

**A.** 2Q
**B.** 4Q
**C.** 8Q
**D.** 16Q

Questions **64-66** are based on the following:

A pressurized cylindrical tank is 5 m in diameter. Water exits from the pipe at point C with a velocity of 13 m/s. Point A is 10 m above point B, and point C is 3 m above point B. The cross-sectional area of the pipe at point B is 0.08 m$^2$, and the pipe narrows to a cross-sectional area of 0.04 m$^2$ at point C. Assume an ideal fluid in laminar flow. The density of water is 1,000 kg/m$^3$. (Use the acceleration due to gravity $g = 9.8$ m/s$^2$)

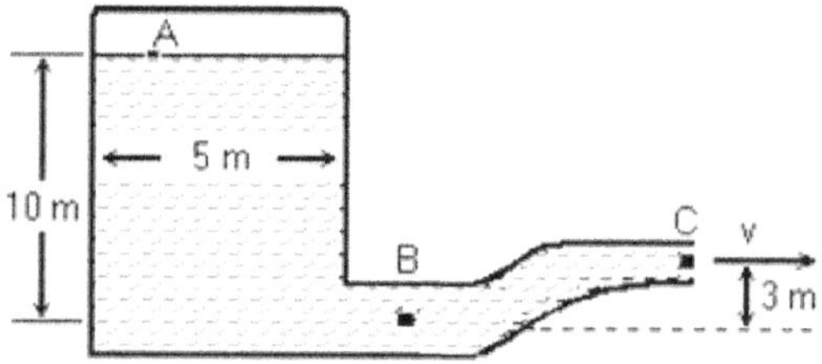

**64.** What is the mass flow rate in the pipe at point C?

**A.** 320 kg/s
**B.** 520 kg/s
**C.** 570 kg/s
**D.** 610 kg/s

**65.** What is the rate at which the water is falling in the tank?

**A.** 12 mm/s
**B.** 15 mm/s
**C.** 26 mm/s
**D.** 44 mm/s

**66.** What is the gauge pressure in the pipe at point B?

**A.** 82 kPa
**B.** 167 kPa
**C.** 71 kPa
**D.** 98 kPa

**67.** A closed cubical chamber resting on the floor contains oil and a piston. If the piston is pushed down hard enough to increase the pressure just below the piston by an amount $\Delta P$, which of the following statements is correct?

**A.** The pressure at the top of the oil increases by less than $\Delta P$
**B.** The increase in the force on the top of the chamber equals the increase in the force on the bottom of the chamber
**C.** The pressure in the oil increases by less than $\Delta P$
**D.** The pressure at the bottom of the oil increases by more than $\Delta P$

**68.** The tires support the weight of a stationary car. If one tire has a slow leak, the air pressure within the tire [ ], the surface area between the tire and the road [ ] and the net force the tire exerts on the road [ ]. (Assume the car is nearly level and remains nearly level, and that initially, each tire supports the same force)

**A.** decreases … increases … remains constant
**B.** increases … increases … increases
**C.** decreases … increases … increases
**D.** decreases … increases … decreases

**69.** Jack is breathing through a snorkel as he swims in the Caribbean Sea. He has trouble breathing when his chest is submerged about 1 meter under water. Which expression gives the force that his muscles must exert to expand his chest?

**A.** (atmospheric pressure) × (area of his chest)
**B.** (atmospheric pressure) × (area of snorkel hole + area of his chest)
**C.** (gauge pressure of the water) × (area of his chest)
**D.** (gauge pressure of the water) × (area of snorkel hole)

**70.** What is the mass of a cylindrical rod with a length of 14 cm and a diameter of 2 cm that just barely floats in water? (Use the density of water $\rho$ = 1,000 kg/m$^3$)

**A.** 44 g
**B.** 70 g
**C.** 140 g
**D.** 28 g

**71.** A silver necklace with a mass of 60 grams and a volume of 5.7 cm3 is lowered into a container of water and tied to a string connected to a force meter. What is the reading on the force meter? (Use the density of water = 1 g/cm$^3$ and acceleration due to gravity $g$ = 9.8 m/s$^2$)

**A.** 0.53 N
**B.** 0.22 N
**C.** 0.62 N
**D.** 0.38 N

**72.** A water tank is filled to a depth of 6 m, and the bottom of the tank is 22 m above ground. A water-filled hose that is 2 cm in diameter extends from the bottom of the tank to the ground, but no water is flowing in the hose. What is the gauge water pressure at ground level in the hose? (Use the density of water $\rho$ = 1,000 kg/m$^3$ and acceleration due to gravity $g$ = 9.8 m/s$^2$)

**A.** $2.7 \times 10^5$ N/m$^2$
**B.** $5.3 \times 10^4$ N/m$^2$
**C.** 8.7 N/m$^2$
**D.** Requires the cross-sectional area of the tank

**73.** When a dam began to leak, Mike placed his finger in the hole to stop the flow. The dam is 20 m high and 100 km long and sits on top of a lake, another 980 m deep, 100 km wide, and 100 km long. The hole that Mike blocked is a square 0.01 m by 0.01 m located 1 m below the surface of the water. Assuming that the viscosity of the water is negligible, what force does Mike have to exert to prevent water from leaking? (Use atmospheric pressure $P_{atm} = 10^5$ Pa, the density of water $\rho = 10^3$ kg/m³ and the acceleration due to gravity $g = 10$ m/s²)

**A.** 10 N
**B.** 1 N
**C.** 1,000 N
**D.** 0.1 N

**74.** What is the density of an object if it weighs 7.86 N when it is in air and 6.92 N when it is immersed in water? (Use the acceleration due to gravity $g = 9.8$ m/s² and the density of water $\rho = 1{,}000$ kg/m³)

**A.** 6,042 kg/m³
**B.** 7,286 kg/m³
**C.** 8,333 kg/m³
**D.** 9,240 kg/m³

**75.** As a cubical block of marble is lowered at a steady rate into the ocean by a crane, the top and bottom faces are kept horizontal. Which graph depicts the total pressure (P) on the bottom of the block as a function of time ($t$) as the block just enters the water at $t = 0$ s?

**A.**
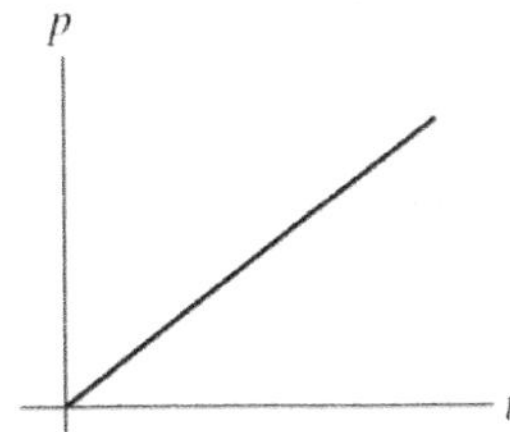


**C.**
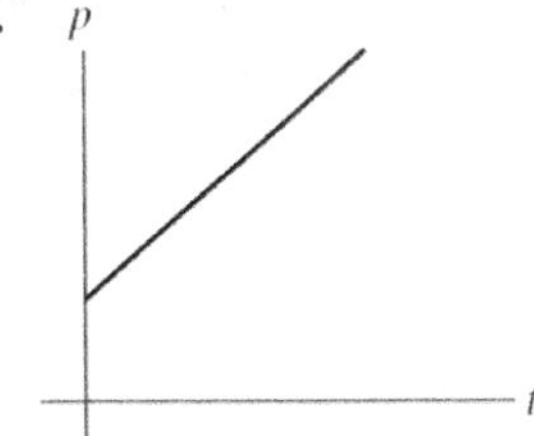


**B.**
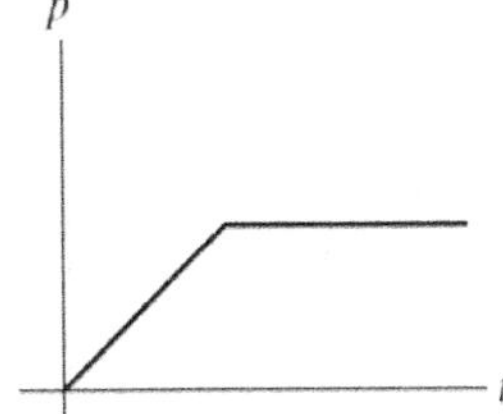


**D.**
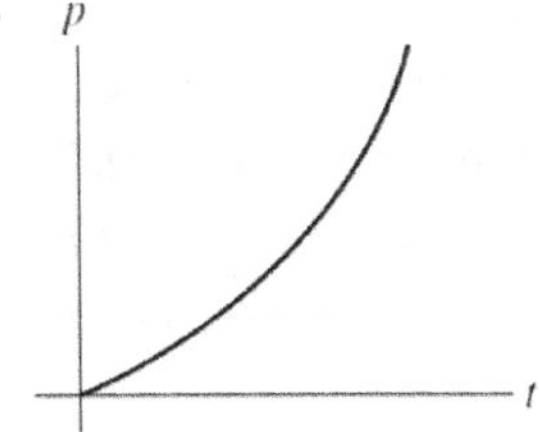

**76.** A spherical inflated balloon is submerged in a swimming pool. How is the buoyant force affected if the balloon is inflated to double its radius?

**A.** 8 times larger
**B.** 2 times larger
**C.** 4 times larger
**D.** 2 times smaller

**77.** A 4.2 m steel wire has a diameter of 1.8 mm. The wire stretches 1.8 mm when it bears a load. What is the mass of the load? (Use Young's modulus for steel = $2 \times 10^{11}$ N/m$^2$ and acceleration due to gravity $g = 9.8$ m/s$^2$)

**A.** 22 kg
**B.** 26 kg
**C.** 30 kg
**D.** 16 kg

**78.** A plastic container in the shape of a cube with 0.2 m sides is suspended in a vacuum. The container is filled to 3 atm with 10 g of $N_2$ gas. What is the force that the $N_2$ gas exerts on one face of the cube? (Use the ideal gas constant $R = 0.0821$ L atm/K mol and 1 atm = $1.01 \times 10^5$ Pa)

**A.** $2.1 \times 10^3$ N
**B.** $5.6 \times 10^4$ N
**C.** $1.2 \times 10^4$ N
**D.** $4.2 \times 10^4$ N

**79.** If the amount of fluid flowing through a tube remains constant, by what factor does the speed of the fluid change when the radius of the tube decreases from 16 cm to 4 cm?

**A.** Increases by $\sqrt{2}$
**B.** Increases by 16
**C.** Decreases by 16
**D.** Decreases by 4

**80.** Ice has a lower density than water because ice:

**A.** molecules vibrate at lower rates than water molecules
**B.** is made of open-structured, hexagonal crystals
**C.** is denser and therefore sinks when in liquid water
**D.** molecules are more compact in the solid state

 

*Notes for active learning*

*Notes for active learning*

 

# 5 – Chemical Bonding

## Practice Set 1: Questions 1–20

**1.** What is the number of valence electrons in tin (Sn)?

**A.** 14 **B.** 8 **C.** 2 **D.** 4

**2.** Unhybridized *p* orbitals participate in π bonds as double and triple bonds. How many distinct and degenerate *p* orbitals exist in the second electron shell, where n = 2?

**A.** 3 **B.** 2 **C.** 1 **D.** 0

**3.** What is the number of valence electrons in a sulfite ion, $SO_3^{2-}$?

**A.** 22 **B.** 24 **C.** 26 **D.** 34

**4.** Which type of attractive force occurs in molecules regardless of the atoms they possess?

**A.** Dipole–ion interactions
**B.** Ion–ion interactions
**C.** Dipole–dipole attractions
**D.** London dispersion forces

**5.** Given the structure of glucose, which statement explains the hydrogen bonding between glucose and water?

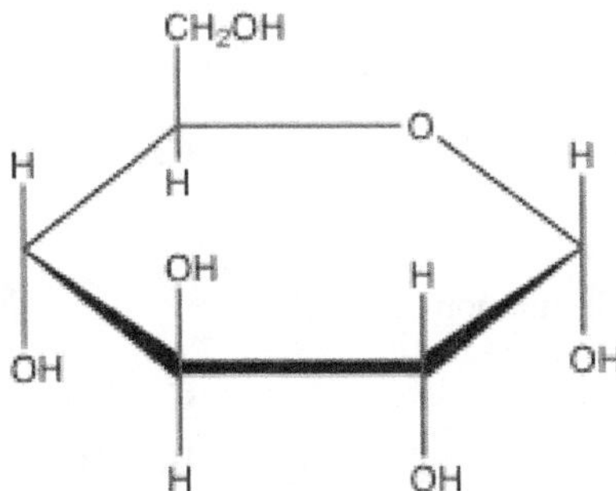


**A.** Due to the cyclic structure of glucose, there is no H–bonding with water
**B.** Each glucose molecule could H–bond with up to 17 water molecules
**C.** H–bonds form, with water being the H–bond donor
**D.** H–bonds form, with glucose being the H–bond donor

**6.** How many valence electrons are in the Lewis dot structure of $C_2H_6$?

**A.** 2 **B.** 14 **C.** 6 **D.** 8

**7.** The term for a bond where the electrons are shared unequally is:

**A.** polar covalent
**B.** nonpolar covalent
**C.** coordinate covalent
**D.** nonpolar ionic

**8.** What is the name for the weak forces of attraction between nonpolar molecules due to temporary dipoles between adjacent nonpolar molecules?

**A.** van der Waals forces
**B.** hydrophobic forces
**C.** hydrogen bonding forces
**D.** nonpolar covalent forces

**9.** Nitrogen has five valence electrons; which of the following bonding is/are possible?

I. one single and one double bond
II. three single bonds
III. one triple bond

**A.** I only
**B.** II only
**C.** I and III only
**D.** I, II and III

**10.** What is the number of valence electrons in antimony (Sb)?

**A.** 1 **B.** 2 **C.** 5 **D.** 4

**11.** How many resonance structures, if any, can be drawn for a nitrite ion?

**A.** 1 **B.** 2 **C.** 3 **D.** 4

**12.** What is the formula for the ammonium ion?

**A.** $NH_4^-$
**B.** $N_4H^+$
**C.** $NH_4^{2-}$
**D.** $NH_4^+$

**13.** What type of bond forms between oppositely charged ions?

**A.** dipole
**B.** covalent
**C.** ionic
**D.** induced dipole

**14.** Which elements are arranged in order of increasing electronegativity?

**A.** Fr, Mg, Si, O
**B.** Br, Cl, S, P
**C.** F, B, O, Li
**D.** Cl, S, Se, Te

**15.** The attraction due to London dispersion forces between molecules depends on what two factors?

**A.** Volatility and shape
**B.** Molar mass and volatility
**C.** Vapor pressure and size
**D.** Molar mass and shape

**16.** Which of the substances below would have the largest dipole?

**A.** $CO_2$
**B.** $SO_2$
**C.** $H_2O$
**D.** $CCl_4$

**17.** What is the name for the attraction between $H_2O$ molecules?

**A.** adhesion
**B.** polarity
**C.** cohesion
**D.** van der Waals

**18.** Which compound contains only covalent bonds?

**A.** $HC_2H_3O_2$
**B.** NaCl
**C.** $NH_4OH$
**D.** $Ca_3(PO_4)_2$

**19.** The distance between two atomic nuclei in a chemical bond is determined by the:

**A.** size of the valence electrons
**B.** size of the nucleus
**C.** size of the protons
**D.** balance between the repulsion of the nuclei and the attraction of the nuclei for the bonding electrons

**20.** Carbonic acid has the chemical formula of $H_2CO_3$. The carbonate ion has the molecular formula of $CO_3^{2-}$. From the Lewis structure for the $CO_3^{2-}$ ion, what is the number of reasonable resonance structures for the anion?

**A.** original structure only
**B.** 2
**C.** 3
**D.** 4

---

**Practice Set 2: Questions 21–40**

---

**21.** Which of the following molecules would contain a dipole?

**A.** F–F
**B.** H–H
**C.** Cl–Cl
**D.** H–F

**22.** Which is the correct formula for the ionic compound formed between Ca and I?

**A.** $Ca_3I_2$
**B.** $Ca_2I_3$
**C.** $CaI_2$
**D.** $Ca_2I$

**23.** Which bonding is NOT possible for a carbon atom with four valence electrons?

**A.** 1 single and 1 triple bond
**B.** 1 double and 1 triple bond
**C.** 4 single bonds
**D.** 2 single and 1 double bond

**24.** During strenuous exercise, why does perspiration on a person's skin form droplets?

**A.** Ability of $H_2O$ to dissipate heat
**B.** High specific heat of $H_2O$
**C.** Adhesive properties of $H_2O$
**D.** Cohesive properties of $H_2O$

**25.** Why does water dissolve an ionic compound if an ionic bond is stronger than a dipole-dipole interaction?

**A.** Ions do not overcome their interatomic attraction and, therefore, are not soluble
**B.** Ion–dipole interaction causes the ions to heat up and vibrate free of the crystal
**C.** Ionic bond is weakened by the ion–dipole interactions, and ionic repulsion ejects the ions from the crystal
**D.** Ion–dipole interactions of several water molecules aggregate with the ionic bond and dissociate it into the solution

**26.** Which one of these molecules can act as a hydrogen bond acceptor but not a donor?

**A.** $CH_3NH_2$
**B.** $CH_3–O–CH_3$
**C.** $H_2O$
**D.** $C_2H_5OH$

 

**27.** When NaCl dissolves in water, what is the force of attraction between $Na^+$ and $H_2O$?

**A.** ion–dipole
**B.** hydrogen bonding
**C.** ion–ion
**D.** dipole–dipole

**28.** Based on the Lewis structure, how many polar and nonpolar bonds are in $H_2CO$?

**A.** 3 polar bonds and 0 nonpolar bonds
**B.** 2 polar bonds and 1 nonpolar bond
**C.** 1 polar bond and 2 nonpolar bonds
**D.** 0 polar bonds and 3 nonpolar bonds

**29.** How many more electrons can fit within the valence shell of a hydrogen atom?

**A.** 1 **B.** 2 **C.** 7 **D.** 0

**30.** Which element likely forms a cation with a +2 charge?

**A.** Na **B.** S **C.** Si **D.** Mg

**31.** Which statement accurately describes the structure of the ionic compound NaCl?

**A.** Alternating rows of $Na^+$ and $Cl^-$ ions are present
**B.** Each ion present is surrounded by six ions of opposite charge
**C.** Alternating layers of Na and Cl atoms are present
**D.** Alternating layers of $Na^+$ and $Cl^-$ ions are present

**32.** What is the force holding two atoms in a chemical bond?

**A.** gravitational force
**B.** strong nuclear force
**C.** electrostatic force
**D.** weak nuclear force

**33.** Which of the following diatomic molecules contains the bond of greatest polarity?

**A.** $CH_4$ **B.** Te–F **C.** Cl–F **D.** $P_4$

**34.** Why does $H_2O$ have an unusually high boiling point compared to $H_2S$?

**A.** Hydrogen bonding
**B.** Van der Waals forces
**C.** $H_2O$ molecules pack more closely than $H_2S$
**D.** Covalent bonds are stronger in $H_2O$

**35.** What is the shape of a molecule in which the central atom has 2 bonding electron pairs and 2 nonbonding electron pairs?

**A.** trigonal planar
**B.** trigonal pyramidal
**C.** linear
**D.** bent

**36.** Which of the following represents the breaking of a noncovalent interaction?

**A.** Ionization of water
**B.** Decomposition of hydrogen peroxide
**C.** Hydrolysis of an ester
**D.** Dissolving salt crystals

**37.** Which pair of elements is most likely to form an ionic compound when reacted?

**A.** C and Cl
**B.** K and I
**C.** Ga and Si
**D.** Fe and Mg

**38.** Which of the following statements concerning coordinate covalent bonds is correct?

**A.** Once formed, they are indistinguishable from other covalent bonds
**B.** They are always single bonds
**C.** One of the atoms involved must be a metal and the other a nonmetal
**D.** Both atoms involved in the bond contribute an equal number of electrons to the bond

**39.** The greatest dipole moment within a bond is when:

**A.** both bonding elements have a low electronegativity
**B.** one bonding element has a high electronegativity, and the other has a low electronegativity
**C.** both bonding elements have a high electronegativity
**D.** one bonding element has a high electronegativity, and the other has a moderate electronegativity

**40.** What must occur for an atom to obtain the noble gas configuration?

**A.** lose, gain, or share an electron
**B.** loss or gain an electron
**C.** lose an electron
**D.** share an electron

## Practice Set 3: Questions 41–60

**41.** Based on the Lewis structure for hydrogen peroxide, $H_2O_2$, how many polar and nonpolar bonds are present?

**A.** 3 polar and 0 nonpolar bonds
**B.** 2 polar and 2 nonpolar bonds
**C.** 2 polar and 1 nonpolar bond
**D.** 0 polar and 3 nonpolar bonds

**42.** To form an octet, an atom of selenium must:

**A.** gain 2 electrons
**B.** lose 2 electrons
**C.** gain 6 electrons
**D.** lose 6 electrons

**43.** Which of the following is an example of a chemical reaction?

**A.** two solids mix to form a heterogeneous mixture
**B.** two liquids mix to form a homogeneous mixture
**C.** one or more new compounds are formed by rearranging atoms
**D.** a new element is formed by rearranging nucleons

**44.** Which compound is NOT correctly matched with the predominant intermolecular force associated with that compound in the liquid state?

| Compound | Intermolecular force |
|---|---|
| **A.** $CH_3OH$ | hydrogen bonding |
| **B.** HF | hydrogen bonding |
| **C.** $Cl_2O$ | dipole–dipole interactions |
| **D.** HBr | van der Waals interactions |

**45.** Which element forms an ion with the greatest positive charge?

**A.** Mg **B.** Ca **C.** Al **D.** Na

**46.** What is the difference between a dipole–dipole and an ion–dipole interaction?

**A.** One interaction involves dipole attraction between neutral molecules, while the other involves dipole interactions with ions
**B.** One interaction involves ionic molecules interacting with other ionic molecules while the other involves polar molecules
**C.** One interaction involves salts and water, while the other does not involve water
**D.** One interaction involves hydrogen bonding, while the others do not

**47.** Which is a true statement about $H_2O$ as it begins to freeze?

**A.** Hydrogen bonds break
**B.** The number of hydrogen bonds increases
**C.** Covalent bond strength increases
**D.** Molecules move closer

**48.** In the process of forming sodium nitride ($Na_3N$) from its elements, what happens to the electrons of each sodium and the electrons of each nitrogen, respectively?

**A.** one lost; three gained
**B.** three lost; three gained
**C.** three lost; one gained
**D.** one lost; two gained

**49.** Which species below has the least valence electrons in its Lewis symbol?

**A.** $S^{2-}$ **B.** $Ga^+$ **C.** $Ar^+$ **D.** $Mg^{2+}$

**50.** Which of the following occur(s) naturally as nonpolar diatomic molecules?

I. sulfur II. chlorine III. argon

**A.** I only
**B.** II only
**C.** I and III only
**D.** I and II only

**51.** In a chemical reaction, the bonds being formed are:

**A.** more energetic than the ones broken
**B.** less energetic than the ones broken
**C.** the same as the bonds broken
**D.** different from the ones broken

**52.** The charge on a sulfide ion is:

**A.** +1 **B.** +2 **C.** 0 **D.** –2

 

**53.** Which formula for an ionic compound is NOT correct?

**A.** $Al_2(CO_3)_3$
**B.** $Li_2SO_4$
**C.** $Na_2S$
**D.** $MgHCO_3$

**54.** What term best describes the smallest whole number repeating ratio of ions in an ionic compound?

**A.** lattice
**B.** unit cell
**C.** formula unit
**D.** covalent unit

**55.** In the nitrogen monoxide molecule, the dipole moment is 0.16 D, and the bond length is 115 pm. What is the sign and magnitude of the charge on the oxygen atom? (Use the conversion factor of 1 D = $3.34 \times 10^{-30}$ C·m and the charge of 1 electron = $1.602 \times 10^{-19}$ C)

**A.** $-0.098\ e$
**B.** $-0.71\ e$
**C.** $-1.3\ e$
**D.** $-0.029\ e$

**56.** From the electronegativity below, which single covalent bond is the most polar?

| Element: | H | C | N | O |
|---|---|---|---|---|
| Electronegativity | 2.1 | 2.5 | 3.0 | 3.5 |

**A.** O–C
**B.** O–N
**C.** N–C
**D.** C–H

**57.** Which of the following pairs is NOT correctly matched?

| Formula | Molecular Geometry |
|---|---|
| **A.** $CH_4$ | tetrahedral |
| **B.** $OF_2$ | bent |
| **C.** $PCl_3$ | trigonal planar |
| **D.** $Cl_2CO$ | trigonal planar |

**58.** Why are adjacent water molecules attracted to each other?

**A.** Ionic bonding between the hydrogens of $H_2O$
**B.** Covalent bonding between adjacent oxygens
**C.** Electrostatic attraction between the H of one $H_2O$ and the O of another
**D.** Covalent bonding between the H of one $H_2O$ and the O of another

**59.** What is the chemical formula for a compound that contains $K^+$ and $CO_3^{2-}$ ions?

**A.** $K(CO_3)_3$
**B.** $K_2CO_3$
**C.** $K_3(CO_3)_3$
**D.** $KCO_3$

**60.** Under what conditions is graphite converted to diamond?

**A.** low temperature, high-pressure
**B.** high temperature, low-pressure
**C.** high temperature, high-pressure
**D.** low temperature, low-pressure

---

**Practice Set 4: Questions 60–88**

---

**61.** Which of the following molecules is a Lewis acid?

**A.** $BH_3$
**B.** $NH_3$
**C.** $NH_4^+$
**D.** $CH_3COOH$

**62.** Which molecule(s) is/are most likely to show a dipole-dipole interaction?

I. H–C≡C–H
II. $CH_4$
III. $CH_3SH$
IV. $CH_3CH_2OH$

**A.** I only
**B.** II only
**C.** III only
**D.** III and IV only

**63.** C=C, C=O, C=N and N=N bonds are observed in many organic compounds. However, C=S, C=P, C=Si, and other similar bonds are not often found. What is the most probable explanation for this observation?

**A.** The comparative sizes of 3*p* atomic orbitals make an effective overlap between them less likely than between two 2*p* orbitals
**B.** S, P, and Si do not undergo hybridization of orbitals
**C.** S, P, and Si do not form π bonds due to the lack of occupied *p* orbitals in their ground state electron configurations
**D.** Carbon does not combine with elements found below the second row of the periodic table

**64.** Write the formula for the ionic compound formed from magnesium and sulfur:

**A.** $Mg_2S$
**B.** MgS
**C.** $MgS_2$
**D.** $MgS_3$

**65.** Explain why chlorine, $Cl_2$, is a gas at room temperature, while bromine, $Br_2$, is a liquid.

**A.** Bromine ions are held by ionic bonds
**B.** Chlorine molecules are smaller and, therefore, pack tighter in their physical orientation
**C.** Bromine atoms are larger, which forms a stronger induced dipole–induced dipole attraction
**D.** Chlorine atoms are larger, which forms a stronger induced dipole–induced dipole attraction

**66.** What is the major intermolecular force in $(CH_3)_2NH$?

**A.** hydrogen bonding
**B.** dipole–dipole attractions
**C.** London dispersion forces
**D.** ion–dipole attractions

**67.** Which of the following correctly describes the molecule potassium oxide and its bond?

**A.** It is a weak electrolyte with an ionic bond
**B.** It is a strong electrolyte with an ionic bond
**C.** It is a non-electrolyte with a covalent bond
**D.** It is a strong electrolyte with a covalent bond

**68.** What charge does an ion with an atomic number of 34 and 36 electrons have?

**A.** +2 **B.** –36 **C.** +34 **D.** –2

**69.** Based on the Lewis structure for $H_3C–NH_2$, the formal charge on N is:

**A.** –1 **B.** 0 **C.** +2 **D.** +1

**70.** The name of $S^{2-}$ is:

**A.** sulfite ion
**B.** sulfide ion
**C.** sulfur
**D.** sulfate ion

**71.** An ionic bond forms between two atoms when:

**A.** protons are transferred from the nucleus of the nonmetal to the nucleus of the metal
**B.** each atom acquires a negative charge
**C.** electrons are transferred from metallic to nonmetallic atoms
**D.** four electrons are shared

**72.** Which of the following pairings of ions is NOT consistent with the formula?

**A.** $Co_2S_3$ ($Co^{3+}$ and $S^{2-}$)
**B.** $K_2O$ ($K^+$ and $O^-$)
**C.** $Na_3P$ ($Na^+$ and $P^{3-}$)
**D.** $BaF_2$ ($Ba^{2+}$ and $F^-$)

**73.** Which of the following solids is likely to have the smallest exothermic lattice energy?

**A.** $Al_2O_3$ **B.** KF **C.** NaCl **D.** LiF

 

**74.** In chlorine monoxide, chlorine has a charge of +0.167 *e*. If the bond length is 154 pm, what is the molecule's dipole moment? (Use the conversion 1 meter = $1 \times 10^{-12}$ m/pm and 1 electron = $1.602 \times 10^{-19}$ C)

**A.** 2.30 D
**B.** 0.167 D
**C.** 1.24 D
**D.** 3.11 D

**75.** All the following are examples of polar molecules, EXCEPT:

**A.** $H_2O$
**B.** $CCl_4$
**C.** $CH_2Cl_2$
**D.** HF

**76.** Which is the most likely noncovalent interaction between an alcohol and a carboxylic acid at pH = 10?

**A.** formation of an anhydride bond
**B.** dipole–dipole interaction
**C.** dipole–charge interaction
**D.** charge–charge interaction

**77.** Which electron geometry is characteristic of an $sp^2$ hybridized atom?

**A.** trigonal planar
**B.** tetrahedral
**C.** trigonal bipyramidal
**D.** bent

**78.** Which of the following statements about noble gases is NOT correct?

**A.** They have very stable electron arrangements
**B.** They are the most reactive of all gases
**C.** They exist in nature as individual atoms rather than in the molecular form
**D.** They have 8 valence electrons

**79.** How are intermolecular forces and solubility related?

**A.** Solubility is a measure of how weak the intermolecular forces in the solute are
**B.** Solubility is a measure of how strong a solvent's intermolecular forces are
**C.** Solubility depends on the solute's ability to overcome the intermolecular forces in the solvent
**D.** Solubility depends on the solvent's ability to overcome the intermolecular forces in a solute

**80.** In a bond between two of the following atoms, the bonding electrons would be most strongly attracted to:

**A.** I **B.** He **C.** Cs **D.** Cl

**81.** Which choice lists the intermolecular forces in increasing order for small molecules of comparable molecular weight?

**A.** dipole–dipole forces < hydrogen bonds < London forces
**B.** London forces < hydrogen bonds < dipole–dipole forces
**C.** London forces < dipole–dipole forces < hydrogen bonds
**D.** hydrogen bonds < dipole–dipole forces < London forces

**82.** How many electrons are in the outermost electron shell of $O_2$ with an atomic number of 8?

**A.** 2 **B.** 6 **C.** 8 **D.** 16

**83.** Based on the Lewis structure, how many resonance structures are reasonable for the nitrate ion?

**A.** 1 **B.** 2 **C.** 3 **D.** 4

**84.** Which of the compounds is most likely to be ionic?

**A.** $SrBr_2$
**B.** $H_2O$
**C.** $CH_2Cl_2$
**D.** $NO_2$

**85.** The ability of an atom in a molecule to attract electrons to itself is:

**A.** ionization energy
**B.** paramagnetism
**C.** electronegativity
**D.** electron affinity

**86.** The "octet rule" relates to the number eight because:

**A.** all orbitals can hold 8 electrons
**B.** electron arrangements involving 8 valence electrons are highly stable
**C.** all atoms have 8 valence electrons
**D.** only atoms with 8 valence electrons undergo a chemical reaction

 

**87.** Which of the following molecules is polar?

**A.** $SO_3$ **B.** $SO_2$ **C.** $CO_2$ **D.** $CH_4$

**88.** Which force is intermolecular?

**A.** polar covalent bond
**B.** ionic bond
**C.** dipole–dipole interactions
**D.** nonpolar covalent bond

*Notes for active learning*

# 6 – Stoichiometry

## Practice Set 1: Questions 1–20

**1.** What is the volume of three moles of $O_2$ at STP?

**A.** 11.20 L
**B.** 22.71 L
**C.** 68.13 L
**D.** 32.00 L

**2.** In the following reaction, which of the following describes $H_2SO_4$?

$H_2SO_4 + HI \rightarrow I_2 + SO_2 + H_2O$

**A.** reducing agent and is reduced
**B.** reducing agent and is oxidized
**C.** oxidizing agent and is reduced
**D.** oxidizing agent and is oxidized

**3.** Select the balanced chemical equation for the reaction: $C_6H_{14} + O_2 \rightarrow CO_2 + H_2O$

**A.** $3\ C_6H_{14} + O_2 \rightarrow 18\ CO_2 + 22\ H_2O$
**B.** $2\ C_6H_{14} + 12\ O_2 \rightarrow 12\ CO_2 + 14\ H_2O$
**C.** $2\ C_6H_{14} + 19\ O_2 \rightarrow 12\ CO_2 + 14\ H_2O$
**D.** $2\ C_6H_{14} + 9\ O_2 \rightarrow 12\ CO_2 + 7\ H_2O$

**4.** An oxidation number is the [ ] that an atom [ ] when the electrons in each bond are assigned to the [ ] electronegative of the two atoms involved in the bond.

**A.** number of protons … definitely has … more
**B.** charge … definitely has … less
**C.** number of electrons … definitely has … more
**D.** charge … appears to have … more

**5.** What is the empirical formula of acetic acid $CH_3COOH$?

**A.** $CH_3COOH$
**B.** $C_2H_4O_2$
**C.** $CH_2O$
**D.** $CO_2H_2$

**6.** In which of the following compounds does Cl have an oxidation number of +7?

**A.** $NaClO_2$
**B.** $Al(ClO_4)_3$
**C.** $Ca(ClO_3)_2$
**D.** $LiClO_3$

**7.** What is the formula mass of a molecule of $CO_2$?

**A.** 44 amu
**B.** 52 amu
**C.** 56.5 amu
**D.** 112 amu

**8.** What is the product of heating cadmium metal and powdered sulfur?

**A.** $CdS_2$
**B.** $Cd_2S_3$
**C.** $CdS$
**D.** $Cd_2S$

**9.** After balancing the following redox reaction, what is the coefficient of NaCl?

$$Cl_2\,(g) + NaI\,(aq) \rightarrow I_2\,(s) + NaCl\,(aq)$$

**A.** 1
**B.** 2
**C.** 3
**D.** 5

**10.** Given the following spontaneous redox reaction, Which substance listed below is the strongest reducing agent?

$$FeCl_3\,(aq) + NaI\,(aq) \rightarrow I_2\,(s) + FeCl_2\,(aq) + NaCl\,(aq)$$

**A.** $FeCl_2$
**B.** $I_2$
**C.** NaI
**D.** $FeCl_3$

**11.** 14.5 moles of $N_2$ gas is mixed with 34 moles of $H_2$ gas in the following reaction:

$$N_2\,(g) + 3\,H_2\,(g) \rightarrow 2\,NH_3\,(g)$$

How many moles of $N_2$ gas remain if the reaction produces 18 moles of $NH_3$ gas when performed at 600 K?

**A.** 0.6 moles
**B.** 1.4 moles
**C.** 5.5 moles
**D.** 7.4 moles

**12.** Which equation is NOT correctly classified by the type of chemical reaction?

**A.** $AgNO_3 + NaCl \rightarrow AgCl + NaNO_3$ (double-replacement/non-redox)
**B.** $Cl_2 + F_2 \rightarrow 2\ ClF$ (synthesis/redox)
**C.** $H_2O + SO_2 \rightarrow H_2SO_3$ (synthesis/non-redox)
**D.** $CaCO_3 \rightarrow CaO + CO_2$ (decomposition/redox)

**13.** How many grams of $H_2O$ can be formed from a reaction between 10 grams of oxygen and 1 gram of hydrogen?

**A.** 11 grams of $H_2O$ are formed since mass must be conserved
**B.** 10 grams of $H_2O$ are formed since the mass of water produced cannot be greater than the amount of oxygen reacting
**C.** 9 grams of $H_2O$ are formed because oxygen and hydrogen react in an 8:1 mass ratio
**D.** No $H_2O$ is formed because there is insufficient hydrogen to react with oxygen

**14.** How many grams of $Ba^{2+}$ ions are in an aqueous solution of $BaCl_2$ that contains $6.8 \times 10^{22}$ Cl ions?

**A.** $3.2 \times 10^{48}$ g
**B.** 12 g
**C.** 14.5 g
**D.** 7.8 g

**15.** What is the oxidation number of Br in $NaBrO_3$?

**A.** –1
**B.** +1
**C.** +3
**D.** +5

**16.** When an aluminum metal reacts with ferric oxide ($Fe_2O_3$), a displacement reaction yields two products, one being metallic iron. What is the sum of the coefficients of the products of the balanced reaction?

**A.** 3
**B.** 6
**C.** 2
**D.** 5

**17.** Which of the following reactions is NOT correctly classified?

**A.** $AgNO_3\ (aq) + KOH\ (aq) \rightarrow KNO_3\ (aq) + AgOH\ (s)$ : non-redox / double-replacement
**B.** $2\ H_2O_2\ (s) \rightarrow 2\ H_2O\ (l) + O_2\ (g)$ : non-redox / decomposition
**C.** $Pb(NO_3)_2\ (aq) + 2\ Na\ (s) \rightarrow Pb\ (s) + 2\ NaNO_3\ (aq)$ : redox / single-replacement
**D.** $HNO_3\ (aq) + LiOH\ (aq) \rightarrow LiNO_3\ (aq) + H_2O\ (l)$ : non-redox / double-replacement

**18.** Calculate the number of $O_2$ molecules if a 15.0 L cylinder was filled with $O_2$ gas at STP. (Use the conversion factor of 1 mole of $O_2 = 6.02 \times 10^{23}$ $O_2$ molecules)

**A.** 443 molecules
**B.** $6.59 \times 10^{24}$ molecules
**C.** $4.03 \times 10^{23}$ molecules
**D.** $2.77 \times 10^{22}$ molecules

**19.** Which of the following is a guideline for balancing redox equations by the oxidation number method?

**A.** Verify that the number of atoms and the ionic charge is the same for reactants and products
**B.** In front of the substance reduced, place a coefficient that corresponds to the number of electrons lost by the substance oxidized
**C.** In front of the oxidized substance, place a coefficient that corresponds to the number of electrons gained by the substance-reduced
**D.** All the above

**20.** Which of the following represents the oxidation of $Co^{2+}$?

**A.** $Co \rightarrow Co^{2+} + 2\ e^-$
**B.** $Co^{3+} + e^- \rightarrow Co^{2+}$
**C.** $Co^{2+} + 2\ e^- \rightarrow Co$
**D.** $Co^{2+} \rightarrow Co^{3+} + e^-$

---

**Practice Set 2: Questions 21–40**

---

**21.** How many moles of phosphorous trichloride are required to produce 365 grams of HCl when the reaction yields 75%?

$$PCl_3\ (g) + 3\ NH_3\ (g) \rightarrow P(NH_2)_3 + 3\ HCl\ (g)$$

**A.** 1 mol
**B.** 2.5 mol
**C.** 3.5 mol
**D.** 4.5 mol

**22.** What is the oxidation number of liquid bromine in the elemental state?

**A.** 0
**B.** –1
**C.** –2
**D.** –3

**23.** Campers use propane burners for cooking. What volume of $H_2O$ is produced by complete combustion, as shown in the unbalanced equation, of 2.6 L of propane ($C_3H_8$) gas when measured at the same temperature and pressure?

$$C_3H_8\ (g) + O_2\ (g) \rightarrow CO_2\ (g) + H_2O\ (g)$$

**A.** 0.65 L
**B.** 10.4 L
**C.** 5.2 L
**D.** 2.6 L

**24.** Which substance is reduced in the following redox reaction?

$$HgCl_2\ (aq) + Sn^{2+}\ (aq) \rightarrow Sn^{4+}\ (aq) + Hg_2Cl_2\ (s) + Cl^-\ (aq)$$

**A.** $Sn^{4+}$
**B.** $Hg_2Cl_2$
**C.** $HgCl_2$
**D.** $Sn^{2+}$

**25.** What is the coefficient ($n$) of P for the balanced equation: $nP\ (s) + nO_2\ (g) \rightarrow nP_2O_5\ (s)$?

**A.** 1
**B.** 2
**C.** 4
**D.** 5

**26.** In a basic solution, which of the following are guidelines for balancing a redox equation using the half-reaction method?

I. Add the two half-reactions and cancel identical species on each side of the equation

II. Multiply each half-reaction by a whole number so that the number of electrons lost by the substance oxidized is equal to the electrons gained by the substance reduced

III. Write a balanced half-reaction for the substance oxidized and the substance reduced

**A.** II only
**B.** III only
**C.** I and III only
**D.** I, II and III

**27.** What is the molecular formula of galactose if the empirical formula is $CH_2O$ and the approximate molar mass is 180 g/mol?

**A.** $C_6H_{12}O_6$
**B.** $CH_2O_6$
**C.** $CH_2O$
**D.** CHO

**28.** How many formula units of lithium iodide (LiI) have a mass equal to 6.45 g? (Use the molecular mass of LiI = 133.85 g)

**A.** $3.45 \times 10^{23}$ formula units
**B.** $2.90 \times 10^{22}$ formula units
**C.** $1.65 \times 10^{24}$ formula units
**D.** $7.74 \times 10^{25}$ formula units

**29.** What is/are the product(s) of the reaction of $N_2$ and $O_2$ gases in a combustion engine?

I. NO II. $NO_2$ III. $N_2O$

**A.** I only
**B.** II only
**C.** III only
**D.** I, II and III

**30.** What is the coefficient ($n$) of $O_2$ gas for the balanced equation?

$$nP\ (s) + nO_2\ (g) \rightarrow nP_2O_3\ (s)$$

**A.** 1
**B.** 2
**C.** 3
**D.** 5

**31.** Given the spontaneous redox reaction, which substance is the weakest reducing agent?

$Mg\ (s) + Sn^{2+}\ (aq) \rightarrow Mg^{2+}\ (aq) + Sn\ (s)$

**A.** Sn
**B.** $Mg^{2+}$
**C.** $Sn^{2+}$
**D.** Mg

**32.** After balancing the following redox reaction in an acidic solution, what is the coefficient of $H^+$?

$Mg\ (s) + NO_3^-\ (aq) \rightarrow Mg^{2+}\ (aq) + NO_2\ (aq)$

**A.** 1
**B.** 2
**C.** 4
**D.** 6

**33.** Which substance contains the greatest number of moles in a 10 g sample?

**A.** $SiO_2$
**B.** $CH_4$
**C.** $CBr_4$
**D.** $CO_2$

**34.** Which chemistry law is illustrated when ethyl alcohol contains 52% carbon, 13% hydrogen, and 35% oxygen by mass?

**A.** law of constant composition
**B.** law of constant percentages
**C.** law of multiple proportions
**D.** law of conservation of mass

**35.** Which could NOT be true for the following reaction: N2 $(g) + 3\ H_2\ (g) \rightarrow 2\ NH_3\ (g)$?

**A.** 25 grams of $N_2$ gas react with 75 grams of $H_2$ gas to form 50 grams of $NH_3$ gas
**B.** 28 grams of $N_2$ gas react with 6 grams of $H_2$ gas to form 34 grams of $NH_3$ gas
**C.** 15 moles of $N_2$ gas react with 45 moles of $H_2$ gas to form 30 moles of $NH_3$ gas
**D.** 5 molecules of $N_2$ gas react with 15 molecules of $H_2$ gas to form 10 molecules of $NH_3$ gas

**36.** From the following reaction, if 0.2 moles of Al are allowed to react with 0.4 moles of $Fe_2O_3$, how many grams of aluminum oxide are produced?

$2\ Al + Fe_2O_3 \rightarrow 2\ Fe + Al_2O_3$

**A.** 2.8 g
**B.** 5.1 g
**C.** 10.2 g
**D.** 14.2 g

**37.** In the following compounds, the oxidation number of hydrogen is +1, EXCEPT:

**A.** $NH_3$ **B.** $HClO_2$ **C.** $H_2SO_4$ **D.** NaH

**38.** Which equation is NOT correctly classified by the type of chemical reaction?

**A.** $PbO + C \rightarrow Pb + CO$ : single-replacement/non-redox
**B.** $2\ Na + 2\ HCl \rightarrow 2\ NaCl + H_2$ : single-replacement/redox
**C.** $NaHCO_3 + HCl \rightarrow NaCl + H_2O + CO_2$ : double-replacement/non-redox
**D.** $2\ Na + H_2 \rightarrow 2\ NaH$ : synthesis/redox

**39.** What are the oxidation states of sulfur in $H_2SO_4$ and $H_2SO_3$, respectively?

**A.** +4 and +4
**B.** +6 and +4
**C.** +2 and +4
**D.** +4 and +2

**40.** A latex balloon has a volume of 500 mL filled with gas at a pressure of 780 torrs and a temperature of 320 K. How many moles of gas does the balloon contain? (Use the ideal gas constant $R = 0.08206\ L \cdot atm\ K^{-1}\ mol^{-1}$)

**A.** 0.0195
**B.** 0.822
**C.** 3.156
**D.** 18.87

 

---

**Practice Set 3: Questions 41–60**

---

**41.** What is the term for a substance that causes the reduction of another substance in a redox reaction?

**A.** oxidizing agent
**B.** reducing agent
**C.** anode
**D.** cathode

**42.** Which of the following is a method for balancing a redox equation in an acidic solution by the half-reaction method?

**A.** Multiply each half-reaction by a whole number so that the number of electrons lost by the substance oxidized is equal to the electrons gained by the substance reduced
**B.** Add the two half-reactions and cancel identical species from each side of the equation
**C.** Write a half-reaction for the substance oxidized and the substance reduced
**D.** All the above

**43.** What is the formula mass of a molecule of $C_6H_{12}O_6$?

**A.** 148 amu
**B.** 27 amu
**C.** 91 amu
**D.** 180 amu

**44.** Is it possible to have a macroscopic sample of oxygen that has a mass of 12 atomic mass units?

**A.** No, because this is less than the mass of a single oxygen atom
**B.** Yes, because it would have the same density as nitrogen
**C.** No, because this is less than a macroscopic quantity
**D.** Yes, but it would need to be made of isotopes of oxygen atoms

**45.** What is the coefficient for $O_2$ when balanced with the lowest whole number coefficients?

$$C_2H_6 + O_2 \rightarrow CO_2 + H_2O$$

**A.** 3 **B.** 4 **C.** 6 **D.** 7

 

**46.** Which element is reduced in the following redox reaction?

$BaSO_4 + 4\ C \rightarrow BaS + 4\ CO$

**A.** O in CO
**B.** S in $BaSO_4$
**C.** S in BaS
**D.** C in CO

**47.** Ethanol ($C_2H_5OH$) is blended with gasoline as a fuel additive. If the combustion of ethanol produces carbon dioxide and water, what is the coefficient of oxygen in the balanced equation?

$$\_\_C_2H_5OH\ (g) + \_\_O_2\ (g) \xrightarrow{\textit{Spark}} \_\_CO_2\ (g) + \_\_H_2O\ (g)$$

**A.** 1 **B.** 2 **C.** 3 **D.** 6

**48.** How many moles of C are in a 4.50 g sample if the atomic mass of C is 12.011 amu?

**A.** 0.375 moles
**B.** 2.67 moles
**C.** 1.00 moles
**D.** 0.54 moles

**49.** A 6.84 g hydrocarbon sample yielded 8.98 grams of carbon dioxide upon combustion analysis. The percent by mass of carbon in the hydrocarbon is:

**A.** 18.6%
**B.** 23.7%
**C.** 35.8%
**D.** 11.4%

**50.** Select the balanced chemical equation:

**A.** $2\ C_2H_5OH + 2\ Na_2Cr_2O_7 + 8\ H_2SO_4 \rightarrow 2\ HC_2H_3O_2 + 2\ Cr_2(SO_4)_3 + 4\ Na_2SO_4 + 11\ H_2O$
**B.** $3\ C_2H_5OH + 2\ Na_2Cr_2O_7 + 8\ H_2SO_4 \rightarrow 3\ HC_2H_3O_2 + 2\ Cr_2(SO_4)_3 + 2\ Na_2SO_4 + 11\ H_2O$
**C.** $C_2H_5OH + 2\ Na_2Cr_2O_7 + 8\ H_2SO_4 \rightarrow HC_2H_3O_2 + 2\ Cr_2(SO_4)_3 + 2\ Na_2SO_4 + 11\ H_2O$
**D.** $C_2H_5OH + Na_2Cr_2O_7 + 2\ H_2SO_4 \rightarrow HC_2H_3O_2 + Cr_2(SO_4)_3 + 2\ Na_2SO_4 + 11\ H_2O$

**51.** Under acidic conditions, what is the sum of the coefficients in the balanced reaction below?

$Fe^{2+} + Cr_2O_7^{2-} \rightarrow Fe^{3+} + Cr^{3+}$

**A.** 8 **B.** 14 **C.** 17 **D.** 36

 

**52.** Assuming STP, if 49 g of $H_2SO_4$ are produced in the following reaction, what volume of $O_2$ must be used in the reaction?

$$RuS\ (s) + O_2 + H_2O \rightarrow Ru_2O_3\ (s) + H_2SO_4$$

**A.** 25.2 liters
**B.** 31.2 liters
**C.** 28.3 liters
**D.** 29.1 liters

**53.** From the following reaction, if 0.20 mole of Al is allowed to react with 0.40 mole of $Fe_2O_3$, how many moles of iron are produced?

$$2\ Al + Fe_2O_3 \rightarrow 2\ Fe + Al_2O_3$$

**A.** 0.05 mole
**B.** 0.075 mole
**C.** 0.20 mole
**D.** 0.10 mole

**54.** What is the oxidation number of sulfur in the $S_2O_8^{2-}$ ion?

**A.** –1 **B.** +7 **C.** +2 **D.** +6

**55.** What are the oxidation numbers for the elements in $Na_2CrO_4$?

**A.** +2 for Na, +5 for Cr and –6 for O
**B.** +2 for Na, +3 for Cr and –2 for O
**C.** +1 for Na, +4 for Cr and –6 for O
**D.** +1 for Na, +6 for Cr and –2 for O

**56.** What is the oxidation state of sulfur in sulfuric acid?

**A.** +8 **B.** +6 **C.** –2 **D.** –6

**57.** Which substance is oxidized in the following redox reaction?

$$HgCl_2\ (aq) + Sn^{2+}\ (aq) \rightarrow Sn^{4+}\ (aq) + Hg_2Cl_2\ (s) + Cl^-\ (aq)$$

**A.** $Hg_2Cl_2$
**B.** $Sn^{4+}$
**C.** $Sn^{2+}$
**D.** $HgCl_2$

 

**58.** Which of the following statements is NOT true with respect to the balanced equation?

$$Na_2SO_4\ (aq) + BaCl_2\ (aq) \rightarrow 2\ NaCl\ (aq) + BaSO_4\ (s)$$

**A.** Barium sulfate and sodium chloride are products
**B.** Barium chloride is dissolved in water
**C.** Barium sulfate is a solid
**D.** 2 NaCl (*aq*) could be written as $Na_2Cl_2$ (*aq*)

**59.** Which of the following represents 1 mol of phosphine gas ($PH_3$)?

I. 22.71 L phosphine gas at STP
II. 34.00 g phosphine gas
III. $6.02 \times 10^{23}$ phosphine molecules

**A.** I only
**B.** II only
**C.** III only
**D.** II and III only

**60.** Which of the following represents the oxidation of $Co^{2+}$?

**A.** $Co \rightarrow Co^{2+} + 2\ e^-$
**B.** $Co^{3+} + e^- \rightarrow Co^{2+}$
**C.** $Co^{2+} + 2\ e^- \rightarrow Co$
**D.** $Co^{2+} \rightarrow Co^{3+} + e^-$

---

**Practice Set 4: Questions 61–80**

---

**61.** What is the term for a substance that causes oxidation in a redox reaction?

**A.** oxidized
**B.** oxidizing agent
**C.** anode
**D.** cathode

**62.** Carbon tetrachloride ($CCl_4$) is a potent hepatotoxin (toxic to the liver) commonly used in the past in fire extinguishers and as a refrigerant. What is the percent by mass of Cl in carbon tetrachloride?

**A.** 25% **B.** 66% **C.** 78% **D.** 92%

**63.** Which of the following is true for a redox reaction to be balanced?

I. Total ionic charge of reactants must equal the ionic charge of the products
II. Atoms of each reactant must equal atoms of the product
III. Electron gain must equal electron loss

**A.** I only
**B.** II only
**C.** I and II only
**D.** I, II and III

**64.** The mass percent of a compound is approximately 71.8% Cl, 24.2% C, and 4.0% H. If the compound's molecular weight is 99 g/mol, what is the molecular formula of the compound?

**A.** $Cl_2C_3H_6$
**B.** $Cl_2C_2H_4$
**C.** $ClCH_3$
**D.** $ClC_2H_2$

**65.** In the early 1980s, benzene was used as a solvent for waxes and oils until the EPA listed it as a carcinogen. What is the molecular formula of benzene if the empirical formula is $C_1H_1$ and the approximate molar mass is 78 g/mol?

**A.** $CH_{12}$
**B.** $C_{12}H_{12}$
**C.** $C_6H_6$
**D.** $CH_6$

**66.** How many O atoms are in the formula unit $GaO(NO_3)_2$?

**A.** 3 **B.** 4 **C.** 5 **D.** 7

**67.** Which reaction represents the balanced reaction for the combustion of ethanol?

**A.** $4\ C_2H_5OH + 13\ O_2 \rightarrow 8\ CO_2 + 10\ H_2$
**B.** $C_2H_5OH + 3\ O_2 \rightarrow 2\ CO_2 + 3\ H_2O$
**C.** $C_2H_5OH + 2\ O_2 \rightarrow 2\ CO_2 + 2\ H_2O$
**D.** $C_2H_5OH + O_2 \rightarrow CO_2 + H_2O$

**68.** What is the coefficient for $O_2$ when the following equation is balanced with the lowest whole number coefficients?

$$\_\_C_3H_7OH + \_\_O_2 \rightarrow \_\_CO_2 + \_\_H_2O$$

**A.** 3 **B.** 6 **C.** 9 **D.** 13/2

**69.** After balancing the following redox reaction, what is the coefficient of $CO_2$?

$$\_\_Co_2O_3\ (s) + \_\_CO\ (g) \rightarrow \_\_Co\ (s) + \_\_CO_2\ (g)$$

**A.** 1 **B.** 2 **C.** 3 **D.** 4

**70.** What is the term for the volume occupied by 1 mol of gas at STP?

**A.** STP volume
**B.** molar volume
**C.** standard volume
**D.** Avogadro's volume

**71.** Which substance listed below is the weakest oxidizing agent given the following spontaneous redox reaction?

$$Mg\ (s) + Sn^{2+}\ (aq) \rightarrow Mg^{2+}\ (aq) + Sn\ (s)$$

**A.** $Mg^{2+}$
**B.** Sn
**C.** Mg
**D.** $Sn^{2+}$

**72.** In the following reaction performed at 500 K, 18.0 moles of $N_2$ gas is mixed with 24.0 moles of $H_2$ gas. What is the percent yield of $NH_3$ if the reaction produces 13.5 moles of $NH_3$?

$$N_2\ (g) + 3\ H_2\ (g) \rightarrow 2\ NH_3\ (g)$$

**A.** 16%
**B.** 66%
**C.** 72%
**D.** 84%

 

**73.** How many grams of $H_2O$ can be produced from the reaction of 25.0 grams of $H_2$ and 225 grams of $O_2$?

**A.** 266 grams
**B.** 223 grams
**C.** 184 grams
**D.** 27 grams

**74.** What is the oxidation number of Cl in $LiClO_2$?

**A.** −1
**B.** +1
**C.** +3
**D.** +5

**75.** In acidic conditions, what is the sum of the coefficients in the products of the balanced reaction?

$$MnO_4^- + C_3H_7OH \rightarrow Mn^{2+} + C_2H_5COOH$$

**A.** 12
**B.** 16
**C.** 18
**D.** 20

**76.** Which reaction is NOT correctly classified?

**A.** $PbO\ (s) + C\ (s) \rightarrow Pb\ (s) + CO\ (g)$ : (double-replacement)
**B.** $CaO\ (s) + H_2O\ (l) \rightarrow Ca(OH)_2\ (aq)$ : (synthesis)
**C.** $Pb(NO_3)_2\ (aq) + 2\ LiCl\ (aq) \rightarrow 2\ LiNO_3\ (aq) + PbCl_2\ (s)$ : (double-replacement)
**D.** $Mg\ (s) + 2\ HCl\ (aq) \rightarrow MgCl_2\ (aq) + H_2\ (g)$ : (single-replacement)

**77.** The reactants for this chemical reaction are:

$$C_6H_{12}O_6 + 6\ H_2O + 6\ O_2 \rightarrow 6\ CO_2 + 12\ H_2O$$

**A.** $C_6H_{12}O_6$, $H_2O$, $O_2$ and $CO_2$
**B.** $C_6H_{12}O_6$ and $H_2O$
**C.** $C_6H_{12}O_6$
**D.** $C_6H_{12}O_6$, $H_2O$ and $O_2$

**78.** What is the charge of the electrons in 4 grams of He? (Use the Faraday constant $F$ = 96,500 C/mol)

**A.** 48,250 C
**B.** 96,500 C
**C.** 193,000 C
**D.** 386,000 C

**79.** What is the oxidation number of iron in the compound $FeBr_3$?

**A.** –2 **B.** +1 **C.** +2 **D.** +3

**80.** What is the term for the amount of substance that contains $6.02 \times 10^{23}$ particles?

**A.** molar mass

**B.** mole

**C.** Avogadro's number

**D.** formula mass

**Practice Set 5: Questions 81–100**

**81.** Which of the following species undergoes oxidation in 2 CuBr → 2 Cu + $Br_2$?

**A.** $Cu^+$
**B.** $Br^-$
**C.** Cu
**D.** CuBr

**82.** How many electrons are lost or gained by each formula unit of $CuBr_2$ in this reaction?

$$Zn + CuBr_2 \rightarrow ZnBr_2 + Cu$$

**A.** loses 1 electron
**B.** gains 6 electrons
**C.** gains 2 electrons
**D.** loses 2 electrons

**83.** How many molecules of $CO_2$ are in 168.0 grams of $CO_2$?

**A.** $3.96 \times 10^{23}$
**B.** $2.30 \times 10^{24}$
**C.** $4.24 \times 10^{22}$
**D.** $6.82 \times 10^{24}$

**84.** What is the empirical formula of a compound that, by mass, contains 64% silver, 8% nitrogen, and 28% oxygen?

**A.** $AgNO_3$
**B.** $Ag_3NO_3$
**C.** $AgNO_2$
**D.** $Ag_3NO_2$

**85.** What is the mass of one mole of a gas with a density of 1.34 g/L at STP?

**A.** 48.0 g
**B.** 56.4 g
**C.** 18.3 g
**D.** 30.1 g

**86.** Propane ($C_3H_8$) is flammable and used as a substitute for natural gas. What is the coefficient of oxygen in the balanced equation for the combustion of propane?

$$\_\_C_3H_8\,(g) + \_\_O_2\,(g) \xrightarrow{\textit{Spark}} \_\_CO_2\,(g) + \_\_H_2O\,(g)$$

**A.** 1
**B.** 7
**C.** 5
**D.** 10

**87.** How many atoms of oxygen does this reaction yield?

$C_6H_{12}O_6 + 6\ H_2O + 6\ O_2 \rightarrow 6\ CO_2 + 12\ H_2O$

**A.** 3 **B.** 12 **C.** 14 **D.** 24

**88.** After balancing the following redox reaction, what is the coefficient of $O_2$?

$\_Al_2O_3\ (s) + \_Cl_2\ (g) \rightarrow \_AlCl_3\ (aq) + \_O_2\ (g)$

**A.** 1 **B.** 2 **C.** 3 **D.** 5

**89.** Which reaction is the correctly balanced half-reaction (in acid solution) for the process below?

$Cr_2O_7^{2-}\ (aq) \rightarrow Cr^{3+}\ (aq)$

**A.** $8\ H^+ + Cr_2O_7 \rightarrow 2\ Cr^{3+} + 4\ H_2O + 3\ e^-$
**B.** $12\ H^+ + Cr_2O_7^{2-} + 3\ e^- \rightarrow 2\ Cr^{3+} + 6\ H_2O$
**C.** $14\ H^+ + Cr_2O_7^{2-} + 6\ e^- \rightarrow 2\ Cr^{3+} + 7\ H_2O$
**D.** $8\ H^+ + Cr_2O_7 + 3\ e^- \rightarrow 2\ Cr^{3+} + 4\ H_2O$

**90.** Determine the oxidation number of C in $NaHCO_3$:

**A.** +5 **B.** +4 **C.** +12 **D.** +6

**91.** What principle states that equal volumes of gases at the same temperature and pressure contain equal numbers of molecules?

**A.** Dalton's theory
**B.** Charles' theory
**C.** Boyle's theory
**D.** Avogadro's theory

**92.** What are the products for this double-replacement reaction?

$BaCl_2\ (aq) + K_2SO_4\ (aq) \rightarrow$

**A.** $BaSO_3$ and $KClO_4$
**B.** $BaSO_4$ and 2 KCl
**C.** BaS and $KClO_4$
**D.** $BaSO_3$ and KCl

**93.** The formula for mustard gas used in chemical warfare is $C_4H_8SCl_2$. What is the percentage by mass of chlorine in mustard gas? (Use the molecular mass of mustard gas = 159.09 g/mol)

**A.** 18.4%
**B.** 44.6%
**C.** 14.6%
**D.** 31.2%

**94.** Which substance listed is the weakest oxidizing agent given the following *spontaneous* redox reaction?

$$FeCl_3\ (aq) + NaI\ (aq) \rightarrow I_2\ (s) + FeCl_2\ (aq) + NaCl\ (aq)$$

**A.** $I_2$
**B.** $FeCl_2$
**C.** $FeCl_3$
**D.** NaI

**95.** If the reaction below is run at STP with excess $H_2O$, and 22.4 liters of $O_2$ react with 67 g of RuS, how many grams of $H_2SO_4$ are produced?

$$RuS\ (s) + O_2 + H_2O \rightarrow Ru_2O_3\ (s) + H_2SO_4$$

**A.** 28 g
**B.** 32 g
**C.** 44 g
**D.** 54 g

**96.** What is the oxidation number of sulfur in calcium sulfate, $CaSO_4$?

**A.** +6
**B.** +4
**C.** +2
**D.** 0

**97.** How many moles of $O_2$ gas are required for combustion with one mole of $C_6H_{12}O_6$ in the unbalanced reaction?

$$C_6H_{12}O_6\ (s) + O_2\ (g) \rightarrow CO_2\ (g) + H_2O\ (g)$$

**A.** 1
**B.** 2.5
**C.** 6
**D.** 10

**98.** Which reaction below is a *synthesis* reaction?

**A.** $3\ CuSO_4 + 2\ Al \rightarrow Al_2(SO_4)_3 + 3\ Cu$
**B.** $SO_3 + H_2O \rightarrow H_2SO_4$
**C.** $2\ NaHCO_3 \rightarrow Na_2CO_3 + CO_2 + H_2O$
**D.** $C_3H_8 + 5\ O_2 \rightarrow 3\ CO_2 + 4\ H_2O$

**99.** Which statement is true regarding the coefficients in a chemical equation?

I. They appear before the chemical formulas
II. They appear as subscripts
III. Coefficients of reactants always sum to those of the products

**A.** I only
**B.** II only
**C.** I and II only
**D.** I and III only

**100.** What volume is occupied by 17.0 g of NO gas at STP?

**A.** 23.4 L
**B.** 58.4 L
**C.** 0.58 L
**D.** 12. 7 L

 

---

**Practice Set 6: Questions 101–120**

---

**101.** Which of the following is/are likely to act as an oxidizing agent?

I. $Cl^-$ II. $Cl_2$ III. $Na^+$

**A.** I only
**B.** II only
**C.** III only
**D.** I and II only

**102.** When a substance loses electrons, it is [ ] while the substance itself acts as [ ] agent.

**A.** reduced… a reducing
**B.** oxidized… a reducing
**C.** reduced… an oxidizing
**D.** oxidized… an oxidizing

**103.** What is the compound's chemical formula expressing the number of atoms of each element in a molecule?

**A.** molecular formula
**B.** empirical formula
**C.** elemental formula
**D.** atomic formula

**104.** Which substance is the weakest reducing agent, given the spontaneous redox reaction?

$FeCl_3$ (*aq*) + NaI (*aq*) → $I_2$ (*s*) + $FeCl_2$ (*aq*) + NaCl (*aq*)

**A.** NaCl
**B.** $I_2$
**C.** NaI
**D.** $FeCl_2$

**105.** What is the mass of 3.5 moles of glucose with a molecular formula of $C_6H_{12}O_6$?

**A.** 180 g
**B.** 90 g
**C.** 630 g
**D.** 51.4 g

**106.** Butane ($C_4H_{10}$) is flammable and used in butane lighters. What is the coefficient of oxygen in the balanced equation for the combustion of butane?

*Spark*

__$C_4H_{10}$ (*g*) + __$O_2$ (*g*) → __$CO_2$ (*g*) + __$H_2O$ (*g*)

**A.** 9
**B.** 13
**C.** 18
**D.** 26

**107.** Which of the following is an accurate statement for Avogadro's number?

I. Equal to the number of atoms in 1 mole of an element
II. Equal to the number of molecules in a compound
III. Equal to approximately $6.022 \times 10^{23}$

**A.** I only
**B.** II only
**C.** III only
**D.** I and III only

**108.** What is the sum of the coefficients in the balanced reaction (no fractional coefficients)?

$$\_\_RuS\ (s) + \_\_O_2 + \_\_H_2O \rightarrow \_\_Ru_2O_3\ (s) + \_\_H_2SO_4$$

**A.** 23
**B.** 18
**C.** 13
**D.** 25

**109.** Which is sufficient for determining the molecular formula of a compound?

I. The % by mass of a compound
II. The molecular weight of a compound
III. The amount of a compound in a sample

**A.** I only
**B.** II only
**C.** I and II only
**D.** I and III only

**110.** Which equation(s) is/are balanced?

I. $Mg\ (s) + 2\ HCl\ (aq) \rightarrow MgCl_2\ (aq) + H_2\ (g)$
II. $3\ Al\ (s) + 3\ Br_2\ (l) \rightarrow Al_2Br_3\ (s)$
III. $2\ HgO\ (s) \rightarrow 2\ Hg\ (l) + O_2\ (g)$

**A.** I only
**B.** II only
**C.** I and III only
**D.** I and II only

**111.** How many moles are in 60.2 g of $MgSO_4$?

**A.** 0.25 mole
**B.** 0.5 mole
**C.** 0.45 mole
**D.** 0.65 mole

**112.** How many atoms of Mg are in a solid 72.9 g sample of magnesium? (Use the molecular mass of Mg = 24.3 g/mol)

**A.** $1.81 \times 10^{24}$ atoms
**B.** 4,800 atom
**C.** $3.01 \times 10^{23}$ atoms
**D.** $2.42 \times 10^{25}$ atoms

**113.** The formula for the illegal drug cocaine is $C_{17}H_{21}NO_4$ (303.39 g/mol). What is the percentage by mass of oxygen in cocaine?

**A.** 21.1%
**B.** 6.5%
**C.** 457%
**D.** 4.4%

**114.** If 7 moles of RuS are used in the following reaction, what is the maximum number of moles of $Ru_2O_3$ that can be produced?

$$RuS\ (s) + O_2 + H_2O \rightarrow Ru_2O_3\ (s) + H_2SO_4$$

**A.** 3.5 moles
**B.** 1.8 moles
**C.** 5.8 moles
**D.** 2.2 moles

**115.** In which sequence are sulfur-containing species arranged by *decreasing* oxidation numbers for S?

**A.** $SO_4^{2-}$, $S^{2-}$, $S_2O_3^{2-}$
**B.** $S_2O_3^{2-}$, $SO_3^{2-}$, $S^{2-}$
**C.** $SO_4^{2-}$, $S_2O_3^{2-}$, $S^{2-}$
**D.** $SO_3^{2-}$, $SO_4^{2-}$, $S^{2-}$

**116.** In all the following compounds, the oxidation number of oxygen is −2, EXCEPT:

**A.** $NaClO_2$
**B.** $Li_2O_2$
**C.** $Ba(OH)_2$
**D.** $Na_2SO_4$

**117.** Which of the following is a *double-replacement* reaction?

**A.** $2\ HI \rightarrow H_2 + I_2$
**B.** $SO_2 + H_2O \rightarrow H_2SO_3$
**C.** $HBr + KOH \rightarrow H_2O + KBr$
**D.** $CuO + H_2 \rightarrow Cu + H_2O$

**118.** Which of the following substances has the lowest density?

**A.** A mass of 750 g and a volume of 70 dL
**B.** A mass of 5 μg and a volume of 25 μL
**C.** A mass of 1.5 kg and a volume of 1.2 L
**D.** A mass of 25 g and a volume of 20 mL

**119.** How many moles of $O_2$ gas is required for combustion with 2 moles of hexane in the unbalanced reaction: $C_6H_{14}$ (*g*) + $O_2$ (*g*) → $CO_2$ (*g*) + $H_2O$ (*g*)

**A.** 11 **B.** 14 **C.** 12 **D.** 19

**120.** How is Avogadro's number related to the numbers on the periodic table?

**A.** The atomic mass listed is the mass of Avogadro's number of atoms
**B.** The periodic table provides the mass of one atom, while Avogadro's number is the number of moles
**C.** The masses are divisible by Avogadro's number, which provides the weight of one mole
**D.** The periodic table only provides atomic numbers, not atomic mass

 

---

**Practice Set 7: Questions 121–140**

---

**121.** Which of the following is/are likely to act as an oxidizing agent?

I. $Cl_2$ II. $Cl^-$ III. $Na^+$

**A.** I only
**B.** II only
**C.** III only
**D.** I and II only

**122.** In a redox reaction, the substance that is reduced:

**A.** gains electrons
**B.** loses electrons
**C.** contains an element that increases in oxidation number
**D.** is the reducing agent

**123.** What is the term for a chemical formula that expresses the simplest whole-number ratio of atoms of each element in a molecule?

**A.** empirical formula
**B.** molecular formula
**C.** atomic formula
**D.** elemental formula

**124.** How many H atoms are in the molecule $C_6H_3(C_3H_7)_2(C_2H_5)$?

**A.** 27 **B.** 15 **C.** 22 **D.** 20

**125.** Which scientific principle is the basis for balancing chemical equations?

**A.** Law of Conservation of Mass and Energy
**B.** Law of Definite Proportions
**C.** Law of Conservation of Energy
**D.** Law of Conservation of Mass

**126.** What coefficient is needed for the $O_2$ molecule to balance the following equation?

$$\_C_4H_{10}\,(g) + \_O_2\,(g) \xrightarrow{\textit{Spark}} \_CO_2\,(g) + \_H_2O\,(g)$$

**A.** 5 **B.** 1 **C.** 8 **D.** 13

**127.** Which substance is reduced in the following redox reaction?

$F_2\ (g) + 2\ Br^-\ (aq) \rightarrow 2\ F^-\ (aq) + Br_2\ (l)$

**A.** $F^-$ **B.** $Br_2$ **C.** $F_2$ **D.** $Br^-$

**128.** What is the total number of coefficients when balanced with the lowest whole number of coefficients?

$N_2H_4 + H_2O_2 \rightarrow N_2 + H_2O$

**A.** 4 **B.** 8 **C.** 10 **D.** 12

**129.** How many molecules of $CH_4$ gas have a mass equal to 6.40 g?

**A.** $3.33 \times 10^{23}$ molecules
**B.** $2.40 \times 10^{23}$ molecules
**C.** $1.15 \times 10^{24}$ molecules
**D.** $3.75 \times 10^{24}$ molecules

**130.** If 0.2 moles of Al are allowed to react with 0.4 moles of $Fe_2O_3$, what is the limiting reactant in the following reaction?

$2\ Al + Fe_2O_3 \rightarrow 2\ Fe + Al_2O_3$

**A.** $Fe_2O_3$ **B.** $Al_2O_3$ **C.** Al **D.** Fe

**131.** Balance the equation: __$H_2\ (g)$ + __$N_2\ (g)$ → __$NH_3\ (g)$

**A.** 3, 2, 2
**B.** 3, 1, 2
**C.** 2, 2, 3
**D.** 2, 2, 5

**132.** How many oxygen atoms are in 5.40 g of fructose ($C_6H_{12}O_6$)?

**A.** $2.34 \times 10^{22}$
**B.** $1.08 \times 10^{23}$
**C.** $3.34 \times 10^{24}$
**D.** $3.16 \times 10^{23}$

**133.** Which is the properly balanced equation for the reaction occurring when solid iron (III) oxide is reduced with carbon to produce carbon dioxide and molten iron?

**A.** $2\ Fe_2O_3 + 3\ C\ (s) \rightarrow 4\ Fe\ (l) + 3\ CO_2\ (g)$
**B.** $4\ Fe_2O_3 + 6\ C\ (s) \rightarrow 8\ Fe\ (l) + 6\ CO_2\ (g)$
**C.** $2\ FeO_3 + 3\ C\ (s) \rightarrow 2\ Fe\ (l) + 3\ CO_2\ (g)$
**D.** $2\ Fe_3O + C\ (s) \rightarrow 6\ Fe\ (l) + CO_2\ (g)$

 

**134.** How many moles of $Sn^{4+}$ are produced from one mole of $Sn^{2+}$ and excess $O_2$ in the following reaction?

$Sn^{2+} + O_2 \rightarrow Sn^{4+}$

**A.** 2 moles
**B.** 2.5 moles
**C.** 1 mole
**D.** ½ mole

**135.** The oxidation number +7 is for the element:

**A.** Mn in $KMnO_4$
**B.** Br in $NaBrO_3$
**C.** C in $MgC_2O_4$
**D.** S in $H_2SO_4$

**136.** Which of the following reactions is NOT correctly classified?

**A.** $AgNO_3$ (*aq*) + KOH (*aq*) → $KNO_3$ (*aq*) + AgOH (*s*) : non-redox / double replacement
**B.** 2 $H_2O_2$ (*s*) → 2 $H_2O$ (*l*) + $O_2$ (*g*) : non-redox / decomposition
**C.** $Pb(NO_3)_2$ (*aq*) + 2 Na (*s*) → Pb (*s*) + 2 $NaNO_3$ (*aq*) : redox / single-replacement
**D.** $HNO_3$ (*aq*) + LiOH (*aq*) → $LiNO_3$ (*aq*) + $H_2O$ (*l*) : non-redox / double-replacement

**137.** What is the molecular formula of a compound with an empirical formula of CHCl with a molar mass of 194 g/mol?

**A.** $C_4H_4Cl_4$
**B.** $C_2H_4Cl_3$
**C.** $C_3H_5Cl_3$
**D.** CHCl

**138.** How many moles of Al are in an 8.52 g sample if the atomic mass of Al is 26.98 amu?

**A.** 0.316 moles
**B.** 0.233 moles
**C.** 1.46 moles
**D.** 4.34 moles

**139.** What is the oxidation number of Cr in the compound $HCr_2O_4Cl$?

**A.** +3 **B.** +2 **C.** +4 **D.** +5

**140.** What general term refers to the mass of 1 mol of any substance?

**A.** gram-formula mass
**B.** molar mass
**C.** gram-atomic mass
**D.** gram-molecular mass

*Notes or active learning*

# 7 – Chemical Kinetics and Equilibria

---

**Practice Set 1: Questions 1–20**

---

**1.** What is the general equilibrium constant ($K_{eq}$) expression for the following reversible reaction?

$2\ A + 3\ B \leftrightarrow C$

**A.** $K_{eq} = [C] / [A]^2 \cdot [B]^3$

**B.** $K_{eq} = [C] / [A] \cdot [B]$

**C.** $K_{eq} = [A] \cdot [B] / [C]$

**D.** $K_{eq} = [A]^2 \cdot [B]^3 / [C]$

**2.** What are gases A and B likely to be if gas A has twice the average velocity of gas B in a mixture of these two gases?

**A.** Ar and Kr

**B.** N and Fe

**C.** Ne and Ar

**D.** Mg and K

**3.** For a hypothetical reaction, A + B → C, predict which reaction occurs at the slowest rate from the following reaction conditions.

| Reaction | Activation energy | Temperature |
|---|---|---|
| 1 | 103 kJ/mol | 15 °C |
| 2 | 46 kJ/mol | 22 °C |
| 3 | 103 kJ/mol | 24 °C |
| 4 | 46 kJ/mol | 30 °C |

**A.** 1 **B.** 2 **C.** 3 **D.** 4

**4.** From the data below, what is the order of the reaction with respect to reactant A?

Determining Rate Law from Experimental Data: A + B → Products

| Exp. | Initial [A] | Initial [B] | Initial Rate M/s |
|---|---|---|---|
| 1 | 0.015 | 0.022 | 0.125 |
| 2 | 0.030 | 0.044 | 0.500 |
| 3 | 0.060 | 0.044 | 0.500 |
| 4 | 0.060 | 0.066 | 1.125 |
| 5 | 0.085 | 0.088 | ? |

**A.** Zero **B.** First **C.** Second **D.** Third

**5.** For the combustion of ethanol ($C_2H_6O$) to form carbon dioxide and water, what is the rate at which carbon dioxide is produced if the ethanol is consumed at 4.0 M $s^{-1}$?

**A.** 1.5 M $s^{-1}$ **B.** 12.0 M $s^{-1}$ **C.** 8.0 M $s^{-1}$ **D.** 9.0 M $s^{-1}$

**6.** Which is the correct equilibrium constant ($K_{eq}$) expression for the following reaction?

$2\ Ag\ (s) + Cl_2\ (g) \leftrightarrow 2\ AgCl\ (s)$

**A.** $K_{eq} = [AgCl] / [Ag]^2 \times [Cl_2]$

**B.** $K_{eq} = [2\ AgCl] / [2\ Ag] \times [Cl_2]$

**C.** $K_{eq} = [AgCl]^2 / [Ag]^2 \times [Cl_2]$

**D.** $K_{eq} = 1 / [Cl_2]$

**7.** The equilibrium position for a system where $K_{eq} = 6.3 \times 10^{-14}$ is favored for [ ], and the product concentration is relatively [ ].

**A.** the left; large

**B.** the left; small

**C.** the right; large

**D.** the right; small

**8.** What can be deduced about the activation energy of a reaction that takes billions of years to complete and a reaction that takes only a fraction of a millisecond?

**A.** The slow reaction has high activation energy, while the fast reaction has low activation energy

**B.** The slow reaction has low activation energy, while the fast reaction has a high activation energy

**C.** The activation energy of both reactions is very low

**D.** The activation energy of both reactions is very high

 

**9.** Which influences the rate of a first-order chemical reaction?

I. catalyst II. temperature III. concentration

**A.** I only
**B.** I and II only
**C.** I and III only
**D.** I, II and III

Questions **10** through **14** are based on the following:

Energy profiles for four reactions (with the same scale).

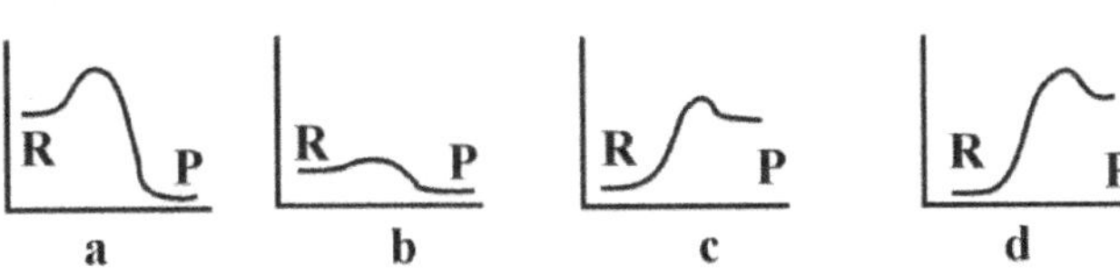


R = reactants P = products

**10.** Which reaction requires the most energy?

**A.** a **B.** b **C.** c **D.** d

**11.** Which reaction has the highest activation energy?

**A.** a **B.** b **C.** c **D.** d

**12.** Which reaction has the lowest activation energy?

**A.** a **B.** b **C.** c **D.** d

**13.** Which reaction proceeds the slowest?

**A.** a **B.** b **C.** c **D.** d

**14.** If the graphs are for the same reaction, which most likely has a catalyst?

**A.** a **B.** b **C.** c **D.** d

**15.** What is an explanation for the observation that the reaction stops before all reactants are converted to products in the following reaction? (Use the notation of [ ] for concentration)

$NH_3\ (aq) + HC_2H_3O_2\ (aq) \rightarrow NH_4^+\ (aq) + C_2H_3O_2^-\ (aq)$

**A.** The catalyst is depleted
**B.** The reverse rate increases while the forward rate decreases until they are equal
**C.** As [products] increase, the acetic acid begins to dissociate, stopping the reaction
**D.** As [reactants] decrease, $NH_3$ and $HC_2H_3O_2$ molecules stop colliding

**16.** What is the term for the principle that the rate of reaction is regulated by the frequency, energy, and orientation of molecules striking each other?

**A.** orientation theory
**B.** frequency theory
**C.** energy theory
**D.** collision theory

**17.** What is the effect on the energy of the activated complex and the reaction rate when a catalyst is added to a chemical reaction?

**A.** The energy of the activated complex increases, and the reaction rate decreases
**B.** The energy of the activated complex decreases, and the reaction rate increases
**C.** The energy of the activated complex and the reaction rate increase
**D.** The energy of the activated complex and the reaction rate decrease

**18.** Which change shifts the equilibrium to the right for the reversible reaction in an aqueous solution?

$HNO_2\ (aq) \leftrightarrow H^+\ (aq) + NO_2^-\ (aq)$

I. Add solid NaOH
II. Decrease $[NO_2^-]$
III. Decrease $[H^+]$
IV. Increase $[HNO_2]$

**A.** II and III only
**B.** I and IV only
**C.** II, III and IV only
**D.** I, II, III and IV

**19.** Which conditions would favor driving the reaction to completion?

$$2\ N_2\ (g) + 6\ H_2O\ (g) + \textit{heat} \leftrightarrow 4\ NH_3\ (g) + 3\ O_2\ (g)$$

**A.** Increasing the reaction temperature
**B.** Continual addition of $NH_3$ gas to the reaction mixture
**C.** Decreasing the pressure on the reaction vessel
**D.** Continual removal of $N_2$ gas

**20.** The system, $H_2\ (g) + X_2\ (g) \leftrightarrow 2\ HX\ (g)$ has a value of 24.4 for $K_c$. A catalyst was introduced into a reaction within a 4.0-liters reactor containing 0.20 moles of $H_2$, 0.20 moles of $X_2$ and 0.800 moles of $HX$. The reaction proceeds in which direction?

**A.** to the right, $Q > K_c$
**B.** to the left, $Q > K_c$
**C.** to the right, $Q < K_c$
**D.** to the left, $Q < K_c$

---

**Practice Set 2: Questions 21–40**

---

**21.** Which is the $K_c$ equilibrium expression for the following reaction?

$$4\ CuO\ (s) + CH_4\ (g) \leftrightarrow CO_2\ (g) + 4\ Cu\ (s) + 2\ H_2O\ (g)$$

**A.** $[Cu]^4 / [CuO]^4$
**B.** $[CO_2]·[H_2O]^2 / [CH_4]$
**C.** $[CH_4] / [CO_2]·[H_2O]^2$
**D.** $[CuO]^4 / [Cu]^4$

**22.** Which of the following concentrations of $CH_2Cl_2$ should be used in the rate law for Step 2 if $CH_2Cl_2$ is a product of the fast (first) step and a reactant of the slow (second) step?

**A.** $[CH_2Cl_2]$ at equilibrium
**B.** $[CH_2Cl_2]$ in Step 2 cannot be predicted because Step 1 is the fast step
**C.** Zero moles per liter
**D.** $[CH_2Cl_2]$ after Step 1 is completed

**23.** What is the term for a substance that allows a reaction to proceed faster by lowering activation energy?

**A.** rate barrier
**B.** energy barrier
**C.** collision energy
**D.** catalyst

**24.** Which statement is true for the grams of products present after a chemical reaction reaches equilibrium?

**A.** Must equal the grams of the initial reactants
**B.** May be less than, equal to, or greater than the grams of reactants present, depending upon the chemical reaction
**C.** Must be greater than the grams of the initial reactants
**D.** Must be less than the grams of the initial reactants

 

**25.** What is the rate law when rates were measured at different concentrations for the dissociation of hydrogen gas: $H_2\ (g) \rightarrow 2\ H\ (g)$?

| $[H_2]$ | Rate M/s $s^{-1}$ |
|---|---|
| 1.0 | $1.3 \times 10^5$ |
| 1.5 | $2.6 \times 10^5$ |
| 2.0 | $5.2 \times 10^5$ |

**A.** rate = $k^2[H] / [H_2]$

**B.** rate = $k[H]^2 / [H_2]$

**C.** rate = $k[H_2]^2$

**D.** rate = $k[H_2] / [H]^2$

**26.** Which change to this reaction system causes the equilibrium to shift to the right?

$N_2\ (g) + 3\ H_2\ (g) \leftrightarrow 2\ NH_3\ (g) + heat$

**A.** Heating the system

**B.** Removal of $H_2\ (g)$

**C.** Addition of $NH_3\ (g)$

**D.** Lowering the temperature

**27.** Which statement is NOT correct for $aA + bB \rightarrow dD + eE$ whereby rate = $k[A]^q \cdot [B]^r$?

**A.** The overall order of the reaction is $q + r$

**B.** The exponents $q$ and $r$ are equal to the coefficients $a$ and $b$, respectively

**C.** The exponents $q$ and $r$ must be determined experimentally

**D.** The exponents $q$ and $r$ are often integers

**28.** Which is the correct equilibrium constant ($K_{eq}$) expression for the following reaction?

$CO\ (g) + 2\ H_2\ (g) \leftrightarrow CH_3OH\ (l)$

**A.** $K_{eq} = 1 / [CO] \cdot [H_2]^2$

**B.** $K_{eq} = [CO] \cdot [H_2]^2$

**C.** $K_{eq} = [CH_3OH] / [CO] \cdot [H_2]^2$

**D.** $K_{eq} = [CH_3OH] / [CO] \cdot [H_2]$

Questions **29-32** are based on the following graph and net reaction:

The reaction proceeds in two consecutive steps.

$$XY + Z \leftrightarrow XYZ \leftrightarrow X + YZ$$

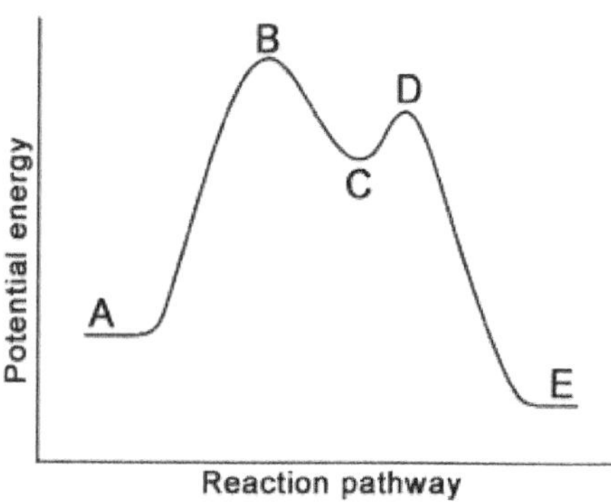


**29.** Where is the activated complex for this reaction?

**A.** A
**B.** A and C
**C.** C
**D.** B and D

**30.** The activation energy of the slow step for the forward reaction is given by:

**A.** A $\rightarrow$ B
**B.** A $\rightarrow$ C
**C.** B $\rightarrow$ C
**D.** C $\rightarrow$ E

**31.** The activation energy of the slow step for the reverse reaction is given by:

**A.** C $\rightarrow$ A
**B.** E $\rightarrow$ C
**C.** C $\rightarrow$ B
**D.** E $\rightarrow$ D

**32.** The change in energy ($\Delta$E) of the overall reaction is given by the difference between:

**A.** A and B
**B.** A and E
**C.** A and C
**D.** B and D

**33.** Which of the following increases the collision energy of gaseous molecules?

I. Increasing the temperature
II. Adding a catalyst
III. Increasing the concentration

**A.** I only
**B.** II only
**C.** III only
**D.** I and II only

**34.** Which of the following conditions favors the formation of NO (*g*) in a closed container?

$N_2 (g) + O_2 (g) \leftrightarrow 2 NO (g)$ (Use $\Delta H = +181$ kJ/mol)

**A.** Increasing the temperature
**B.** Decreasing the temperature
**C.** Increasing the pressure
**D.** Decreasing the pressure

**35.** A chemical system is considered to have reached dynamic equilibrium when the:

**A.** activation energy of the forward reaction equals the activation energy of the reverse reaction
**B.** rate of production of each of the products equals the rate of their consumption by the reverse reaction
**C.** frequency of collisions between the reactant molecules equals the frequency of collisions between the product molecules
**D.** sum of the concentrations of each of the reactant species equals the sum of the concentrations of each of the product species

**36.** Chemical equilibrium is reached in a system when:

**A.** complete conversion of reactants to products has occurred
**B.** product molecules begin reacting with each other
**C.** reactant concentrations steadily decrease
**D.** product and reactant concentrations remain constant

**37.** At a given temperature, $K = 46.0$ for the reaction:

$4 HCl (g) + O_2 (g) \leftrightarrow 2 H_2O (g) + 2 Cl_2 (g)$

At equilibrium, $[HCl] = 0.150$, $[O_2] = 0.395$ and $[H_2O] = 0.625$. What is the concentration of $Cl_2$ at equilibrium?

**A.** 0.153 M
**B.** 0.444 M
**C.** 1.14 M
**D.** 0.00547 M\

**38.** If at equilibrium, reactant concentrations are slightly smaller than product concentrations, the equilibrium constant would be:

**A.** slightly greater than 1
**B.** slightly lower than 1
**C.** much lower than 1
**D.** much greater than 1

**39.** Hydrogen gas reacts with iron (III) oxide to form iron metal (which produces steel), as shown in the reaction below. Which statement is NOT correct concerning the equilibrium system?

$$Fe_2O_3\ (s) + 3\ H_2\ (g) + heat \leftrightarrow 2\ Fe\ (s) + 3\ H_2O\ (g)$$

**A.** Continually removing water from the reaction chamber increases the yield of iron
**B.** Decreasing the volume of hydrogen gas reduces the yield of iron
**C.** Lowering the reaction temperature increases the concentration of hydrogen gas
**D.** Increasing the pressure on the reaction chamber increases the formation of products

**40.** Which changes shift the equilibrium to the right for the following reversible reaction?

$$CO\ (g) + H_2O\ (g) \leftrightarrow CO_2\ (g) + H_2\ (g) + heat$$

**A.** increasing volume
**B.** increasing temperature
**C.** increasing [CO]
**D.** adding a catalyst

---

**Practice Set 3: Questions 41–60**

---

**41.** Which is the correct equilibrium constant ($K_{eq}$) expression for the following reaction?

$$4\ NH_3\ (g) + 5\ O_2\ (g) \leftrightarrow 4\ NO\ (g) + 6\ H_2O\ (g)$$

**A.** $K_{eq} = [NO]^4 \times [H_2O]^6 / [NH_3]^4 \times [O_2]^5$

**B.** $K_{eq} = [NH_3]^4 \times [O_2]^5 / [NO]^4 \times [H_2O]^6$

**C.** $K_{eq} = [NO] \times [H_2O] / [NH_3] \times [O_2]$

**D.** $K_{eq} = [NH_3] \times [O_2] / [NO] \times [H_2O]$

**42.** Carbonic acid equilibrium in blood:

$$CO_2\ (g) + H_2O\ (l) \leftrightarrow H_2CO_3\ (aq) \leftrightarrow H^+\ (aq) + HCO_3^-\ (aq)$$

If a person hyperventilates, rapid breathing expels carbon dioxide gas. Which of the following decreases when a person hyperventilates?

I. $[HCO_3^-]$ II. $[H^+]$ III. $[H_2CO_3]$

**A.** I only

**B.** II only

**C.** III only

**D.** I, II and III

**43.** What is the overall order of the reaction if the units of the rate constant for a particular reaction are $min^{-1}$?

**A.** Zero

**B.** First

**C.** Second

**D.** Third

**44.** Heat is often added to chemical reactions performed in the laboratory to:

**A.** compensate for the natural tendency of energy to disperse

**B.** increase the rate at which reactants collide

**C.** allow a greater number of reactants to overcome the barrier of the activation energy

**D.** all the above

**45.** Predict which reaction occurs faster for a hypothetical reaction X + Y → W + Z.

| Reaction | Activation energy | Temperature |
|---|---|---|
| 1 | low | low |
| 2 | low | high |
| 3 | high | high |
| 4 | high | low |

**A.** 1 **B.** 2 **C.** 3 **D.** 4

**46.** For the reaction, 2 *XO* + $O_2$ → 2 $XO_2$, data obtained from measurement of the initial rate of reaction at varying concentrations are:

| Experiment | [*XO*] | [$O_2$] | Rate (mmol $L^{-1}$ $s^{-1}$) |
|---|---|---|---|
| 1 | 0.010 | 0.010 | 2.5 |
| 2 | 0.010 | 0.020 | 5.0 |
| 3 | 0.030 | 0.020 | 45.0 |

What is the expression for the rate law?

**A.** rate = $k[XO]\cdot[O_2]$
**B.** rate = $k[XO]^2\cdot[O_2]^2$
**C.** rate = $k[XO]^2\cdot[O_2]$
**D.** rate = $k[XO]\cdot[O_2]^2$

**47.** What is the equilibrium ($K_{eq}$) expression for the following reaction?

$CaO\ (s) + CO_2\ (g) \leftrightarrow CaCO_3\ (s)$

**A.** $K_{eq} = [CaCO_3] / [CaO]$
**B.** $K_{eq} = 1 / [CO_2]$
**C.** $K_{eq} = [CaCO_3] / [CaO]\cdot[CO_2]$
**D.** $K_{eq} = [CO_2]$

**48.** If a reaction does not occur extensively and gives a low concentration of products at equilibrium, which of the following is true?

**A.** The rate of the forward reaction is greater than the reverse reaction
**B.** The rate of the reverse reaction is greater than the forward reaction
**C.** The equilibrium constant is greater than one; that is, $K_{eq}$ is larger than 1
**D.** The equilibrium constant is less than one; that is, $K_{eq}$ is smaller than 1

 

**49.** Which of the following changes most likely decreases the rate of a reaction?

**A.** Increasing the reaction temperature
**B.** Increasing the concentration of a reactant
**C.** Increasing the activation energy for the reaction
**D.** Decreasing the activation energy for the reaction

**50.** Which factors would increase the rate of a reversible chemical reaction?

I. Increasing the temperature of the reaction
II. Removing products as they form
III. Adding a catalyst to the reaction vessel

**A.** I only
**B.** II only
**C.** I and III only
**D.** I, II and III

**51.** What is the ionization equilibrium constant ($K_i$) expression for the following weak acid?

$H_2S\ (aq) \leftrightarrow H^+\ (aq) + HS^-\ (aq)$

**A.** $K_i = [H^+]^2 \cdot [S^{2-}] / [H_2S]$
**B.** $K_i = [H_2S] / [H^+] \cdot [HS^-]$
**C.** $K_i = [H^+] \cdot [HS^-] / [H_2S]$
**D.** $K_i = [H^+]^2 \cdot [HS^-] / [H_2S]$

**52.** Increasing the temperature of a chemical reaction:

**A.** increases the reaction rate by lowering the activation energy
**B.** increases the reaction rate by increasing reactant collisions per unit time
**C.** increases the activation energy, thus increasing the reaction rate
**D.** raises the activation energy, thus decreasing the reaction rate

**53.** Reaction rates are determined by the following factors EXCEPT:

**A.** orientation of collisions between molecules
**B.** spontaneity of the reaction
**C.** force of collisions between molecules
**D.** number of collisions between molecules

**54.** Coal-burning plants release sulfur dioxide into the atmosphere, while nitrogen monoxide is released via industrial processes and combustion engines. Sulfur dioxide can also be produced in the atmosphere by the following equilibrium reaction:

$$SO_3\,(g) + NO\,(g) + heat \leftrightarrow SO_2\,(g) + NO_2\,(g)$$

Which of the following does NOT shift the equilibrium to the right?

**A.** $[NO_2]$ decrease
**B.** [NO] increase
**C.** Decrease the reaction chamber volume
**D.** Temperature increase

**55.** Which of the following statements can be assumed true about how reactions occur?

**A.** Reactant particles must collide with each other
**B.** Energy must be released as the reaction proceeds
**C.** Catalysts must be present in the reaction
**D.** Energy must be absorbed as the reaction proceeds

**56.** At equilibrium, increasing the temperature of an exothermic reaction likely:

**A.** increases the heat of reaction
**B.** decreases the heat of reaction
**C.** increases the forward reaction
**D.** decreases the forward reaction

**57.** Which shifts the equilibrium to the left for the reversible reaction in an aqueous solution?

$$HC_2H_3O_2\,(aq) \leftrightarrow H^+\,(aq) + C_2H_3O_2^-\,(aq)$$

I. increase pH
II. increase $[HC_2H_3O_2]$
III. add solid $KC_2H_3O_2$

**A.** I only
**B.** I and III only
**C.** III only
**D.** II and III only

 

**58.** Which of the following statements is true concerning the equilibrium system, whereby S combines with $H_2$ to form hydrogen sulfide, a toxic gas from the decay of organic material? (Use the equilibrium constant, $K_{eq} = 2.8 \times 10^{-21}$)

$$S\ (g) + H_2\ (g) \leftrightarrow H_2S\ (g)$$

**A.** Almost all the starting molecules are converted to product
**B.** Little hydrogen sulfide gas is present in the equilibrium
**C.** Decreasing $[H_2]$ shifts the equilibrium to the right
**D.** Increasing the volume of the sealed reaction container shifts the equilibrium to the right

**59.** Which of the following conditions characterizes a system in a state of chemical equilibrium?

**A.** Product concentrations are greater than reactant concentrations
**B.** Reactant molecules no longer react with each other
**C.** Concentrations of reactants and products are equal
**D.** Reactants are being consumed at the same rate they are being produced

**60.** Which statement regarding an equilibrium constant for a particular reaction is NOT true?

**A.** It does not change as the product is removed
**B.** It does not change as an additional quantity of a reactant is added
**C.** It changes when a catalyst is added
**D.** It changes as the temperature increases

**Practice Set 4: Questions 61–80**

Questions **61** through **63** refer to the rate data for the conversion of reactants W, X, and Y to product Z.

| Trial number | Concentration (moles/L) | | | Rate of formation of Z |
|---|---|---|---|---|
| | W | X | Y | (moles/l·s) |
| 1 | 0.01 | 0.05 | 0.04 | 0.04 |
| 2 | 0.015 | 0.07 | 0.06 | 0.08 |
| 3 | 0.01 | 0.15 | 0.04 | 0.36 |
| 4 | 0.03 | 0.07 | 0.06 | 0.08 |
| 5 | 0.01 | 0.05 | 0.16 | 0.08 |

**61.** From the above data, what is the overall order of the reaction?

**A.** 2½ **B.** 4 **C.** 3 **D.** 2

**62.** From the above data, the order with respect to W suggests that the rate of formation of Z is:

**A.** dependent on [W]
**B.** independent of [W]
**C.** semi-dependent on [W]
**D.** unable to be determined

**63.** From the above data, the magnitude of *k* for trial 1 is:

**A.** 20 **B.** 40 **C.** 60 **D.** 80

 

**64.** Which changes shift the equilibrium to the left for the given reversible reaction?

$SO_3\,(g) + NO\,(g) + heat \leftrightarrow SO_2\,(g) + NO_2\,(g)$

**A.** Decrease temperature
**B.** Decrease volume
**C.** Increase [NO]
**D.** Decrease $[SO_2]$

**65.** What is the ionization equilibrium constant ($K_i$) expression for the following weak acid?

$H_3PO_4\,(aq) \leftrightarrow H^+\,(aq) + H_2PO_4^-\,(aq)$

**A.** $K_i = [H_3PO_4] / [H^+]\cdot[H_2PO_4^-]$
**B.** $K_i = [H^+]^3\cdot[PO_4^{3-}] / [H_3PO_4]$
**C.** $K_i = [H^+]^3\cdot[H_2PO_4^-] / [H_3PO_4]$
**D.** $K_i = [H^+]\cdot[H_2PO_4^-] / [H_3PO_4]$

**66.** What effect does a catalyst have on equilibrium?

**A.** It increases the rate of the forward reaction
**B.** It shifts the reaction to the right
**C.** It increases the rate at which equilibrium is reached without changing ΔG
**D.** It increases the rate at which equilibrium is reached and lowers ΔG

**67.** What is the correct ionization equilibrium constant ($K_i$) expression for the following weak acid?

$H_2SO_3\,(aq) \leftrightarrow H^+\,(aq) + HSO_3^-\,(aq)$

**A.** $K_i = [H_2SO_3] / [H^+]\cdot[HSO_3^-]$
**B.** $K_i = [H^+]^2\cdot[SO_3^{2-}] / [H_2SO_3]$
**C.** $K_i = [H^+]^2\cdot[HSO_3^-] / [H_2SO_3]$
**D.** $K_i = [H^+]\cdot[HSO_3^-] / [H_2SO_3]$

**68.** Which of the following is true if a reaction occurs extensively and yields a high concentration of products at equilibrium?

**A.** The rate of the reverse reaction is greater than the forward reaction
**B.** The rate of the forward reaction is greater than the reverse reaction
**C.** The equilibrium constant is less than one; $K_{eq}$ is much smaller than 1
**D.** The equilibrium constant is greater than one; $K_{eq}$ is much larger than 1

**69.** The minimum combined kinetic energy reactants must possess for collisions to result in a reaction is:

**A.** orientation energy
**B.** activation energy
**C.** collision energy
**D.** dissociation energy

**70.** Which factors decrease the rate of a reaction?

I. Lowering the temperature
II. Increasing the concentration of reactants
III. Adding a catalyst to the reaction vessel

**A.** I only
**B.** II only
**C.** III only
**D.** I and II only

**71.** For a collision between molecules to result in a reaction, the molecules must possess a favorable orientation relative to each other and:

**A.** be in the gaseous state
**B.** have a specific minimum energy
**C.** adhere for at least 2 nanoseconds
**D.** exchange electrons

**72.** Most reactions are carried out in liquid solution or the gaseous phase because in such situations:

**A.** kinetic energies of reactants are lower
**B.** reactant collisions occur more frequently
**C.** activation energies are higher
**D.** reactant activation energies are lower

**73.** Find the reaction rate for A + B → C:

| Trial | $[A]_{t=0}$ | $[B]_{t=0}$ | Initial rate (M/s) |
|---|---|---|---|
| 1 | 0.05 M | 1.0 M | $1.0 \times 10^{-3}$ |
| 2 | 0.05 M | 4.0 M | $16.0 \times 10^{-3}$ |
| 3 | 0.15 M | 1.0 M | $3.0 \times 10^{-3}$ |

**A.** rate = $k[A]^2 \cdot [B]^2$
**B.** rate = $k[A] \cdot [B]^2$
**C.** rate = $k[A]^2 \cdot [B]$
**D.** rate = $k[A] \cdot [B]$

**74.** Why does a glowing splint of wood burn only slowly in the air but rapidly in a burst of flames when placed in pure oxygen?

**A.** A glowing wood splint is extinguished from pure oxygen because $O_2$ inhibits the smoke
**B.** Pure oxygen can absorb carbon dioxide at a faster rate
**C.** Oxygen is a flammable gas
**D.** There is an increased number of collisions between wood and oxygen molecules

**75.** Which changes shift the equilibrium to the right for the following system at equilibrium?

$N_2 (g) + 3 H_2 (g) \leftrightarrow 2 NH_3 (g) + 92.94$ kJ

I. Removing $NH_3$
II. Adding $NH_3$
III. Removing $N_2$
IV. Adding $N_2$

**A.** I and III
**B.** II and III
**C.** II and IV
**D.** I and IV

**76.** Which changes do not affect equilibrium for reversible reactions in an aqueous solution?

$HC_2H_3O_2 (aq) \leftrightarrow H^+ (aq) + C_2H_3O_2^- (aq)$

**A.** Adding solid $NaC_2H_3O_2$
**B.** Adding solid $NaNO_3$
**C.** Increasing $[HC_2H_3O_2]$
**D.** Increasing $[H^+]$

**77.** Consider the following reaction: $H_2 (g) + I_2 (g) \rightarrow 2 HI (g)$

At 160 K, this reaction has an equilibrium constant of 35. If at 160 K, the concentration of hydrogen gas is 0.4 M, iodine gas is 0.6 M, and hydrogen iodide gas is 3 M:

**A.** system is at equilibrium
**B.** [iodine] decreases
**C.** [hydrogen iodide] increases
**D.** [hydrogen iodide] decreases

**78.** If the concentration of reactants decreases, which of the following is true?

I. The amount of products increases
II. The heat of reaction decreases
III. The rate of reaction decreases

**A.** I only
**B.** II only
**C.** III only
**D.** II and III only

**79.** What does a chemical equilibrium expression of a reaction depend on?

I. mechanism
II. stoichiometry
III. rate

**A.** I only
**B.** II only
**C.** III only
**D.** I and II only

**80.** For the following reaction where $\Delta H < 0$, which factor decreases the magnitude of the equilibrium constant *K*?

$$CO\,(g) + 2\,H_2O\,(g) \leftrightarrow CH_3OH\,(g)$$

**A.** Decreasing the temperature of this system

**B.** Decreasing volume

**C.** Decreasing the pressure of this system

**D.** None of the above

 

**Practice Set 5: Questions 81–100**

**81.** Which changes shift the equilibrium to the product side for the following reaction at equilibrium?

$SO_2Cl_2\ (g) \leftrightarrow SO_2\ (g) + Cl_2\ (g)$

I. Addition of $SO_2Cl_2$
II. Addition of $SO_2$
III. Removal of $SO_2Cl_2$
IV. Removal of $Cl_2$

**A.** I, II and III
**B.** I and IV
**C.** I and II
**D.** III and IV

**82.** Calculate a value for $K_c$ for the following reaction:

$NOCl\ (g) + \frac{1}{2}\ O_2\ (g) \leftrightarrow NO_2\ (g) + \frac{1}{2}\ Cl_2\ (g)$

Use the data: $2\ NO\ (g) + Cl_2\ (g) \leftrightarrow 2\ NOCl\ (g)$ $\quad K_c = 3.20 \times 10^{-3}$

$2\ NO_2\ (g) \leftrightarrow 2\ NO\ (g) + O_2\ (g)$ $\quad K_c = 15.5$

**A.** 4.49
**B.** 0.343
**C.** $4.32 \times 10^{-4}$
**D.** $1.33 \times 10^{-5}$

**83.** Assuming vessels a, b and c are drawn to relative proportions, which of the following reactions proceeds the fastest? (Assume equal temperatures)

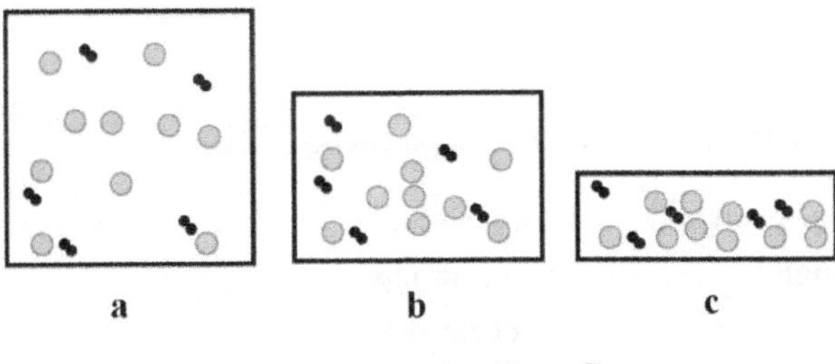


$A + B \rightarrow C$

**A.** a
**B.** b
**C.** c
**D.** All proceed at same rate

 

**84.** The reaction rate is:

**A.** ratio of the masses of products and reactants
**B.** ratio of the molecular masses of the elements in a given compound
**C.** speed at which reactants are consumed, or product is formed
**D.** balanced chemical formula that relates the number of product molecules to the reactant molecules

**85.** What is the equilibrium constant ($K_{eq}$) expression for the reversible reaction below?

$2\,A \leftrightarrow B + 3\,C$

**A.** $K_{eq} = [B] \times [C]^3 / [A]^2$
**B.** $K_{eq} = [B] \times [C] / [A]$
**C.** $K_{eq} = [A]^2 / [B] \times [C]^3$
**D.** $K_{eq} = [A] / [B] \times [C]$

**86.** With respect to A, what is the order of the reaction if the rate law = $k[A]^3[B]^6$?

**A.** 2
**B.** 3
**C.** 4
**D.** 5

**87.** Nitrogen monoxide reacts with bromine at elevated temperatures according to the equation:

$2\,NO\,(g) + Br_2\,(g) \rightarrow 2\,NOBr\,(g)$

What is the rate of $Br_2$ (*g*) consumption if, in a specific reaction mixture, the NOBr (g) formation rate was $4.50 \times 10^{-4}$ mol $L^{-1}$ $s^{-1}$?

**A.** $3.12 \times 10^{-4}$ mol $L^{-1}$ $s^{-1}$
**B.** $8.00 \times 10^{-4}$ mol $L^{-1}$ $s^{-1}$
**C.** $2.25 \times 10^{-4}$ mol $L^{-1}$ $s^{-1}$
**D.** $4.50 \times 10^{-4}$ mol $L^{-1}$ $s^{-1}$

**88.** Which of the following statements about "activation energy" is correct?

**A.** Activation energy is the energy given off when reactants collide
**B.** Activation energy is high for reactions that occur rapidly
**C.** Activation energy is low for reactions that occur rapidly
**D.** Activation energy is the maximum energy a reacting molecule may possess

**89.** Which is the correct equilibrium constant ($K_{eq}$) expression for the following reaction?

$A + 2B \leftrightarrow 2C + D$

**A.** $[C]^2 \cdot [D] / [A] \cdot [B]^2$
**B.** $[A]^2 \cdot [B]^2 / 2[C]^2 \cdot [D]$
**C.** $[A] \cdot 2[B] / 2[C] \cdot [D]$
**D.** $2[C] \cdot [D] / [A] \cdot 2[B]$

**90.** What is the effect on the equilibrium after adding $H_2O$ to the equilibrium mixture if $CO_2$ and $H_2$ react until equilibrium is established?

$CO_2 (g) + H_2 (g) \leftrightarrow H_2O (g) + CO (g)$

**A.** $[H_2]$ decreases and $[H_2O]$ increases
**B.** $[CO]$ and $[CO_2]$ increase
**C.** $[H_2]$ decreases and $[CO_2]$ increases
**D.** Equilibrium shifts to the left

**91.** For a reaction that has an equilibrium constant of $4.3 \times 10^{-17}$ at 25 °C, the relative position of equilibrium is described as:

**A.** equal amounts of reactants and products
**B.** significant amounts of reactants and products
**C.** mostly products
**D.** mostly reactants

**92.** What is the term for the energy necessary for reactants to achieve the transition state and form products?

**A.** heat of reaction
**B.** activation energy
**C.** rate barrier
**D.** collision energy

**93.** What is the equilibrium constant $K_c$ value if, at equilibrium, the concentrations of $[NH_3] = 0.40$ M, $[H_2] = 0.12$ M and $[N_2] = 0.040$ M?

$2 NH_3 (g) \leftrightarrow N_2 (g) + 3 H_2 (g)$

**A.** $6.3 \times 10^{12}$
**B.** $7.1 \times 10^{-7}$
**C.** $4.3 \times 10^{-4}$
**D.** $3.9 \times 10^{-3}$

**94.** Which of the following increases the collision frequency of molecules?

I. Increasing the concentration
II. Adding a catalyst
III. Decreasing the temperature

**A.** I only
**B.** II only
**C.** III only
**D.** I and II only

**95.** For the following reaction with $\Delta H < 0$, which factor increases the equilibrium yield for methanol?

$CO\ (g) + 2\ H_2O\ (g) \leftrightarrow CH_3OH\ (g)$

I. Decreasing the volume
II. Decreasing the pressure
III. Lowering the temperature of the system

**A.** I only
**B.** II only
**C.** I and II only
**D.** I and III only

**96.** If there is too much chlorine in the water, swimmers complain that their eyes burn. Consider the equilibrium found in swimming pools. Predict which increases the chlorine concentration.

$Cl_2\ (g) + H_2O\ (l) \leftrightarrow HClO\ (aq) \leftrightarrow H^+\ (aq) + ClO^-\ (aq)$

**A.** Decreasing the pH
**B.** Adding hydrochloric acid, HCl (*aq*)
**C.** Adding hypochlorous acid, HClO (*aq*)
**D.** All the above

**97.** What is the ratio of the diffusion rates of $H_2$ gas to $O_2$ gas?

**A.** 4:1
**B.** 2:1
**C.** 1:3
**D.** 1:4

**98.** Which of the following increases the product formed from a reaction?

I. Using a UV light catalyst
II. Adding an acid catalyst
III. Adding a metal catalyst

**A.** I only
**B.** II only
**C.** III only
**D.** None of the above

**99.** A mixture of 1.40 moles of A and 2.30 moles of B reacted. At equilibrium, 0.90 moles of A are present. How many moles of C are present at equilibrium?

$$3\text{ A }(g) + 2\text{ B }(g) \rightarrow 4\text{ C }(g)$$

**A.** 2.7 moles
**B.** 1.3 moles
**C.** 0.67 moles
**D.** 1.8 moles

**100.** What is the equilibrium constant $K_c$ for the following reaction?

$$PCl_5\ (g) + 2\ NO\ (g) \leftrightarrow PCl_3\ (g) + 2\ NOCl\ (g)$$

$K_1 = PCl_3\ (g) + Cl_2\ (g) \leftrightarrow PCl_5\ (g)$

$K_2 = 2\ NO\ (g) + Cl_2\ (g) \leftrightarrow 2\ NOCl\ (g)$

**A.** $K_1 / K_2$
**B.** $(K_1K_2)^{-1}$
**C.** $K_1 \times K_2$
**D.** $K_2 / K_1$

---

**Practice Set 6: Questions 101–120**

---

**101.** Write the mass action ($K_c$) expression for the following reaction.

$$4\ Cr\ (s) + 3\ CCl_4\ (g) \leftrightarrow 4\ CrCl_3\ (g) + 4\ C\ (s)$$

**A.** $K_c = [C]\cdot[CrCl_3] / [Cr]\cdot[CCl_4]$

**B.** $K_c = [C]^4\cdot[CrCl_3]^4 / [Cr]^4\cdot[CCl_4]^3$

**C.** $K_c = [CrCl_3]^4 / [CCl_4]^3$

**D.** $K_c = [CrCl_3] / [CCl_4]$

**102.** Which changes shift the equilibrium to the left for the reversible reaction in an aqueous solution?

$$HNO_2\ (aq) \leftrightarrow H^+\ (aq) + NO_2^-\ (aq)$$

**A.** Adding solid $KNO_2$

**B.** Adding solid KCl

**C.** Increasing $[HNO_2]$

**D.** Increasing pH

**103.** Given the equation $x$A + $y$B → $z$, the rate expression reaction in terms of the rate of change of the concentration with respect to time is:

**A.** $k[A]^x[B]^{yz}$

**B.** $k[A]^y[B]^x$

**C.** $k[A]^x[B]^y$

**D.** cannot be determined

**104.** Given the experimental data, what is the rate law for reaction 3 D + E → F + 2 G?

| Experiment | [D] | [E] | Rate (mol L$^{-1}$ s$^{-1}$) |
|---|---|---|---|
| 1 | 0.100 | 0.250 | 0.000250 |
| 2 | 0.200 | 0.250 | 0.000500 |
| 3 | 0.100 | 0.500 | 0.00100 |

**A.** rate = $k[D]^2\cdot[E]^2$

**B.** rate = $k[D]^2\cdot[E]$

**C.** rate = $k[D]^3\cdot[E]$

**D.** rate = $k[D]\cdot[E]^2$

**105.** A 10-mm cube of copper metal is placed in 400 mL of 10 M nitric acid at 27 °C, and the reaction below occurs:

$$Cu(s) + 4\ H^+\ (aq) + 2\ NO_3^-\ (aq) \rightarrow Cu^{2+}\ (aq) + 2\ NO_2\ (g) + 2\ H_2O\ (l)$$

If nitrogen dioxide is produced at the rate of 2.8 × 10–4 M/min, what is the rate at which hydrogen ions are consumed?

**A.** $3.2 \times 10^{-3}$ M/min
**B.** $1.8 \times 10^{-2}$ M/min
**C.** $1.1 \times 10^{-4}$ M/min
**D.** $5.6 \times 10^{-4}$ M/min

**106.** What is the equilibrium constant ($K_{eq}$) expression for the following reaction?

$$2\ CO\ (g) + O_2\ (g) \leftrightarrow 2\ CO_2\ (g)$$

**A.** $K_{eq} = 2[CO_2] / 2[CO] \times [O_2]$
**B.** $K_{eq} = [CO] \times [O_2] / [CO_2]$
**C.** $K_{eq} = [CO_2]^2 / [CO]^2 \times [O_2]$
**D.** $K_{eq} = [CO_2] / [CO] + [O_2]$

**107.** In the reaction A + B → AB, which does NOT increase the reaction rate?

I. increasing the temperature
II. decreasing the temperature
III. adding a catalyst

**A.** I only
**B.** II only
**C.** III only
**D.** I and III only

**108.** If $K_{eq} = 6.1 \times 10^{-11}$, which statement is true?

**A.** Slightly more products are present
**B.** The amount of reactants equals products
**C.** Mostly products are present
**D.** Mostly reactants are present

**109.** By convention, what is the equilibrium constant for step 1 in a reaction?

**A.** $k_1 / k_{-1}$
**B.** $k_1 + k_{+2}$
**C.** $k_{-1}$
**D.** $k_1$

**110.** A catalyst changes which of the following?

**A.** ΔS
**B.** ΔG
**C.** ΔH
**D.** $E_{activation}$

**111.** Which changes shift(s) the equilibrium to the right for the reversible reaction in an aqueous solution?

$HC_2H_3O_2$ (*aq*) ↔ $H^+$ (*aq*) + $C_2H_3O_2^-$ (*aq*)

I. Decreasing [$C_2H_3O_2^-$]

II. Decreasing [$H^+$]

III. Increasing [$HC_2H_3O_2$]

IV. Decreasing [$HC_2H_3O_2$]

**A.** I and II only

**B.** I and III only

**C.** II and III only

**D.** I, II and III only

**112.** For the reaction A + B → C, which proceeds the slowest? (Assume equal temperatures)

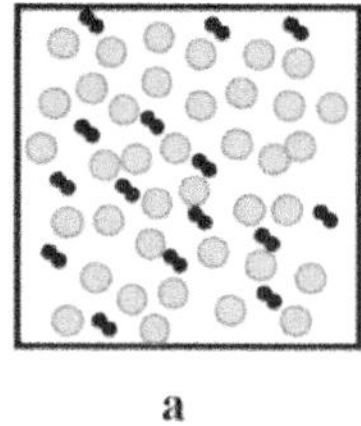
a

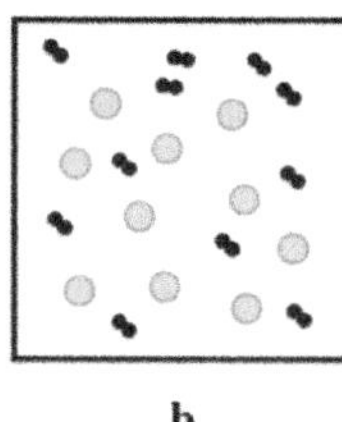
b

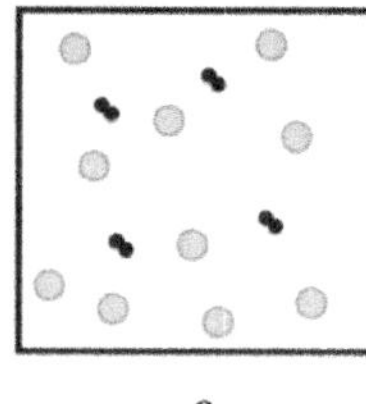
c

**A.** a

**B.** b

**C.** c

**D.** All proceed at the same rate

**113.** Why might increasing the concentration of a set of reactants increase the reaction rate?

**A.** The rate of reaction depends only on the mass of the atoms and, therefore, increases as the mass of the reactants increase

**B.** There is an increased probability that two reactant molecules collide and react

**C.** There is an increased ratio of reactants to products

**D.** The concentration of reactants is unrelated to the rate of reaction

**114.** For a chemical reaction to occur, all the following must happen EXCEPT:

**A.** reactant particles must collide with the correct orientation

**B.** a large enough number of collisions must occur

**C.** chemical bonds must break or form

**D.** reactant particles must collide with enough energy for change to occur

 

**115.** If a catalyst is added to the reaction, which direction does the equilibrium shift toward?

$CO + H_2O + \textit{heat} \leftrightarrow CO_2 + H_2$

**A.** To the left
**B.** To the right
**C.** No effect
**D.** Not enough information

**116.** For a chemical reaction at equilibrium, which of the following decreases the concentration of the products?

**A.** Decreasing the pressure
**B.** Increasing the temperature and decreasing the temperature
**C.** Increasing the temperature
**D.** Decreasing the concentration of a gaseous or aqueous reactant

**117.** Which factor does NOT describe activated complexes?

**A.** May be chemically isolated
**B.** Decompose rapidly
**C.** Have specific geometry
**D.** Are extremely reactive

**118.** When a reaction system is at equilibrium:

**A.** rate in the forward and reverse directions is equal
**B.** rate in the forward direction is at a maximum
**C.** amounts of reactants and products are equal
**D.** rate in the reverse direction is at a minimum

**119.** A reaction vessel contains $NH_3$, $N_2$, and $H_2$ at equilibrium with $[NH_3] = 0.1$ M, $[N_2] =$ 0.2 M, and $[H_2] = 0.3$ M. For decomposition, what is $K$ for $NH_3 \rightarrow N_2$ and $H_2$?

**A.** $K = (0.2)\cdot(0.3)^3 / (0.1)^2$
**B.** $K = (0.1) / (0.2)^2\cdot(1.5)^3$
**C.** $K = (0.1)^2 / (0.2)\cdot(0.3)^3$
**D.** $K = (0.2)\cdot(0.3)^2 / (0.1)$

**120.** The data provides the rate of a reaction as affected by the concentration of the reactants. What is the order of the reaction rate?

| Experiment | [X] | [Y] | [Z] | Rate (mol $L^{-1}$ $hr^{-1}$) |
|---|---|---|---|---|
| 1 | 0.200 M | 0.100 M | 0.600 M | 5.0 |
| 2 | 0.200 M | 0.400 M | 0.400 M | 80.0 |
| 3 | 0.600 M | 0.100 M | 0.200 M | 15.0 |
| 4 | 0.200 M | 0.100 M | 0.200 M | 5.0 |
| 5 | 0.200 M | 0.200 M | 0.400 M | 20.0 |

**A.** Zero order with respect to X

**B.** Order for X is minus one (rate proportional to 1 / [X])

**C.** First order with respect to X

**D.** Second order with respect to X

*Notes for active learning*

*Notes for active learning*

 

# 8 – Solution Chemistry

## Practice Set 1: Questions 1–20

**1.** If the solubility of nitrogen in blood is 1.90 cc/100 cc at 1.00 atm, what is the solubility of nitrogen in a scuba diver's blood at a depth of 125 feet where the pressure is 4.5 atm?

**A.** 1.90 cc/100 cc
**B.** 8.55 cc/100 cc
**C.** 4.5 cc/100 cc
**D.** 0.236 cc/100 cc

**2.** All the statements about molarity are correct, EXCEPT:

**A.** volume = moles/molarity
**B.** moles = molarity × volume
**C.** molarity equals moles of solute per mole of solvent
**D.** abbreviation is M

**3.** Which of the following molecules is expected to be most soluble in water?

**A.** NaCl
**B.** $CH_3CH_2CH_2COOH$
**C.** $CH_3CH_2CH_2$ OH
**D.** $Al(OH)_3$

**4.** The equation for the reaction shown below can be written as an ionic equation.

$$BaCl_2\ (aq) + K_2CrO_4\ (aq) \rightarrow BaCrO_4\ (s) + 2\ KCl\ (aq)$$

In the ionic equation, the spectator ions are:

**A.** $K^+$ and $Cl^-$
**B.** $Ba^{2+}$ and $CrO_4^{2-}$
**C.** $Ba^{2+}$ and $K^+$
**D.** $K^+$ and $CrO_4^{2-}$

**5.** Carbon dioxide gas is injected into soda in commercially prepared soft drinks. Under what conditions are carbon dioxide gas most soluble?

**A.** High temperature, high-pressure
**B.** High temperature, low-pressure
**C.** Low temperature, low-pressure
**D.** Low temperature, high-pressure

**6.** A solute is a:

**A.** substance that dissolves into a solvent
**B.** substance containing a solid, liquid, or gas
**C.** solid substance that does not dissolve into water
**D.** solid substance that does not dissolve at a given temperature

**7.** How many ions are produced in solution by dissociating one formula unit of $Co(NO_3)_2 \cdot 6H_2O$?

**A.** 2 **B.** 3 **C.** 4 **D.** 6

**8.** 15 grams of an unknown substance is dissolved in 60 grams of water. When the solution is transferred to another container, it weighs 78 grams. Which of the following is a possible explanation?

**A.** The solution reacted with the second container, forming a precipitate
**B.** Some of the solution remained in the first container
**C.** The reaction was endothermic, which increased the average molecular speed
**D.** Solution reacted with the first container, and some byproducts were transferred with the solution

**9.** Which of the following is NOT soluble in $H_2O$?

**A.** Iron (III) hydroxide
**B.** Iron (III) nitrate
**C.** Potassium sulfate
**D.** Ammonium sulfate

**10.** What is the v/v% concentration of a solution made by adding 25 mL of acetone to 75 mL of water?

**A.** 33% v/v
**B.** 0.33% v/v
**C.** 25% v/v
**D.** 2.5% v/v

**11.** Why is octane less soluble in $H_2O$ than in benzene?

**A.** Bonds between benzene and octane are much stronger than the bonds between $H_2O$ and octane
**B.** Octane cannot dissociate in the presence of $H_2O$
**C.** Bonds between $H_2O$ and octane are weaker than the bonds between $H_2O$ molecules
**D.** Octane and benzene have similar molecular weights

**12.** What is the $K_{sp}$ for slightly soluble copper (II) phosphate in an aqueous solution?

$Cu_3(PO_4)_2$ (*s*) ↔ 3 $Cu^{2+}$ (*aq*) + 2 $PO_4^{3-}$ (*aq*)

**A.** $K_{sp} = [Cu^{2+}]^3 \cdot [PO_4^{3-}]^2$

**B.** $K_{sp} = [Cu^{2+}] \cdot [PO_4^{3-}]^2$

**C.** $K_{sp} = [Cu^{2+}]^3 \cdot [PO_4^{3-}]$

**D.** $K_{sp} = [Cu^{2+}] \cdot [PO_4^{3-}]$

**13.** Apply the *like dissolves like* rule to predict which liquids are miscible with water.

I. carbon tetrachloride, $CCl_4$

II. toluene, $C_7H_8$

III. ethanol, $C_2H_5OH$

**A.** I only

**B.** II only

**C.** III only

**D.** I and II only

**14.** Which of the following pairs of substances would have both species in the pair be written in the molecular form in a net ionic equation?

I. $CO_2$ and $H_2SO_4$ II. LiOH and $H_2$ III. HF and $CO_2$

**A.** I only

**B.** II only

**C.** III only

**D.** I, II and III

**15.** Which statement best describes a supersaturated solution?

**A.** It contains dissolved solute in equilibrium with undissolved solid

**B.** It rapidly precipitates if a seed crystal is added

**C.** It contains as much solvent as it can accommodate

**D.** It contains no double bonds

**16.** What is the mass of a 7.50% urine sample that contains 122 g of dissolved solute?

**A.** 1,250 g

**B.** 935 g

**C.** 49.35 g

**D.** 1,627 g

**17.** Calculate the molarity of a solution prepared by dissolving 15.0 g of $NH_3$ in 250 g of water with a final density of 0.974 g/mL.

**A.** 36.2 M

**B.** 3.23 M

**C.** 0.0462 M

**D.** 0.664 M

**18.** Which of the following compounds has the highest boiling point?

**A.** 0.2 M $Al(NO_3)_3$
**B.** 0.2 M $MgCl_2$
**C.** 0.2 M glucose ($C_6H_{12}O_6$)
**D.** 0.2 M $Na_2SO_4$

**19.** Which compound produces four ions per formula unit by dissociation when dissolved in water?

**A.** $Li_3PO_4$
**B.** $Ca(NO_3)_2$
**C.** $MgSO_4$
**D.** $(NH_4)_2SO_4$

**20.** Which of the following would be a weak electrolyte in a solution?

**A.** HBr (*aq*)
**B.** KCl
**C.** KOH
**D.** $HC_2H_3O_2$

---

**Practice Set 2: Questions 21–40**

---

**21.** The ions $Ca^{2+}$, $Mg^{2+}$, $Fe^{2+}$, and $Fe^{3+}$, which are present in groundwater, can be removed by pretreating the water with:

**A.** $PbSO_4$
**B.** $Na_2CO_3 \cdot 10H_2O$
**C.** $KNO_3$
**D.** $CaCl_2$

**22.** Choose the spectator ions: $Pb(NO_3)_2$ (*aq*) + $H_2SO_4$ (*aq*) → ?

**A.** $NO_3^-$ and $H^+$
**B.** $H^+$ and $SO_4^{2-}$
**C.** $Pb^{2+}$ and $H^+$
**D.** $Pb^{2+}$ and $NO_3^-$

**23.** From the *like dissolves like* rule, predict which vitamin is soluble in water:

**A.** α-tocopherol ($C_{29}H_{50}O_2$)
**B.** calciferol ($C_{27}H_{44}O$)
**C.** ascorbic acid ($C_6H_8O_6$)
**D.** retinol ($C_{20}H_{30}O$)

**24.** How much water must be added when 125 mL of a 2.00 M solution of HCl is diluted to a final concentration of 0.400 M?

**A.** 150 mL
**B.** 500 mL
**C.** 625 mL
**D.** 750 mL

**25.** Which of the following is the sulfate ion?

**A.** $SO_4^{2-}$
**B.** $S^{2-}$
**C.** $CO_3^{2-}$
**D.** $PO_4^{3-}$

**26.** Which of the following explains why bubbles form inside a pot of water when water is heated?

**A.** As temperature increases, the vapor pressure increases
**B.** As temperature increases, the atmospheric pressure decreases
**C.** As temperature increases, the solubility of air decreases
**D.** As temperature increases, the kinetic energy decreases

**27.** A solution in which the rate of crystallization is equal to the rate of dissolution is:

**A.** saturated
**B.** supersaturated
**C.** dilute
**D.** unsaturated

**28.** What is the term that refers to liquids that do not dissolve in one another and separate into two layers?

**A.** Soluble
**B.** Miscible
**C.** Insoluble
**D.** Immiscible

**29.** Which is a correctly balanced hydration equation for the hydration of $Na_2SO_4$?

**A.** $Na_2SO_4\ (s) \xrightarrow{H_2O} Na^+\ (aq) + 2SO_4^{2-}\ (aq)$

**B.** $Na_2SO_4\ (s) \xrightarrow{H_2O} 2\ Na^{2+}\ (aq) + S^{2-}\ (aq) + O_4^{2-}\ (aq)$

**C.** $Na_2SO_4\ (s) \xrightarrow{H_2O} Na_2^{2+}\ (aq) + SO_4^{2-}\ (aq)$

**D.** $Na_2SO_4\ (s) \xrightarrow{H_2O} 2\ Na^+\ (aq) + SO_4^{2-}\ (aq)$

**30.** Soft drinks are carbonated by injection with carbon dioxide gas. Under what conditions is carbon dioxide gas least soluble?

**A.** High temperature, low-pressure
**B.** High temperature, high-pressure
**C.** Low temperature, high-pressure
**D.** Low temperature, low-pressure

**31.** Which type of compound is likely to dissolve in $H_2O$?

I. One with hydrogen bonds
II. Highly polar compound
III. Salt

**A.** I only
**B.** **B.** II only
**C.** III only
**D.** I, II and III

**32.** Which of the following might have the best solubility in water?

**A.** $CH_3CH_3$
**B.** $CH_3OH$
**C.** $CCl_4$
**D.** $O_2$

**33.** What is the $K_{sp}$ for slightly soluble gold (III) chloride in an aqueous solution for the reaction shown?

$AuCl_3\ (s) \leftrightarrow Au^{3+}\ (aq) + 3\ Cl^-\ (aq)$

**A.** $K_{sp} = [Au^{3+}]^3 \cdot [Cl^-] / [AuCl_3]$
**B.** $K_{sp} = [Au^{3+}] \cdot [Cl^-]^3$
**C.** $K_{sp} = [Au^{3+}]^3 \cdot [Cl^-]$
**D.** $K_{sp} = [Au^{3+}] \cdot [Cl^-]$

**34.** What is the net ionic equation for the reaction shown?

$CaCO_3 + 2\ HNO_3 \rightarrow Ca(NO_3)_2 + CO_2 + H_2O$

**A.** $CO_3^{2-} + H^+ \rightarrow CO_2$
**B.** $CaCO_3 + 2\ H^+ \rightarrow Ca^{2+} + CO_2 + H_2O$
**C.** $Ca^{2+} + 2\ NO_3^- \rightarrow Ca(NO_3)_2$
**D.** $CaCO_3 + 2\ NO_3^- \rightarrow Ca(NO_3)_2 + CO_3^{2-}$

**35.** What is the volume of a 0.550 M $Fe(NO_3)_3$ solution needed to supply 0.950 moles of nitrate ions?

**A.** 265 mL
**B.** 0.828 mL
**C.** 22.2 mL
**D.** 576 mL

**36.** What is the molarity of a solution that contains 48 mEq $Ca^{2+}$ per liter?

**A.** 0.024 M
**B.** 0.048 M
**C.** 1.8 M
**D.** 2.4 M

**37.** If 36.0 g of LiOH is dissolved in water to make 975 mL of solution, what is the molarity of the LiOH solution? (Use molecular mass of LiOH = 24.0 g/mol)

**A.** 1.54 M
**B.** 2.48 M
**C.** 0. 844 M
**D.** 0.268 M

**38.** What volume of 8.50% (m/v) solution contains 60.0 grams of glucose?

**A.** 170 mL
**B.** 448 mL
**C.** 706 mL
**D.** 344 mL

**39.** In an AgCl solution, if the $K_{sp}$ for AgCl is $A$, and the concentration $Cl^-$ in a container is $B$ molar, what is the concentration of Ag (in moles/liter)?

I. $A$ moles/liter    II. $B$ moles/liter    III. $A/B$ moles/liter

**A.** I only
**B.** II only
**C.** III only
**D.** II and III only

**40.** Which statement below is generally true?

**A.** Bases are strong electrolytes and ionize completely when dissolved in water
**B.** Salts are strong electrolytes and dissociate completely when dissolved in water
**C.** Acids are strong electrolytes and ionize completely when dissolved in water
**D.** Bases are weak electrolytes and ionize completely when dissolved in water

---

**Practice Set 3: Questions 41–60**

---

**41.** Which of the following intermolecular attractions is/are important for forming a solution?

I. solute-solute II. solvent-solute III. solvent-solvent

**A.** I only
**B.** III only
**C.** I and II only
**D.** I, II and III

**42.** What is the concentration of $I^-$ ions in a 0.40 M solution of magnesium iodide?

**A.** 0.05 M
**B.** 0.80 M
**C.** 0.60 M
**D.** 0.20 M

**43.** Which is most likely soluble in $NH_3$?

**A.** $CO_2$
**B.** $SO_2$
**C.** $CCl_4$
**D.** $N_2$

**44.** Which of the following represents the symbol for the chlorite ion?

**A.** $ClO_2^-$
**B.** $ClO^-$
**C.** $ClO_4^-$
**D.** $ClO_3^-$

**45.** Which statement best describes what is happening in a water softening unit?

**A.** Sodium is removed from the water, making the water interact less with the soap molecules
**B.** Ions in the water softener are softened by chemically bonding with sodium
**C.** Hard ions are trapped in the softener, which filters out the ions
**D.** Hard ions in water are exchanged for ions that do not interact as strongly with soaps

**46.** Which of the following compounds are soluble in water?

I. $Mn(OH)_2$ II. $Cr(NO_3)_3$ III. $Ni_3(PO_4)_2$

**A.** I only
**B.** II only
**C.** III only
**D.** I and III only

**47.** The hydration number of an ion is the number of:

**A.** water molecules bonded to an ion in an aqueous solution
**B.** water molecules required to dissolve one mole of ions
**C.** ions bonded to one mole of water molecules
**D.** ions dissolved in one liter of an aqueous solution

**48.** When a solid dissolves, each molecule is removed from the crystal by interaction with the solvent. This process of surrounding each ion with solvent molecules is called:

**A.** hemolysis
**B.** electrolysis
**C.** solvation
**D.** dilution

**49.** The term *miscible* describes which type of solution?

**A.** Solid/solid
**B.** Liquid/gas
**C.** Liquid/solid
**D.** Liquid/liquid

**50.** Apply the *like dissolves like* rule to predict which of the following vitamins is insoluble in water:

**A.** niacinamide ($C_6H_6N_2O$)
**B.** pyridoxine ($C_8H_{11}NO_3$)
**C.** retinol ($C_{20}H_{30}O$)
**D.** thiamine ($C_{12}H_{17}N_4OS$)

**51.** Which species is NOT written as its constituent ions when the equation is expanded into the ionic equation?

$$Mg(OH)_2\ (s) + 2\ HCl\ (aq) \rightarrow MgCl_2\ (aq) + 2\ H_2O\ (l)$$

**A.** $Mg(OH)_2$ only
**B.** $H_2O$ and $Mg(OH)_2$
**C.** HCl
**D.** $MgCl_2$

**52.** If $x$ moles of $PbCl_2$ fully dissociate in 1 liter of $H_2O$, the $K_{sp}$ is equivalent to:

**A.** $x^2$
**B.** $2x^4$
**C.** $4x^3$
**D.** $2x^3$

 

**53.** Which of the following are strong electrolytes?

I. salts II. strong bases III. weak acids

**A.** I only
**B.** I and II only
**C.** III only
**D.** I, II and III

**54.** What are the spectator ions in the reaction between KOH and $HNO_3$?

**A.** $K^+$ and $NO_3^-$
**B.** $H^+$ and $NO_3^-$
**C.** $K^+$ and $H^+$
**D.** $H^+$ and $^-OH$

**55.** What volume of 14 M acid must be diluted with distilled water to prepare 6.0 L of 0.20 M acid?

**A.** 86 mL
**B.** 62 mL
**C.** 0.94 mL
**D.** 6.8 mL

**56.** What is the molarity of the solution obtained by diluting 160 mL of 4.50 M NaOH to 595 mL?

**A.** 0.242 M
**B.** 1.21 M
**C.** 2.42 M
**D.** 1.72 M

**57.** Which compound is more soluble in the nonpolar solvent of benzene than in water?

**A.** $SO_2$
**B.** $CO_2$
**C.** Silver chloride
**D.** $H_2S$

**58.** Which of the following concentrations is dependent on temperature?

**A.** Mole fraction
**B.** Molarity
**C.** Mass percent
**D.** Molality

**59.** Which of the following solutions is the most concentrated?

**A.** One liter of water with 1 gram of sugar
**B.** One liter of water with 2 grams of sugar
**C.** One liter of water with 5 grams of sugar
**D.** One liter of water with 10 grams of sugar

**60.** What mass of NaOH is contained in 75.0 mL of a 5.0% (w/v) NaOH solution?

**A.** 6.50 g
**B.** 15.0 g
**C.** 3.75 g
**D.** 0.65 g

 

## Practice Set 4: Questions 61–80

Questions **61** through **63** are based on the following data:

| | $K_{sp}$ |
|---|---|
| $PbCl_2$ | $1.0 \times 10^{-5}$ |
| AgCl | $1.0 \times 10^{-10}$ |
| $PbCO_3$ | $1.0 \times 10^{-15}$ |

**61.** Consider a saturated solution of $PbCl_2$. The addition of NaCl would:

I. decrease $[Pb^{2+}]$
II. increase the precipitation of $PbCl_2$
III. not affect the precipitation of $PbCl_2$

**A.** I only
**B.** II only
**C.** III only
**D.** I and II only

**62.** What occurs when $AgNO_3$ is added to a saturated solution of $PbCl_2$?

I. AgCl precipitates
II. $Pb(NO_3)_2$ forms a white precipitate
III. More $PbCl_2$ forms

**A.** I only
**B.** II only
**C.** III only
**D.** I, II and III

**63.** Comparing equal volumes of saturated solutions for $PbCl_2$ and AgCl, which solution contains a greater concentration of $Cl^-$?

I. $PbCl_2$ II. AgCl III. Both have the same concentration of $Cl^-$

**A.** I only
**B.** II only
**C.** III only
**D.** I and II only

**64.** Which of the following is the reason why hexane is significantly soluble in octane?

**A.** Entropy increases for the two substances as the dominant factor in the $\Delta G$ when mixed
**B.** Hexane hydrogen bonds with octane
**C.** Intermolecular bonds between hexane-octane are much stronger than hexane-hexane or octane-octane molecular bonds
**D.** $\Delta H$ for hexane-octane is greater than hexane-$H_2O$

**65.** Which of the following are characteristics of an ideally dilute solution?

I. Solute molecules do not interact with each other
II. Solvent molecules do not interact with each other
III. The mole fraction of the solvent approaches 1

**A.** I only
**B.** II only
**C.** I and III only
**D.** I, II and III

**66.** Which principle states that the solubility of a gas in a liquid is proportional to the partial pressure of the gas above the liquid?

**A.** Solubility principle
**B.** Tyndall effect
**C.** Colloid principle
**D.** Henry's law

**67.** Water and methanol are two liquids that dissolve in each other. When the two are mixed, they form one layer because the liquids are:

**A.** unsaturated
**B.** saturated
**C.** miscible
**D.** immiscible

**68.** What is the molarity of a glucose solution that contains 10.0 g of $C_6H_{12}O_6$ dissolved in 100.0 mL of solution? (Use the molecular mass of $C_6H_{12}O_6$ = 180.0 g/mol)

**A.** 1.80 M
**B.** 0.555 M
**C.** 0.0555 M
**D.** 0.00555 M

**69.** Which of the following solid compounds is insoluble in water?

I. $BaSO_4$ II. $Hg_2Cl_2$ III. $PbCl_2$

**A.** I only
**B.** II only
**C.** III only
**D.** I, II and III

**70.** The net ionic equation for the reaction between zinc and a hydrochloric acid solution is:

**A.** $Zn\ (s) + 2\ H^+\ (aq) + 2\ Cl^-\ (aq) \rightarrow Zn^{2+}\ (aq) + 2\ Cl^-\ (aq) + H_2\ (g)$
**B.** $ZnCl_2\ (aq) + H_2\ (g) \rightarrow Zn\ (s) + 2\ HCl\ (aq)$
**C.** $Zn\ (s) + 2\ H^+\ (aq) \rightarrow Zn^{2+}\ (aq) + H_2\ (g)$
**D.** $Zn\ (s) + 2HCl\ (aq) \rightarrow ZnCl_2\ (aq) + H_2\ (g)$

**71.** Apply the *like dissolves like* rule to predict which of the following liquids is/are miscible with water:

I. methyl ethyl ketone, $C_4H_8O$
II. glycerin, $C_3H_5(OH)_3$
III. formic acid, $HCHO_2$

**A.** I only
**B.** II only
**C.** III only
**D.** I, II and III

**72.** What is the $K_{sp}$ for calcium fluoride ($CaF_2$) if the calcium ion concentration in a saturated solution is 0.00021 *M*?

**A.** $K_{sp} = 3.7 \times 10^{-11}$
**B.** $K_{sp} = 2.6 \times 10^{-10}$
**C.** $K_{sp} = 3.6 \times 10^{-9}$
**D.** $K_{sp} = 8.1 \times 10^{-10}$

**73.** A 4 M solution of $H_3A$ is completely dissociated in water. How many equivalents of $H^+$ are in 1/3 liter?

**A.** ¼
**B.** 4
**C.** 1.5
**D.** 3

**74.** Which of the following aqueous solutions is a poor conductor of electricity?

I. sucrose, $C_{12}H_{22}O_{11}$
II. barium nitrate, $Ba(NO_3)_2$
III. calcium bromide, $CaBr_2$

**A.** I only
**B.** II only
**C.** III only
**D.** I and II only

**75.** Which of the following solid compounds is insoluble in water?

**A.** $BaSO_4$
**B.** $Na_2S$
**C.** $(NH_4)_2CO_3$
**D.** $K_2CrO_4$

**76.** Which is true if the ion concentration product of a solution of AgCl is less than the $K_{sp}$?

I. Precipitation occurs
II. The ions are insoluble in water
III. Precipitation does not occur

**A.** I only
**B.** II only
**C.** III only
**D.** I and II only

**77.** Under which conditions is the expected solubility of oxygen gas in water the highest?

**A.** High temperature and high $O_2$ pressure above the solution
**B.** Low temperature and low $O_2$ pressure above the solution
**C.** Low temperature and high $O_2$ pressure above the solution
**D.** High temperature and low $O_2$ pressure above the solution

**78.** What is the molarity of an 8.60 molal methanol ($CH_3OH$) solution with a density of 0.94 g/mL?

**A.** 0.155 M
**B.** 23.5 M
**C.** 6.34 M
**D.** 9.68 M

**79.** Which of the following is NOT a unit factor related to a 15.0% aqueous potassium iodide (KI) solution?

**A.** 100 g solution / 85.0 g water
**B.** 15.0 g KI / 100 g water
**C.** 85.0 g water / 100 g solution
**D.** 15.0 g KI / 85.0 g water

**80.** What is the molar concentration of a solution containing 0.75 mol of solute in 75 $cm^3$ of a solution?

**A.** 0.1 M
**B.** 1.5 M
**C.** 3 M
**D.** 10 M

---

**Practice Set 5: Questions 81–107**

---

**81.** If 25.0 mL of seawater has a mass of 25.88 g and contains 1.35 g of solute, what is the mass/mass percent concentration of solute in the seawater sample?

**A.** 1.14%
**B.** 12.84%
**C.** 2.62%
**D.** 5.22%

**82.** Which of the following statements describing solutions is NOT true?

**A.** Solutions are colorless
**B.** The particles in a solution are atomic or molecular
**C.** Making a solution involves a physical change
**D.** Solutions are homogeneous

**83.** Calculate the solubility product of AgCl if the solubility of AgCl in $H_2O$ is $1.3 \times 10^{-4}$ mol/L?

**A.** $1.3 \times 10^{-4}$
**B.** $1.7 \times 10^{-8}$
**C.** $2.6 \times 10^{-4}$
**D.** $3.9 \times 10^{-5}$

**84.** Which of the following solutions is the most dilute?

**A.** 0.1 liters of $H_2O$ with 1 gram of sugar
**B.** 0.2 liters of $H_2O$ with 2 grams of sugar
**C.** 0.5 liters of $H_2O$ with 5 grams of sugar
**D.** All have the same concentration

**85.** When salt A is dissolved into water to form a 1 molar unsaturated solution, the temperature of the solution decreases. Under these conditions, which statement is accurate when salt A is dissolved in water?

**A.** $\Delta H°$ and $\Delta G°$ are positive
**B.** $\Delta H°$ is positive and $\Delta G°$ is negative
**C.** $\Delta H°$ is negative and $\Delta G°$ is positive
**D.** $\Delta H°$ and $\Delta G°$ are negative

**86.** The heat of a solution measures the energy absorbed during:

I. the formation of solvent-solute bonds
II. the breaking of solute-solute bonds
III. the breaking of solvent-solvent bonds

**A.** I only
**B.** II only
**C.** I and III only
**D.** II and III only

**87.** Why does the reaction proceed if, when solid potassium chloride is dissolved in $H_2O$, the energy of the bonds formed is less than that of the broken bonds?

**A.** The increased disorder due to mixing increases entropy within the system
**B.** The reaction does not take place under standard conditions
**C.** The decreased disorder due to mixing decreases entropy within the system
**D.** Remaining potassium chloride, which does not dissolve, offsets the portion that dissolves

**88.** Why are salts more soluble in $H_2O$ than in benzene?

**A.** Benzene is aromatic and, therefore, very stable
**B.** The dipole moment of $H_2O$ compensates for the loss of ionic bonding when salt dissolves
**C.** Strong intermolecular attractions in benzene must be disrupted to dissolve salt in benzene
**D.** The molecular mass of $H_2O$ is similar to the atomic mass of most ions

**89.** What is the formula of the solid formed when aqueous barium chloride is mixed with aqueous potassium chromate?

**A.** $K_2CrO_4$
**B.** $K_2Ba$
**C.** KCl
**D.** $BaCrO_4$

**90.** Why might sodium carbonate (washing soda, $Na_2CO_3$) be added to hard water for cleaning?

**A.** The soap gets softer due to the added ions
**B.** The ions solubilize the soap due to ion-ion intermolecular attraction, which improves the cleaning ability
**C.** The hard ions in the water are more attracted to the carbonate ions' –2 charge
**D.** The hard ions are dissolved by the added sodium ions

 

**91.** What volume of 0.25 M hydrochloric acid reacts completely with 0.400 g of sodium hydrogen carbonate, $NaHCO_3$? (Use the molar mass of $NaHCO_3$ = 84.0 g/mol)

$$NaHCO_3\ (s) + HCl\ (aq) \rightarrow NaCl\ (aq) + H_2O\ (l) + CO_2\ (g)$$

**A.** 142 mL
**B.** 34.6 mL
**C.** 19.0 mL
**D.** 54.6 mL

**92.** What is the ppm (m/m) concentration if $8.8 \times 10^{-3}$ g of a contaminant is present in 5,246 g of a particular solution?

**A.** 12.40 ppm
**B.** 1.53 ppm
**C.** 4.06 ppm
**D.** 1.68 ppm

**93.** Which of the following structures represents the bicarbonate ion?

**A.** $HCO_3^-$
**B.** $H_2CO_3^{2-}$
**C.** $CO_3^-$
**D.** $CO_3^{2-}$

**94.** Which of the following can serve as the solute in a solution?

I. solid II. liquid III. gas

**A.** I only
**B.** II only
**C.** III only
**D.** I, II and III

**95.** Hydration involves the:

**A.** formation of water–solute bonds
**B.** breaking of water–water bonds
**C.** breaking of water–solute bonds
**D.** breaking of water-water bonds and formation of water-solute bonds

**96.** Which is the name for a substance represented by a formula written as $M_xLO_y \cdot zH_2O$?

**A.** Solvent
**B.** Solute
**C.** Solid hydrate
**D.** Colloid

 

**97.** In the reaction between aqueous silver nitrate and aqueous potassium chromate, what is the identity of the soluble substance formed?

**A.** Potassium nitrate
**B.** Potassium chromate
**C.** Silver nitrate
**D.** Silver chromate

**98.** Apply the *like dissolves like* rule to predict which of the following liquids is miscible with liquid bromine ($Br_2$).

I. benzene ($C_6H_6$) II. hexane ($C_6H_{14}$) III. carbon tetrachloride ($CCl_4$)

**A.** I only
**B.** II only
**C.** III only
**D.** I, II and III

**99.** Hydrochloric acid contains:

**A.** ionic bonds and is a weak electrolyte
**B.** hydrogen bonds and is a weak electrolyte
**C.** covalent bonds and is a strong electrolyte
**D.** ionic bonds and is a non-electrolyte

**100.** At 22 °C, a one-liter sample of pure water has a vapor pressure of 18.5 torr. If 7.5 g of NaCl is added to the sample, what is the result of the vapor pressure of the water?

**A.** It equals the vapor pressure of the NaCl added to the sample
**B.** It remains unchanged at 18.5 torr
**C.** It increases the vapor pressure
**D.** It decreases the vapor pressure

**101.** Barium hydroxide ($Ba(OH)_2$) is dissolved at room temperature in pure water. From the equilibrium concentrations of $Ba^+$ and $^-OH$ ions, which $K_{sp}$ expression could be used to find the solubility product for barium hydroxide?

**A.** $K_{sp} = [Ba^{2+}]\cdot[^-OH]^2$
**B.** $K_{sp} = 2[Ba^{2+}]\cdot[^-OH]^2$
**C.** $K_{sp} = [Ba^{2+}]\cdot[^-OH]$
**D.** $K_{sp} = [Ba^{2+}] \times 2[^-OH]$

**102.** How many grams of $H_3PO_4$ are needed to make 200 mL of a 0.2 M $H_3PO_4$ solution?

**A.** 3.9 g **B.** 5.8 g **C.** 0.34 g **D.** 2.6 g

 

**103.** What is the concentration of $^{-}OH$ if the concentration of $H_3O^+$ is $1 \times 10^{-8}$ M? (Use the expression $[H_3O^+] \times [^{-}OH] = K_w = 1 \times 10^{-14}$)

**A.** $1 \times 10^8$
**B.** $1 \times 10^6$
**C.** $1 \times 10^{-14}$
**D.** $1 \times 10^{-6}$

**104.** What is the molarity of a solution prepared by dissolving 4.50 mol NaCl in enough water to make 1.50 L of solution?

**A.** 6.58 M
**B.** 3.0 M
**C.** 4.65 M
**D.** 7.68 M

**105.** What is the net ionic equation for the reaction between $HNO_3$ and $Na_2SO_3$?

**A.** $H_2O + SO_2\ (g) \rightarrow H^+\ (aq) + HSO_3^-\ (aq)$
**B.** $2\ H^+\ (aq) + SO_3^{2-}\ (aq) \rightarrow H_2O + SO_2\ (g)$
**C.** $HNO_3\ (aq) + Na_2SO_3\ (aq) \rightarrow H_2O + SO_2\ (g) + NO_3^-\ (aq)$
**D.** $H_2SO_3\ (aq) \rightarrow H_2O + SO_2\ (g)$

**106.** Which percentage concentration units are used by chemists?

I. % (v/v)  II. % (m/v)  III. % (m/m)

**A.** I only
**B.** II only
**C.** III only
**D.** I, II and III

**107.** Which of the following aqueous solutions are poor conductors of electricity?

I. ethanol, $CH_3CH_2OH$
II. ammonium chloride, $NH_4Cl$
III. ethylene glycol, $HOCH_2CH_2OH$

**A.** I only
**B.** II only
**C.** III only
**D.** I and III only

*Notes for active learning*

 

# 9 – Acids and Bases

==================================================================

**Practice Set 1: Questions 1–20**

==================================================================

**1.** Which is the conjugate acid–base pair in the reaction?

$CH_3NH_2 + HCl \leftrightarrow CH_3NH_3^+ + Cl^-$

**A.** HCl and $Cl^-$
**B.** $CH_3NH_3^+$ and $Cl^-$
**C.** $CH_3NH_2$ and $Cl^-$
**D.** $CH_3NH_2$ and HCl

**2.** What is the pH of an aqueous solution if the $[H^+] = 0.10$ M?

**A.** 0.0
**B.** 1.0
**C.** 2.0
**D.** 10.0

**3.** In which of the following pairs of substances are both species salts?

**A.** $NH_4F$ and KCl
**B.** $CaCl_2$ and HCN
**C.** LiOH and $K_2CO_3$
**D.** NaOH and $CaCl_2$

**4.** Which reactant is a Brønsted-Lowry acid?

$HCl\ (aq) + KHS\ (aq) \rightarrow KCl\ (aq) + H_2S\ (aq)$

**A.** KCl
**B.** $H_2S$
**C.** HCl
**D.** KHS

**5.** If a light bulb in a conductivity apparatus glows brightly when testing a solution, which of the following must be true about the solution?

**A.** It is highly reactive
**B.** It is slightly reactive
**C.** It is highly ionized
**D.** It is slightly ionized

**6.** What is the term for a substance capable of donating or accepting a proton in an acid–base reaction?

I. Nonprotic II. Aprotic III. Amphoteric

**A.** I only
**B.** II only
**C.** III only
**D.** I and III only

**7.** Which of the following compounds is a strong acid?

I. $HClO_4$ (*aq*) II. $H_2SO_4$ (*aq*) III. $HNO_3$ (*aq*)

**A.** I only
**B.** II only
**C.** II and III only
**D.** I, II and III

**8.** What is the approximate pH of a strong acid solution where $[H_3O^+] = 8.30 \times 10^{-5}$?

**A.** 4 **B.** 11 **C.** 3 **D.** 5

**9.** Which of the following reactions represents the ionization of $H_2O$?

**A.** $H_2O + H_2O \rightarrow 2\ H_2 + O_2$
**B.** $H_2O + H_2O \rightarrow H_3O^+ + {}^-OH$
**C.** $H_2O + H_3O^+ \rightarrow H_3O^+ + H_2O$
**D.** $H_3O^+ + {}^-OH \rightarrow H_2O + H_2O$

**10.** Which set below contains only weak electrolytes?

**A.** $NH_4Cl$ (*aq*), $HClO_2$ (*aq*), HCN (*aq*)
**B.** $NH_3$ (*aq*), $HCO_3^-$ (*aq*), HCN (*aq*)
**C.** KOH (*aq*), $H_3PO_4$ (*aq*), $NaClO_4$ (*aq*)
**D.** $HNO_3$ (*aq*), $H_2SO_4$ (*aq*), HCN (*aq*)

**11.** Which of the following statements describes a neutral solution?

**A.** $[H_3O^+] / [{}^-OH] = 1 \times 10^{-14}$
**B.** $[H_3O^+] / [{}^-OH] = 1$
**C.** $[H_3O^+] < [{}^-OH]$
**D.** $[H_3O^+] > [{}^-OH]$

**12.** Which of the following is an example of an Arrhenius acid?

**A.** $H_2O$ (*l*)
**B.** RbOH (*aq)*
**C.** $Ba(OH)_2$ (*aq*)
**D.** None of the above

 

**13.** In the following reaction, which reactant is a Brønsted-Lowry base?

$H_2CO_3$ (*aq*) + $Na_2HPO_4$ (*aq*) → $NaHCO_3$ (*aq*) + $NaH_2PO_4$ (*aq*)

**A.** $NaHCO_3$
**B.** $NaH_2PO_4$
**C.** $Na_2HPO_4$
**D.** $H_2CO_3$

**14.** Which of the following is the conjugate base of $^-OH$?

**A.** $O_2$
**B.** $O^{2-}$
**C.** $H_2O$
**D.** $O^-$

**15.** Which of the following describes the solution for a vinegar sample at a pH of 5?

**A.** Weakly basic
**B.** Neutral
**C.** Weakly acidic
**D.** Strongly acidic

**16.** What is the pI for glutamic acid that contains two carboxylic acid groups and an amino group? (Use the carboxyl $pK_{a1} = 2.2$, carboxyl $pK_{a2} = 4.2$ and amino $pK_a = 9.7$)

**A.** 3.2
**B.** 1.0
**C.** 6.4
**D.** 5.4

**17.** Which of the following compounds cannot act as an acid?

**A.** $NH_3$
**B.** $H_2SO_4$
**C.** $HSO_4^{1-}$
**D.** $SO_4^{2-}$

**18.** A weak acid is titrated with a strong base. When the concentration of the conjugate base is equal to the concentration of the acid, the titration is at the:

**A.** endpoint
**B.** equivalence point
**C.** buffering region
**D.** diprotic point

**19.** If $[H_3O^+]$ in an aqueous solution is $7.5 \times 10^{-9}$ M, what is the $[^-OH]$?

**A.** $6.4 \times 10^{-5}$ M
**B.** $3.8 \times 10^{+8}$ M
**C.** $7.5 \times 10^{-23}$ M
**D.** $1.3 \times 10^{-6}$ M

**20.** Which species has a $K_a$ of $5.7 \times 10^{-10}$ if $NH_3$ has a $K_b$ of $1.8 \times 10^{-5}$?

**A.** $H^+$
**B.** $NH_2^-$
**C.** $NH_4^+$
**D.** $H_2O$

================================================================

**Practice Set 2: Questions 21–40**

================================================================

**21.** Which of the following are the conjugate bases of $HSO_4^-$, $CH_3OH$, and $H_3O^+$, respectively:

**A.** $SO_4^{2-}$, $CH_2OH^-$ and $^-OH$
**B.** $CH_3O^-$, $SO_4^{2-}$ and $H_2O$
**C.** $SO_4^{2-}$, $CH_3O^-$ and $H_2O$
**D.** $SO_4^-$, $CH_2OH^-$ and $H_2O$

**22.** If 30.0 mL of 0.10 M $Ca(OH)_2$ is titrated with 0.20 M $HNO_3$, what volume of nitric acid is required to neutralize the base according to the following expression?

$$2\ HNO_3\ (aq) + Ca(OH)_2\ (aq) \rightarrow 2\ Ca(NO_3)_2\ (aq) + 2\ H_2O\ (l)$$

**A.** 30.0 mL
**B.** 15.0 mL
**C.** 10.0 mL
**D.** 20.0 mL

**23.** Which of the following expressions describes an acidic solution?

**A.** $[H_3O^+] / [^-OH] = 1 \times 10^{-14}$
**B.** $[H_3O^+] \times [^-OH] \neq 1 \times 10^{-14}$
**C.** $[H_3O^+] < [^-OH]$
**D.** $[H_3O^+] > [^-OH]$

**24.** Which is incorrectly classified as an acid, a base, or an amphoteric species?

**A.** LiOH / base
**B.** $H_2O$ / amphoteric
**C.** $H_2S$ / acid
**D.** $NH_4^+$ / base

**25.** Which of the following is the strongest weak acid?

**A.** $CH_3COOH$; $K_a = 1.8 \times 10^{-5}$
**B.** HF; $K_a = 6.5 \times 10^{-4}$
**C.** HCN; $K_a = 6.3 \times 10^{-10}$
**D.** HClO; $K_a = 3.0 \times 10^{-8}$

**26.** What are the products from the complete neutralization of phosphoric acid with aqueous lithium hydroxide?

**A.** $LiHPO_4$ $(aq)$ and $H_2O$ $(l)$
**B.** $Li_3PO_4$ $(aq)$ and $H_2O$ $(l)$
**C.** $Li_2HPO_4$ $(aq)$ and $H_2O$ $(l)$
**D.** $LiH_2PO_4$ $(aq)$ and $H_2O$ $(l)$

 

**27.** Which of the following compounds is NOT a strong base?

**A.** $Ca(OH)_2$
**B.** $Fe(OH)_3$
**C.** KOH
**D.** NaOH

**28.** What is the $[H^+]$ in stomach acid that registers a pH of 2.0 on a strip of pH paper?

**A.** 0.2 M
**B.** 0.1 M
**C.** 0.02 M
**D.** 0.01 M

**29.** Which statement is true about distinguishing between dissociation and ionization?

**A.** Ionization is the separation of existing charged particles
**B.** Dissociation produces new charged particles
**C.** Ionization involves polar covalent compounds
**D.** Dissociation involves polar covalent compounds

**30.** Which of the following is a general property of a basic solution?

I. Turns litmus paper red
II. Tastes sour
III. Causes the skin of the fingers to feel slippery

**A.** I only
**B.** II only
**C.** III only
**D.** I and II only

**31.** Which of the following compound-classification pairs is incorrectly matched?

**A.** $Ca(OH)_2$ – weak base
**B.** $LiC_2H_3O_2$ –salt
**C.** $NH_3$ – weak base
**D.** HI – strong acid

**32.** What is the term for a substance that releases $H^+$ in $H_2O$?

**A.** Brønsted-Lowry acid
**B.** Brønsted-Lowry base
**C.** Arrhenius acid
**D.** Arrhenius base

**33.** Which molecule is acting as a base in the following reaction?

$$^{-}OH + NH_4^+ \rightarrow H_2O + NH_3$$

**A.** $^{-}OH$
**B.** $NH_4^+$
**C.** $H_2O$
**D.** $NH_3$

**34.** Citric acid is a triprotic acid with three carboxylic acid groups having p$K_a$ values of 3.2, 4.8 and 6.4. At a pH of 5.7, what is the predominant protonation state of citric acid?

**A.** Three carboxylic acid groups are deprotonated
**B.** Three carboxylic acid groups are protonated
**C.** One carboxylic acid group is deprotonated, while two are protonated
**D.** Two carboxylic acid groups are deprotonated, while one is protonated

**35.** When fully neutralized by treatment with barium hydroxide, a phosphoric acid yields $Ba_2P_2O_7$ as one of its products. The parent acid for the anion in this compound is:

**A.** tetraprotic acid
**B.** diprotic acid
**C.** triprotic acid
**D.** hexaprotic acid

**36.** Is a solution more or less acidic when a weak acid solution is added to a concentrated HCl solution?

**A.** Less acidic because the concentration of $OH^-$ increases
**B.** No change in acidity because [HCl] is too high to be changed by the weak solution
**C.** Less acidic because the solution becomes more dilute with a less concentrated solution of $H_3O^+$ being added
**D.** More acidic because more $H_3O^+$ is being added to the solution

**37.** Which of the following is a triprotic acid?

**A.** $HNO_3$
**B.** $H_3PO_4$
**C.** $H_2SO_3$
**D.** $HC_2H_3O_2$

**38.** For which of the following pairs of substances do the two pair members NOT react?

**A.** $Na_3PO_4$ and HCl
**B.** KCl and NaI
**C.** HF and LiOH
**D.** $PbCl_2$ and $H_2SO_4$

**39.** What happens to the pH when sodium acetate is added to an acetic acid solution?

**A.** Decreases due to the common ion effect
**B.** Increases due to the common ion effect
**C.** Remains constant because sodium acetate is a buffer
**D.** Remains constant because sodium acetate is neither acidic nor basic

 

**40.** Which of the following is the acidic anhydride of phosphoric acid ($H_3PO_4$)?

**A.** $P_2O$ **B.** $P_2O_3$ **C.** $PO_3$ **D.** $P_4O_{10}$

================================================================

**Practice Set 3: Questions 41–60**

================================================================

**41.** Which of the following does NOT act as a Brønsted-Lowry acid?

**A.** $CO_3^{2-}$
**B.** $HS^-$
**C.** $HSO_4^-$
**D.** $H_2O$

**42.** Which of the following is the strongest weak acid?

**A.** HF ; $pK_a = 3.17$
**B.** $HCO_3^-$ ; $pK_a = 10.32$
**C.** $H_2PO_4^-$ ; $pK_a = 7.18$
**D.** $NH_4^+$ ; $pK_a = 9.20$

**43.** Why does boiler scale form on the walls of hot water pipes from groundwater?

**A.** Transformation of $H_2PO_4^-$ ions to $PO_4^{3-}$ ions, which precipitate with the "hardness ions," $Ca^{2+}$, $Mg^{2+}$, $Fe^{2+}/Fe^{3+}$

**B.** Transformation of $HSO_4^-$ ions to $SO_4^{2-}$ ions, which precipitate with the "hardness ions," $Ca^{2+}$, $Mg^{2+}$, $Fe^{2+}/Fe^{3+}$

**C.** Transformation of $HSO_3^-$ ions to $SO_3^{2-}$ ions, which precipitate with the "hardness ions," $Ca^{2+}$, $Mg^{2+}$, $Fe^{2+}/Fe^{3+}$

**D.** Transformation of $HCO_3^-$ ions to $CO_3^{2-}$ ions, which precipitate with the "hardness ions," $Ca^{2+}$, $Mg^{2+}$, $Fe^{2+}/Fe^{3+}$

**44.** Which of the following substances, when added to a sulfoxylic acid ($H_2SO_2$) solution, could be used to prepare a buffer solution?

**A.** $H_2O$
**B.** $NaHSO_2$
**C.** KCl
**D.** HCl

**45.** Which of the following statements is NOT correct?

**A.** Acidic salts are formed by partial neutralization of a diprotic acid by a diprotic base
**B.** Acidic salts are formed by partial neutralization of a triprotic acid by a diprotic base
**C.** Acidic salts are formed by partial neutralization of a monoprotic acid by a monoprotic base
**D.** Acidic salts are formed by partial neutralization of a diprotic acid by a monoprotic base

 

**46.** Which of the following is the acid anhydride for $HClO_4$?

**A.** $ClO$
**B.** $ClO_2$
**C.** $ClO_3$
**D.** $Cl_2O_7$

**47.** Identify the acid/base behavior of each substance for the reaction:

$$H_3O^+ + Cl^- \rightleftharpoons H_2O + HCl$$

**A.** $H_3O^+$ acts as an acid, $Cl^-$ acts as a base, $H_2O$ acts as a base, and HCl acts as an acid
**B.** $H_3O^+$ acts as a base, $Cl^-$ acts as an acid, $H_2O$ acts as a base, and HCl acts as an acid
**C.** $H_3O^+$ acts as an acid, $Cl^-$ acts as a base, $H_2O$ acts as an acid, and HCl acts as a base
**D.** $H_3O^+$ acts as a base, $Cl^-$ acts as an acid, $H_2O$ acts as an acid, and HCl acts as a base

**48.** Given the p$K_a$ values for phosphoric acid of 2.15, 6.87, and 12.35, what is the $HPO_4^{2-}$ / $H_2PO_4^-$ ratio in a typical muscle cell when the pH is 7.35?

**A.** $6.32 \times 10^{-6}$
**B.** $1.18 \times 10^5$
**C.** 0.46
**D.** 3.02

**49.** If a light bulb in a conductivity apparatus glows dimly when testing a solution, which of the following must be true about the solution?

I. It is slightly reactive
II. It is slightly ionized
III. It is highly ionized

**A.** I only
**B.** II only
**C.** III only
**D.** I and II only

**50.** Which of the following properties is NOT characteristic of an acid?

**A.** It is neutralized by a base
**B.** It has a slippery feel
**C.** It produces $H^+$ in water
**D.** It tastes sour

**51.** What is the term for a substance that releases hydroxide ions in water?

**A.** Brønsted-Lowry base
**B.** Brønsted-Lowry acid
**C.** Arrhenius base
**D.** Arrhenius acid

 

**52.** For the reaction below, which of the following is the conjugate acid of $C_5H_5N$?

$C_5H_5N + H_2CO_3 \leftrightarrow C_5H_6N^+ + HCO_3^-$

**A.** $C_5H_6N^+$ **B.** $HCO_3^-$ **C.** $C_5H_5N$ **D.** $H_2CO_3$

**53.** Which of the following terms applies to $Cl^-$ in the reaction below?

$HCl\ (aq) \rightarrow H^+ + Cl^-$

**A.** Weak conjugate base
**B.** Strong conjugate base
**C.** Weak conjugate acid
**D.** Strong conjugate acid

**54.** Which of the following compounds is a diprotic acid?

**A.** HCl
**B.** $H_3PO_4$
**C.** $HNO_3$
**D.** $H_2SO_3$

**55.** Lysine contains two amine groups (p$K_a$ = 9.0 and 10.0) and a carboxylic acid group (p$K_a$ = 2.2). In a solution of pH 9.5, which describes the protonation and charge state of lysine?

**A.** Carboxylic acid is deprotonated and negative; amine (p$K_a$ 9.0) is deprotonated and neutral, whereby the amine (p$K_a$ = 10.0) is protonated and positive
**B.** Carboxylic acid is deprotonated and negative; amine (p$K_a$ = 9.0) is protonated and positive, whereby the amine (p$K_a$ = 10.0) is deprotonated and neutral
**C.** Carboxylic acid is deprotonated and neutral; both amines are protonated and positive
**D.** Carboxylic acid is deprotonated and negative; both amines are deprotonated and neutral

**56.** Which of the following is the chemical species present in acidic solutions?

**A.** $H_2O^+\ (aq)$
**B.** $H_3O^+\ (l)$
**C.** $H_3O^+\ (aq)$
**D.** $^-OH\ (aq)$

**57.** Which compound has a value of $K_a$ that is approximately equal to $10^{-5}$?

**A.** $CH_3CH_2CH_2CO_2H$
**B.** KOH
**C.** NaBr
**D.** $HNO_3$

 

**58.** Relative to a pH of 7, a solution with a pH of 4 has:

**A.** 30 times less $[H^+]$
**B.** 300 times less $[H^+]$
**C.** 1,000 times greater $[H^+]$
**D.** 300 times greater $[H^+]$

**59.** What is the pH of this buffer system if the concentration of undissociated weak acid is equal to the concentration of the conjugate base? (Use the $K_a$ of the buffer = $4.6 \times 10^{-4}$)

**A.** 1 and 2
**B.** 3 and 4
**C.** 5 and 6
**D.** 7 and 8

**60.** Which of the following is the ionization constant expression for water?

**A.** $K_w = [H_2O] / [H^+]\cdot[^-OH]$
**B.** $K_w = [H^+]\cdot[^-OH] / [H_2O]$
**C.** $K_w = [H^+]\cdot[^-OH]$
**D.** $K_w = [H_2O]\cdot[H_2O]$

---

**Practice Set 4: Questions 61–80**

---

**61.** Which of the following statements about strong or weak acids is true?

**A.** A weak acid reacts with a strong base
**B.** A strong acid does not react with a strong base
**C.** A weak acid readily forms ions when dissolved in water
**D.** A weak acid and a strong acid at the same concentration are equally corrosive

**62.** What is the value of $K_w$ at 25 °C?

**A.** 1.0
**B.** $1.0 \times 10^{-7}$
**C.** $1.0 \times 10^{-14}$
**D.** $1.0 \times 10^{7}$

**63.** Which of the statements below best describes the following reaction?

$$HNO_3\ (aq) + LiOH\ (aq) \rightarrow LiNO_3\ (aq) + H_2O\ (l)$$

**A.** Nitric acid and lithium hydroxide solutions produce lithium nitrate solution and $H_2O$
**B.** Nitric acid and lithium hydroxide solutions produce lithium nitrate and $H_2O$
**C.** Nitric acid and lithium hydroxide produce lithium nitrate and $H_2O$
**D.** Aqueous solutions of nitric acid and lithium hydroxide produce aqueous lithium nitrate and $H_2O$

**64.** The Brønsted-Lowry acid and base in the following reaction are, respectively:

$$NH_4^+ + CN^- \rightarrow NH_3 + HCN$$

**A.** $NH_4^+$ and $^-CN$
**B.** $^-CN$ and HCN
**C.** $NH_4^+$ and HCN
**D.** $NH_3$ and $^-CN$

**65.** Which would NOT be used to make a buffer solution?

**A.** $H_2SO_4$
**B.** $H_2CO_3$
**C.** $NH_4OH$
**D.** $CH_3COOH$

**66.** Which of the following is a general property of an acidic solution?

**A.** Turns litmus paper blue
**B.** Neutralizes acids
**C.** Tastes bitter
**D.** None of the above

**67.** What is the term for a solution that is a good conductor of electricity?

**A.** Strong electrolyte
**B.** Weak electrolyte
**C.** Non-electrolyte
**D.** Aqueous electrolyte

**68.** Which of the following compounds is an acid?

**A.** HBr
**B.** $C_2H_6$
**C.** KOH
**D.** NaF

**69.** A metal and a salt solution react only if the metal introduced into the solution is:

**A.** below the replaced metal in the activity series
**B.** above the replaced metal in the activity series
**C.** below hydrogen in the activity series
**D.** above hydrogen in the activity series

**70.** Which of the following is an example of an Arrhenius base?

I. NaOH (*aq*) II. $Al(OH)_3$ (*s*) III. $Ca(OH)_2$ (*aq*)

**A.** I only
**B.** II only
**C.** III only
**D.** I and III only

**71.** Which of the following is NOT a conjugate acid/base pair?

**A.** $S^{2-}$ / $H_2S$
**B.** $HSO_4^-$ / $SO_4^{2-}$
**C.** $H_2O$ / $^-OH$
**D.** $PH_4^+$ / $PH_3$

**72.** If a buffer is made with a pH below the $pK_a$ of the weak acid, the [base] / [acid] ratio is:

**A.** equal to 0
**B.** equal to 1
**C.** greater than 1
**D.** less than 1

**73.** Which of the following acids listed below has the strongest conjugate base?

| **Monoprotic Acids** | $K_a$ |
|---|---|
| Acid I | $1.3 \times 10^{-8}$ |
| Acid II | $2.9 \times 10^{-9}$ |
| Acid III | $4.2 \times 10^{-10}$ |
| Acid IV | $3.8 \times 10^{-8}$ |

**A.** I **B.** II **C.** III **D.** IV

**74.** Complete neutralization of phosphoric acid with barium hydroxide, when separated and dried, yields $Ba_3(PO_4)_2$ as one of the products. Therefore, which term describes phosphoric acid?

**A.** Monoprotic acid
**B.** Diprotic acid
**C.** Hexaprotic acid
**D.** Triprotic acid

**75.** Which is the correct net ionic equation for the hydrolysis reaction of $Na_2S$?

**A.** $Na^+ + H_2O \rightarrow NaOH + H_2$
**B.** $S^{2-} + H_2O \rightarrow OH^- + HS^-$
**C.** $S^{2-} + H_2O \rightarrow 2\ HS^- + OH^-$
**D.** $S^{2-} + 2\ H_2O \rightarrow HS^- + H_3O^+$

**76.** Calculate the pH of 0.0765 M $HNO_3$.

**A.** 1.1 **B.** 3.9 **C.** 11.7 **D.** 7. 9

**77.** Which of the following compounds is NOT a strong acid?

**A.** HBr (*aq*)
**B.** $HNO_3$
**C.** $CH_3COOH$
**D.** $H_2SO_4$

**78.** Which reaction produces $NiCr_2O_7$ as a product?

**A.** Nickel (II) hydroxide and dichromic acid
**B.** Nickel (II) hydroxide and chromic acid
**C.** Nickelic acid and chromium (II) hydroxide
**D.** Nickel (II) hydroxide and chromate acid

**79.** Which of the following statements describes a Brønsted-Lowry base?

**A.** Donates protons to other substances
**B.** Accepts protons from other substances
**C.** Produces hydrogen ions in an aqueous solution
**D.** Produces hydroxide ions in an aqueous solution

**80.** When dissolved in water, the Arrhenius acid/bases KOH, $H_2SO_4$ and $HNO_3$ are, respectively:

**A.** base, acid, and base
**B.** base, base, and acid
**C.** base, acid, and acid
**D.** acid, base, and base

---

## Practice Set 5: Questions 81–100

---

Questions **81–85** are based on the following titration graph:

The unknown acid is completely titrated with NaOH, as shown on the following titration curve:

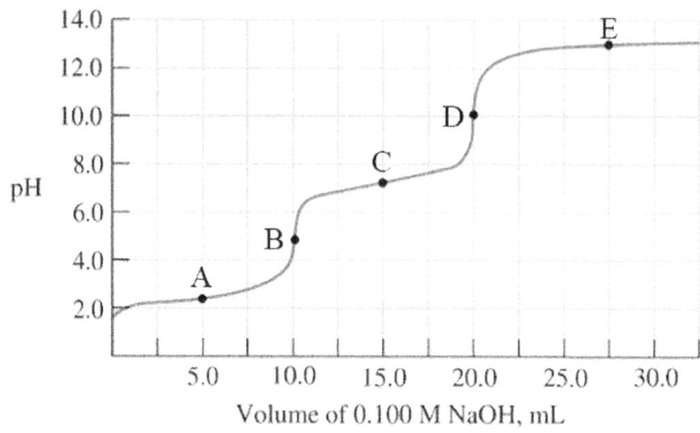


**81.** The unknown acid shown in the graph must be a(n):

**A.** monoprotic acid
**B.** diprotic acid
**C.** triprotic acid
**D.** weak acid

**82.** The p$K_{a2}$ for this acid is located at point:

**A.** A
**B.** B
**C.** C
**D.** D

**83.** At which point does the acid exist as 50% fully protonated and 50% singly deprotonated?

**A.** A
**B.** B
**C.** C
**D.** D

**84.** At which point is the acid 100% singly deprotonated?

**A.** A
**B.** B
**C.** C
**D.** D

**85.** Which points are the best buffer regions?

**A.** A and B
**B.** A and C
**C.** B and D
**D.** C and B

**86.** Sodium hydroxide is a strong base. If a concentrated solution of NaOH spills on a latex glove, it feels like water. Why is it that if the solution were to splash directly on a person's skin, it feels very slippery?

**A.** As a liquid, NaOH is slippery, but this cannot be detected through a latex glove because of the friction between the latex surfaces
**B.** NaOH destroys skin cells on contact, and the remnants of skin cells feel slippery because the cells have been lysed
**C.** NaOH lifts oil directly from skin cells, and the extruded oil causes a slippery sensation
**D.** NaOH reacts with skin oils, transforming them into soap

**87.** Which of the following statements is/are always true for a neutralization reaction?

I. Water is formed
II. It is the reaction of an $^-OH$ with an $H^+$
III. One molecule of acid neutralizes one molecule of base

**A.** I only
**B.** II only
**C.** I and III only
**D.** I and II only

**88.** Which of the following statements is/are true for strong acids?

I. They form positively charged ions when dissolved in $H_2O$
II. They form negatively charged ions when dissolved in $H_2O$
III. They are strongly polar

**A.** I only
**B.** II only
**C.** III only
**D.** I, II and III

**89.** A Brønsted-Lowry base is defined as a substance that:

**A.** increases $[H^+]$ when placed in water
**B.** decreases $[H^+]$ when placed in water
**C.** acts as a proton donor
**D.** acts as a proton acceptor

**90.** What salt-to-acid ratio is needed to prepare a buffer solution with pH = 4.0 and an acid with a $pK_a$ of 3.0?

**A.** 1:1
**B.** 10:1
**C.** 1:100
**D.** 1:5

**91.** Which must be true if an unknown solution is a poor conductor of electricity?

**A.** Solution is slightly reactive
**B.** Solution is highly corrosive
**C.** Solution is highly ionized
**D.** Solution is slightly ionized

**92.** Which of the following is a general property of a basic solution?

I. Turns litmus paper blue  II. Tastes bitter  III. Feels slippery

**A.** I only
**B.** II only
**C.** III only
**D.** I, II and III

**93.** Since $pK_a = -\log K_a$, which of the following is a correct statement?

**A.** Since the $pK_a$ for conversion of the ammonium ion to ammonia is 9.3; ammonia is a weaker base than the ammonium ion
**B.** For carbonic acid with $pK_a$ values of 6.3 and 10.3, the bicarbonate ion is a stronger base than the carbonate ion
**C.** Acetic acid ($pK_a = 4.8$) is a weaker acid than lactic acid ($pK_a = 3.9$)
**D.** Lactic acid ($pK_a = 3.9$) is weaker than other forms of phosphoric acid ($pK_a = 2.1$, 6.9 and 12.4)

**94.** Which of the following compounds is NOT a strong acid?

**A.** $HC_2H_3O_2$
**B.** $HClO_4$
**C.** $HCl$ (*aq*)
**D.** $HNO_3$

**95.** Which of the following is the acid anhydride for $H_3CCOOH$?

**A.** $H_3CCOO^-$
**B.** $H_3CCO_4CCH_3$
**C.** $H_3CCO_3CCH_3$
**D.** $H_3CCO_2CCH_3$

**96.** Acids and bases react to form:

**A.** Brønsted-Lowry acids
**B.** salts
**C.** Lewis acids
**D.** Lewis bases

 

**97.** Which does NOT act as a Brønsted-Lowry acid and base?

**A.** $H_2O$
**B.** $HCO_3^-$
**C.** $H_2PO_4^-$
**D.** $NH_4^+$

**98.** Which of the following is the conjugate acid of hydrogen phosphate, $HPO_4^{2-}$?

**A.** $H_2PO_4^{2-}$
**B.** $H_3PO_4$
**C.** $H_2PO_4^-$
**D.** $H_2PO_3^-$

**99.** A sample of $Mg(OH)_2$ salt is dissolved in water and reaches equilibrium with its dissociated ions. The addition of the strong base NaOH increases the concentration of:

**A.** $H_2O^+$
**B.** $Mg^{2+}$
**C.** undissociated sodium hydroxide
**D.** undissociated magnesium hydroxide

**100.** Which of the following is the weakest acid?

**A.** $HCO_3^-$ ; $pK_a = 10.3$
**B.** $HC_2H_3O_2$ ; $pK_a = 4.8$
**C.** $NH_4^+$ ; $pK_a = 9.2$
**D.** $H_2PO_4^-$ ; $pK_a = 7.2$

---

**Practice Set 6: Questions 101–120**

---

**101.** Which of the following acid pairs are weak acids of both chemical species?

**A.** $HC_2H_3O_2$ and HI
**B.** $H_2CO_3$ and HBr
**C.** HCN and $H_2S$
**D.** $H_3PO_4$ and $H_2SO_4$

**102.** Which indicators are yellow in an acidic solution and blue in a basic solution?

I. methyl red II. phenolphthalein III. bromothymol blue

**A.** I only
**B.** II only
**C.** I and III only
**D.** III only

**103.** Which of the following compounds is a basic anhydride?

**A.** BaO **B.** $O_2$ **C.** $CO_2$ **D.** $SO_2$

**104.** What is the $pK_a$ of an unknown acid if, in a solution at a pH of 7.0, 24% of the acid is in its deprotonated form?

**A.** 6.0 **B.** 6.5 **C.** 7.5 **D.** 8.0

**105.** When acids and bases react, which of the following, other than water, is the product?

**A.** hydronium ion
**B.** salt
**C.** hydrogen ion
**D.** hydroxide ion

**106.** Which of the following acids listed below is the strongest acid?

| Monoprotic Acids | $K_a$ |
|---|---|
| Acid I | $1.0 \times 10^{-8}$ |
| Acid II | $1.8 \times 10^{-9}$ |
| Acid III | $3.7 \times 10^{-10}$ |
| Acid IV | $4.6 \times 10^{-8}$ |

**A.** Acid I
**B.** Acid II
**C.** Acid III
**D.** Acid IV

 

**107.** In a basic solution, the pH is [...], and the $[H_3O^+]$ is [...].

**A.** $< 7$ and $> 1 \times 10^{-7}$ M

**B.** $< 7$ and $< 1 \times 10^{-7}$ M

**C.** $> 7$ and $< 1 \times 10^{-7}$ M

**D.** $< 7$ and $1 \times 10^{-7}$ M

**108.** If a battery acid solution is a strong electrolyte, which of the following must be true of the battery acid?

**A.** It is highly ionized

**B.** It is slightly ionized

**C.** It is highly reactive

**D.** It is slightly reactive

**109.** Which species are formed in the second step of the dissociation of $H_3PO_4$?

**A.** $PO_4^{3-}$

**B.** $H_2PO_4^{2-}$

**C.** $H_2PO_4^-$

**D.** $HPO_4^{2-}$

**110.** Which statement concerning the Arrhenius acid-base theory is NOT correct?

**A.** Neutralization reactions produce $H_2O$, plus a salt

**B.** Acid-base reactions must take place in an aqueous solution

**C.** Arrhenius acids produce $H^+$ in an $H_2O$ solution

**D.** All are correct

**111.** What is the term for a substance that donates a proton in an acid-base reaction?

**A.** Brønsted-Lowry acid

**B.** Brønsted-Lowry base

**C.** Arrhenius acid

**D.** Arrhenius base

**112.** Which of the following does NOT represent a conjugate acid/base pair?

**A.** $HCl$ / $Cl^-$

**B.** $HC_2H_3O_2$ / $^-OH$

**C.** $H_3O^+$ / $H_2O$

**D.** $HCN$ / $^-CN$

**113.** Which of the following statements is correct for a solution of 100% $H_2O$?

**A.** It contains no ions

**B.** It is an electrolyte

**C.** The $[^-OH]$ equals $[H_3O^+]$

**D.** The $[^-OH]$ is greater than $[H_3O^+]$

**114.** The isoelectric point of an amino acid is defined as the pH at which the:

**A.** value is equal to the p$K_a$
**B.** amino acid exists in the acidic form
**C.** amino acid exists in the basic form
**D.** amino acid exists in the zwitterion form

**115.** In an aqueous solution, what is the term for ions that do not participate in a reaction and do not appear in the net ionic equation?

**A.** Zwitterions
**B.** Spectator ions
**C.** Nonelectrolyte ions
**D.** Electrolyte ions

**116.** Salts that result from the reaction of strong acids with strong bases are:

**A.** neutral
**B.** basic
**C.** acidic
**D.** salts

**117.** Which of the following pairs of chemical species contains two polyprotic acids?

**A.** $HNO_3$ and $H_2C_4H_4O_6$
**B.** $HC_2H_3O_2$ and $H_3C_6H_5O_7$
**C.** $H_3PO_4$ and HCN
**D.** $H_2S$ and $H_2CO_3$

**118.** Which of the following pairs of acids and conjugate bases is NOT correctly labeled?

Acid Conjugate Base

**A.** $NH_4^+ \leftrightarrow NH_3$
**B.** $HFO_2 \leftrightarrow HFO_3$
**C.** $H_2SO_4 \leftrightarrow HSO_4^-$
**D.** $HSO_4^- \leftrightarrow SO_4^{2-}$

**119.** What happens to the respective corrosive properties of an acid and base after a neutralization reaction?

**A.** The corrosive properties are doubled because the acid and base are combined in the salt
**B.** The corrosive properties remain the same when the salt is mixed into water
**C.** The corrosive properties are neutralized because the acid and base are transformed
**D.** The corrosive properties are unaffected because salt is a corrosive agent

 

**120.** What is the molarity of a nitric acid solution if 25.00 mL of $HNO_3$ is required to neutralize 0.500 g of calcium carbonate? (Use molecular mass of $CaCO_3$ = 100.09 g/mol)

$$2\ HNO_3\ (aq) + CaCO_3\ (s) \rightarrow Ca(NO_3)_2\ (aq) + H_2O\ (l) + CO_2\ (g)$$

**A.** 0.200 M

**B.** 0.250 M

**C.** 0.400 M

**D.** 0.550 M

================================================================

**Practice Set 7: Questions 121–140**

================================================================

**121.** Which of the following compounds is a weak acid?

**A.** $^{-}OH$ **B.** $NaNH_2$ **C.** $HNO_3$ **D.** HF

**122.** Which indicators are yellow in a basic solution and red in an acidic solution?

I. phenolphthalein
II. bromothymol blue
III. methyl red

**A.** I only
**B.** II only
**C.** III only
**D.** I and II only

**123.** What are the products from the complete neutralization of carbonic acid with aqueous potassium hydroxide?

**A.** $KHCO_4$ (*aq*) and $H_2O$ (*l*)
**B.** $KC_2H_3O_2$ (*aq*) and $H_2O$ (*l*)
**C.** $K_2CO_3$ (*aq*) and $H_2O$ (*l*)
**D.** $KHCO_3$ (*aq*) and $H_2O$ (*l*)

**124.** Which of the following compounds are strong acids?

I. $HNO_3$ (*aq*)
II. $H_2SO_4$ (*aq*)
III. HCl (*aq*)

**A.** I only
**B.** II only
**C.** III only
**D.** I, II and III

**125.** In an acidic solution, what is the $[H_3O^+]$?

**A.** $< 1 \times 10^{-7}$ M
**B.** $> 1 \times 10^{-7}$ M
**C.** $1 \times 10^{-14}$ M
**D.** $1 \times 10^{-7}$ M

**126.** Which of the following substances is NOT an electrolyte?

**A.** $CH_4$ **B.** $NH_4NO_3$ **C.** HBr **D.** KCl

**127.** Which of the following statements about weak acids is correct?

**A.** Weak acids can only be prepared as dilute solutions
**B.** Weak acids have a strong affinity for acidic hydrogens
**C.** Weak acids always contain carbon atoms
**D.** The percentage dissociation for weak acids is usually in the range of 40-60%

**128.** Which of the following is a synonym for hydrogen ion donor?

**A.** Brønsted-Lowry base
**B.** Proton donor
**C.** Amphiprotic
**D.** Arrhenius base

**129.** Which of the following are examples of an Arrhenius acid?

I. $NaC_2H_3O_2$ (*aq*) II. $HC_2H_3O_2$ (*aq*) III. $NH_4C_2H_3O_2$ (*aq*)

**A.** I only
**B.** II only
**C.** III only
**D.** I and II only

**130.** Identify the correct combination of Brønsted-Lowry bases in the following equilibrium reaction:

$H_2PO_4^- + H_2O \leftrightarrow H_3PO_4 + {}^-OH$

**A.** $H_2O + {}^-OH$
**B.** $H_2O + H_3PO_4$
**C.** $H_2PO_4^- + H_3PO_4$
**D.** $H_2PO_4^- + {}^-OH$

**131.** Which of the following is the conjugate acid of water?

**A.** $H^+$ (*aq*) **B.** ${}^-OH$ (*aq*) **C.** $O_2^-$ (*aq*) **D.** $H_3O^+$ (*aq*)

**132.** Calculate the $[H_3O^+]$ and the pH of a 0.045 M $HNO_3$ solution.

**A.** 0.045 M and 1.35
**B.** 4.51 M and –1.88
**C.** $4.5 \times 10^{-13}$ M and 12.55
**D.** $4.5 \times 10^{-13}$ M and 13.55

**133.** At which pH will the net charge on an amino acid be zero if the pI is 6.5?

**A.** 5.5 **B.** 6.0 **C.** 7.0 **D.** 6.5

**134.** What is the term for a substance that changes color according to the pH of the solution?

**A.** Arrhenius acid
**B.** Brønsted-Lowry acid
**C.** Acid–base indicator
**D.** Acid–base signal

**135.** Which of the following acts as a buffer system?

**A.** $NH_3 + H_2O \rightleftharpoons {}^-OH + NH_4^+$
**B.** $H_2PO_4 \rightleftharpoons H^+ + HPO_4^{2-}$
**C.** $HC_2H_3O_2 \rightleftharpoons H^+ + C_2H_3O_2^-$
**D.** All the above

**136.** What is the term for a solution in which concentration has been established accurately, usually to three or four significant digits?

**A.** standard solution
**B.** stock solution
**C.** normal solution
**D.** reference solution

**137.** The hydrogen sulfate ion $HSO_4^-$ is amphoteric. In which of the following equations does it act as an acid?

**A.** $HSO_4^- + {}^-OH \rightarrow H_2SO_4 + O^{2-}$
**B.** $HSO_4^- + H_2O \rightarrow SO_4^{2-} + H_3O^+$
**C.** $HSO_4^- + H_3O^+ \rightarrow SO_3 + 2H_2O$
**D.** $HSO_4^- + H_2O \rightarrow H_2SO_4 + {}^-OH$

**138.** In the following titration curve, what does the inflection point represent?

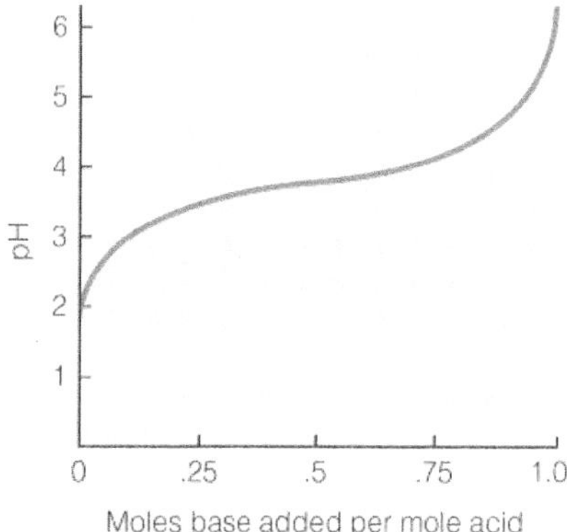


**A.** The weak acid is 50% protonated and 50% deprotonated
**B.** The pH where the solution functions most effectively as a buffer
**C.** Equal concentration of weak acid and conjugate base
**D.** All the above

**139.** What is the sum of the coefficients for the balanced molecular equation of the following acid–base reaction?

$LiOH\ (aq) + H_2SO_4\ (aq) \rightarrow$

**A.** 6 **B.** 7 **C.** 8 **D.** 10

**140.** A typical amino acid has a carboxylic acid and an amine with $pK_a$ values of 2.3 and 9.6, respectively. In a solution of pH 4.5, which describes its protonation and charge state?

**A.** Carboxylic acid is deprotonated and negative; amine is protonated and neutral
**B.** Carboxylic acid is deprotonated and negative; amine is protonated and positive
**C.** Carboxylic acid is protonated and neutral; amine is deprotonated and negative
**D.** Carboxylic acid is protonated and neutral; amine is protonated and neutral

*Notes for active learning*

 

# 10 – Electrochemistry

## Practice Set 1: Questions 1–20

**1.** Which is NOT true regarding the redox reaction occurring in a spontaneous electrochemical cell?

$$Cl_2\ (g) + 2\ Br^-\ (aq) \rightarrow Br_2\ (l) + 2\ Cl^-\ (aq)$$

**A.** Anions flow towards the anode
**B.** Electrons flow from the anode to the cathode
**C.** $Cl_2$ is reduced at the cathode
**D.** $Br^-$ is oxidized at the cathode

**2.** What is the relationship between an element's ionization energy and its ability to function as an oxidizing agent? As ionization energy increases:

**A.** the ability of an element to function as an oxidizing agent remains the same
**B.** the ability of an element to function as an oxidizing agent decreases
**C.** the ability of an element to function as an oxidizing agent increases
**D.** the ability of an element to function as a reducing agent remains the same

**3.** How many electrons are needed to balance the following half-reaction $H_2S \rightarrow S_8$ in an acidic solution?

**A.** 16 electrons to the right side
**B.** 6 electrons to the right side
**C.** 12 electrons to the left side
**D.** 8 electrons to the right side

**4.** Which is NOT true regarding the redox reaction in an electrolytic cell?

$$3\ C + 2\ Co_2O_3 \xrightarrow{\textit{Electricity}} 4\ Co + 3\ CO_2$$

**A.** $CO_2\ (g)$ is produced at the anode
**B.** Co metal is produced at the anode
**C.** Oxidation half-reaction: $C + 2\ O^{2-} \rightarrow CO_2 + 4\ e^-$
**D.** Reduction half-reaction: $Co^{3+} + 3\ e^- \rightarrow Co$

**5.** The anode in a galvanic cell attracts:

**A.** cations
**B.** neutral particles
**C.** anions
**D.** anions and neutral particles

**6.** The electrode with the standard reduction potential of 0 V is assigned as the standard reference electrode and uses the half-reaction:

**A.** $2\ H^+\ (aq) + 2\ e^- \leftrightarrow H_2\ (g)$
**B.** $Ag^+\ (aq) + e^- \leftrightarrow Ag\ (s)$
**C.** $Cu^{2+}\ (aq) + 2\ e^- \leftrightarrow Cu\ (s)$
**D.** $Zn^{2+}\ (aq) + 2\ e^- \leftrightarrow Zn\ (s)$

**7.** Which substance is oxidized in the reaction for a discharging nickel–cadmium (NiCd) battery?

$$Cd\ (s) + NiO_2\ (s) + 2\ H_2O\ (l) \rightarrow Cd(OH)_2\ (s) + Ni(OH)_2\ (s)$$

**A.** $H_2O$
**B.** $Cd(OH)_2$
**C.** Cd
**D.** $NiO_2$

**8.** What happens at the anode if rust forms when Fe is in contact with $H_2O$?

$$4\ Fe + 3\ O_2 \rightarrow 2\ Fe_2O_3$$

**A.** Fe is reduced
**B.** Oxygen is reduced
**C.** Oxygen is oxidized
**D.** Fe is oxidized

**9.** Which is true regarding the redox reaction occurring in the spontaneous electrochemical cell for $Cl_2\ (g) + 2\ Br^-\ (aq) \rightarrow Br_2\ (l) + 2\ Cl^-\ (aq)$?

**A.** Electrons flow from the cathode to the anode
**B.** $Cl_2$ is oxidized at the cathode
**C.** $Br^-$ is reduced at the anode
**D.** Cations in the salt bridge flow from the $Br_2$ half-cell to the $Cl_2$ half-cell

**10.** What is the term for a chemical reaction that involves electron transfer between two reacting substances?

**A.** Reduction reaction
**B.** Redox reaction
**C.** Oxidation reaction
**D.** Half-reaction

 

**11.** Using the following metal ion/metal reaction potentials,

| $Cu^{2+}$ (*aq*)\|Cu (*s*) | $Ag^{+}$ (*aq*)\|Ag (*s*) | $Co^{2+}$ (*aq*)\|Co (*s*) | $Zn^{2+}$ (*aq*)\|Zn (*s*) |
|---|---|---|---|
| +0.34 V | +0.80 V | –0.28 V | –0.76 V |

calculate the standard cell potential for the cell whose reaction is:

$$\text{Co}\ (s) + \text{Cu}^{2+}\ (aq) \rightarrow \text{Co}^{2+}\ (aq) + \text{Cu}\ (s)$$

**A.** +0.62 V
**B.** –0.62 V
**C.** +0.48 V
**D.** –0.48 V

**12.** How are photovoltaic cells different from many other forms of solar energy?

**A.** Light is reflected, and the coolness of the shade is used to provide a temperature differential
**B.** Light is passively converted into heat
**C.** Light is converted directly to electricity
**D.** Light is converted into heat and then into electricity

**13.** What is the term for an electrochemical cell with a single electrode where oxidation or reduction can occur?

**A.** Half-cell
**B.** Voltaic cell
**C.** Dry cell
**D.** Electrolytic cell

**14.** By which method could electrolysis be used to raise the hull of a sunken ship?

**A.** Electrolysis could only be used to raise the hull if the ship is made of iron. If so, the electrolysis of the iron metal might produce sufficient gas to lift the ship
**B.** The gaseous products of the electrolysis of $H_2O$ are collected with bags attached to the hull of the ship and the inflated bags raise the ship
**C.** The electrolysis of the $H_2O$ under the ship's hull boils $H_2O$ and creates upward pressure to raise the ship
**D.** An electric current passed through the hull of the ship produces electrolysis and the gases trapped in compartments of the vessel push it upwards

**15.** What are the products for the single-replacement reaction Zn (*s*) + $CuSO_4$ (*aq*) → ?

**A.** CuO and $ZnSO_4$
**B.** CuO and $ZnSO_3$
**C.** Cu and $ZnSO_4$
**D.** Cu and $ZnSO_3$

**16.** Which of the following is a true statement about the electrochemical reaction for an electrochemical cell with a cell potential of +0.36 V?

**A.** The reaction favors the formation of reactants and would be considered a galvanic cell
**B.** The reaction favors the formation of reactants and would be considered an electrolytic cell
**C.** The reaction is at equilibrium and is a galvanic cell
**D.** The reaction favors the formation of products and would be considered a galvanic cell

**17.** Which of the following is a unit of electrical energy?

**A.** Coulomb
**B.** Joule
**C.** Watt
**D.** Ampere

**18.** Which observation describes the solution near the cathode when an aqueous sodium chloride solution is electrolyzed and hydrogen gas is evolved at the cathode?

**A.** Colored
**B.** Acidic
**C.** Basic
**D.** Frothy

**19.** How many grams of Ag is deposited in the cathode of an electrolytic cell if a current of 3.50 A is applied to a solution of $AgNO_3$ for 12 minutes? (Use the molecular mass of Ag = 107.86 g/mol and the conversion of 1 mol $e^-$ = $9.65 \times 10^4$ C)

**A.** 0.32 g
**B.** 0.86 g
**C.** 2.82 g
**D.** 3.86 g

**20.** In an electrochemical cell, which of the following statements is FALSE?

**A.** The anode is the electrode where oxidation occurs
**B.** A salt bridge provides electrical contact between the half-cells
**C.** The cathode is the electrode where reduction occurs
**D.** All the above are true

 

---

**Practice Set 2: Questions 21–40**

---

**21.** What is the relationship between an element's ionization energy and its ability to function as an oxidizing and reducing agent? Elements with high ionization energy are:

**A.** strong oxidizing and weak reducing agents
**B.** weak oxidizing and weak reducing agents
**C.** strong oxidizing and strong reducing agents
**D.** weak oxidizing and strong reducing agents

**22.** In a basic solution, how many electrons are needed to balance the following half-reaction?

$$C_8H_{10} \rightarrow C_8H_4O_4^{2-}$$

**A.** 8 electrons on the left side
**B.** 12 electrons on the left side
**C.** 4 electrons on the left side
**D.** 12 electrons on the right side

**23.** Which statement regarding the redox reaction occurring in an electrolytic cell is true?

$$3\ C\ (s) + 2\ Co_2O_3\ (l) \xrightarrow{\textit{Electricity}} 4\ Co\ (l) + 3\ CO_2\ (g)$$

**A.** $CO_2$ gas is produced at the anode
**B.** $Co^{3+}$ is produced at the cathode
**C.** Oxidation half-reaction: $Co^{3+} + 3\ e^- \rightarrow Co$
**D.** Reduction half-reaction: $C + 2\ O^{2-} \rightarrow CO_2 + e^-$

**24.** What is true regarding the redox reaction occurring in a spontaneous electrochemical cell?

$$Sn\ (s) + Cu^{2+}\ (aq) \rightarrow Cu\ (s) + Sn^{2+}\ (aq)$$

**A.** Anions in the salt bridge flow from the Cu half-cell to the Sn half-cell
**B.** Electrons flow from the Cu electrode to the Sn electrode
**C.** $Cu^{2+}$ is oxidized at the cathode
**D.** Sn is reduced at the anode

 

**25.** In galvanic cells, reduction occurs at the:

I. salt bridge II. cathode III. anode

**A.** I only
**B.** II only
**C.** III only
**D.** I and II only

**26.** For a battery, what is undergoing reduction in the following oxidation-reduction reaction?

$Mn_2O_3 + ZnO \rightarrow 2\ MnO_2 + Zn$

**A.** $Mn_2O_3$
**B.** $MnO_2$
**C.** ZnO
**D.** Zn

**27.** Given that the following redox reactions go essentially to completion, which of the metals listed below has the greatest tendency to undergo oxidation?

$Ni\ (s) + Ag^+\ (aq) \rightarrow Ag\ (s) + Ni^{2+}\ (aq)$

$Al\ (s) + Cd^{2+}\ (aq) \rightarrow Cd\ (s) + Al^{3+}\ (aq)$

$Cd\ (s) + Ni^{2+}\ (aq) \rightarrow Ni\ (s) + Cd^{2+}\ (aq)$

$Ag\ (s) + H^+\ (aq) \rightarrow$ no reaction

**A.** (H) **B.** Cd **C.** Ni **D.** Al

**28.** What is the purpose of the salt bridge in a voltaic cell?

**A.** Allows for a balance of charge between the two chambers
**B.** Allows the $Fe^{2+}$ and the $Cu^{2+}$ to flow freely between the two chambers
**C.** Allows for the buildup of positively charged ions in one container and negatively charged ions in the other container
**D.** Prevents further migration of electrons through the wire

**29.** If $\Delta G°$ for a cell is positive, the E° is:

**A.** neutral
**B.** negative
**C.** positive
**D.** unable to be determined

 

**30.** In an electrochemical cell, which of the following statements is FALSE?

**A.** Oxidation occurs at the anode
**B.** Reduction occurs at the cathode
**C.** Electrons flow through the salt bridge to complete the cell
**D.** A spontaneous electrochemical cell is called a voltaic cell

**31.** Which of the following statements about electrochemistry is NOT true?

**A.** The study of how protons are transferred from one chemical compound to another
**B.** The use of electrical current to produce an oxidation-reduction reaction
**C.** The use of a set of oxidation-reduction reactions to produce electrical current
**D.** The study of how electrical energy and chemical reactions are related

**32.** In an operating photovoltaic cell, electrons move through the external circuit to the negatively charged p-type silicon wafer. How can the electrons move to the negatively charged silicon wafer if electrons are negatively charged?

**A.** The p-type silicon wafer is positively charged
**B.** The energy of the sunlight moves electrons in a nonspontaneous direction
**C.** Advancements in photovoltaic technology have solved this technological impediment
**D.** An electric current occurs because the energy from the sun reverses the charge of the electrons

**33.** What is the term for the value assigned to an atom in a substance that indicates whether the atom is electron-poor or electron-rich compared to a free atom?

**A.** Reduction number
**B.** Oxidation number
**C.** Cathode number
**D.** Anode number

**34.** What is the term for the relative ability of a substance to undergo reduction?

**A.** Oxidation potential
**B.** Reduction potential
**C.** Anode potential
**D.** Cathode potential

**35.** What is the primary difference between a fuel cell and a battery?

**A.** Fuel cells oxidize to supply electricity, while batteries reduce to supply electricity
**B.** Batteries supply electricity, while fuel cells supply heat
**C.** Batteries can be recharged, while fuel cells cannot
**D.** Fuel cells do not run down because they can be refueled, while batteries run down and need to be recharged

**36.** In balancing the equation for a disproportionation reaction, using the oxidation number method, the substance undergoing disproportionation is:

**A.** initially written twice on the reactant side of the equation
**B.** initially written twice on the product side of the equation
**C.** initially written on the reactant and product sides of the equation
**D.** always assigned an oxidation number of zero

**37.** What is the term for an electrochemical cell in which electrical energy is generated from a spontaneous redox reaction?

**A.** Voltaic cell
**B.** Photoelectric cell
**C.** Dry cell
**D.** Electrolytic cell

**38.** Which observation describes the solution near the anode when an aqueous sodium sulfate solution is electrolyzed and gas is evolved at the anode?

**A.** Colored
**B.** Acidic
**C.** Basic
**D.** Frothy

**39.** Which of the statements below is true regarding the redox reaction occurring in a nonspontaneous electrochemical cell?

$$Cd\ (s) + Zn(NO_3)_2\ (aq) + \text{electricity} \rightarrow Zn\ (s) + Cd(NO_3)_2\ (aq)$$

**A.** Oxidation half-reaction: $Zn^{2+} + 2\ e^- \rightarrow Zn$
**B.** Reduction half-reaction: $Cd \rightarrow Cd^{2+} + 2\ e^-$
**C.** Cd metal is produced at the anode
**D.** Zn metal is produced at the cathode

 

**40.** Which statement is true for electrolysis?

**A.** A spontaneous redox reaction produces electricity
**B.** A nonspontaneous redox reaction is forced to occur by applying an electric current
**C.** Only pure, drinkable water is produced
**D.** There is a cell that reverses the flow of ions

==================================================================

**Practice Set 3: Questions 41–60**

==================================================================

**41.** Which relationship explains an element's electronegativity and ability to act as an oxidizing and reducing agent?

**A.** Atoms with large electronegativity tend to act as strong oxidizing and strong reducing agents
**B.** Atoms with large electronegativity tend to act as weak oxidizing and strong reducing agents
**C.** Atoms with large electronegativity tend to act as strong oxidizing and weak reducing agents
**D.** Atoms with large electronegativity tend to act as weak oxidizing and weak reducing agents

**42.** Which substance is undergoing reduction for a battery if the following two oxidation-reduction reactions take place?

Reaction I: $Zn + 2\ OH^- \rightarrow ZnO + H_2O + 2\ e^-$

Reaction II: $2\ MnO_2 + H_2O + 2\ e^- \rightarrow Mn_2O_3 + 2\ OH^-$

**A.** $MnO_2$ **B.** $H_2O$ **C.** Zn **D.** ZnO

**43.** What is the term for an electrochemical cell in which a nonspontaneous redox reaction forces electricity through the cell?

I. voltaic cell II. dry cell III. electrolytic cell

**A.** I only
**B.** II only
**C.** III only
**D.** I and II only

**44.** Which of the statements is true regarding the following redox reaction occurring in a galvanic cell?

$Cl_2\ (g) + 2\ Br^-\ (aq) \rightarrow Br_2\ (l) + 2\ Cl^-\ (aq)$?

**A.** Cations in the salt bridge flow from the $Cl_2$ half-cell to the $Br_2$ half-cell
**B.** Electrons flow from the anode to the cathode
**C.** $Cl_2$ is reduced at the anode
**D.** $Br^-$ is oxidized at the cathode

 

**45.** Which of the statements below is true regarding the redox reaction occurring in a nonspontaneous electrolytic cell?

$$3\ C\ (s) + 4\ AlCl_3\ (l) \xrightarrow{\text{Electricity}} 4\ Al\ (l) + 3\ CCl_4\ (g)$$

**A.** $Cl^-$ is produced at the anode
**B.** Al metal is produced at the cathode
**C.** Oxidation half-reaction is $Al^{3+} + 3\ e^- \rightarrow Al$
**D.** Reduction half-reaction is $C + 4\ Cl^- \rightarrow CCl_4 + 4\ e^-$

**46.** Which of the following materials is most likely to undergo oxidation?

I. $Cl^-$ II. Na III. $Na^+$

**A.** I only
**B.** II only
**C.** III only
**D.** I and II only

**47.** Which statement regarding the redox reaction occurring in a spontaneous electrochemical cell is NOT true?

$$Zn\ (s) + Cd^{2+}\ (aq) \rightarrow Cd\ (s) + Zn^{2+}\ (aq)$$

**A.** Anions in the salt bridge flow from the Zn half-cell to the Cd half-cell
**B.** Electrons flow from the Zn electrode to the Cd electrode
**C.** $Cd^{2+}$ is reduced at the cathode
**D.** Zn is oxidized at the anode

**48.** Using the provided electrochemical series, determine which is capable of oxidizing Cu (*s*) to $Cu^{2+}$ (*aq*) when added to Cu (*s*) in solution.

**A.** $Al^{3+}$ (*aq*)
**B.** $Ag^+$ (*aq*)
**C.** $K^+$ (*aq*)
**D.** $Pb^{2+}$ (*aq*)

| **Equilibrium** | **E°** |
|---|---|
| $Li^+\ (aq) + e^- \leftrightarrow Li\ (s)$ | –3.03 volts |
| $K^+\ (aq) + e^- \leftrightarrow K\ (s)$ | –2.92 |
| $^*Ca^{2+}\ (aq) + 2\ e^- \leftrightarrow Ca\ (s)$ | –2.87 |
| $^*Na^+\ (aq) + e^- \leftrightarrow Na\ (s)$ | –2.71 |
| $Mg^{2+}\ (aq) + 2\ e^- \leftrightarrow Mg\ (s)$ | –2.37 |
| $Al^{3+}\ (aq) + 3\ e^- \leftrightarrow Al\ (s)$ | –1.66 |
| $Zn^{2+}\ (aq) + 2\ e^- \leftrightarrow Zn\ (s)$ | –0.76 |
| $Fe^{2+}\ (aq) + 2\ e^- \leftrightarrow Fe\ (s)$ | –0.44 |
| $Pb^{2+}\ (aq) + 2\ e^- \leftrightarrow Pb\ (s)$ | –0.13 |
| $2\ H^+\ (aq) + 2\ e^- \leftrightarrow H_2\ (g)$ | 0.0 |
| $Cu^{2+}\ (aq) + 2\ e^- \leftrightarrow Cu\ (s)$ | +0.34 |
| $Ag^+\ (aq) + e^- \leftrightarrow Ag\ (s)$ | +0.80 |
| $Au^{3+}\ (aq) + 3\ e^- \leftrightarrow Au\ (s)$ | +1.50 |

**49.** Why is the anode of a battery indicated with a negative (–) sign?

**A.** Electrons move to the anode to react with $NH_4Cl$ in the battery
**B.** It indicates the electrode where the chemicals are reduced
**C.** Electrons are attracted to the negative electrode
**D.** The electrode is the source of negatively charged electrons

**50.** What is the term for converting chemical energy to electrical energy from redox reactions?

**A.** Redox chemistry
**B.** Electrochemistry
**C.** Cell chemistry
**D.** Battery chemistry

**51.** In electrolysis, E° tends to be:

**A.** negative
**B.** neutral
**C.** positive
**D.** greater than 1

**52.** A major source of chlorine gas is the electrolysis of concentrated salt water, NaCl (*aq*). What is the sign of the electrode where the chlorine gas is formed?

**A.** Neither, since chlorine gas is a neutral molecule and there is no electrode attraction
**B.** Negative, since the chlorine gas needs to deposit electrons to form chloride ions
**C.** Positive, since the chloride ions lose electrons to form chlorine molecules
**D.** Both, since the chloride ions from NaCl (*aq*) are attracted to the positive electrode to form chlorine molecules, while the produced chlorine gas molecules move to deposit electrons at the negative electrode

**53.** Which statement is correct for an electrolytic cell with two electrodes?

**A.** Oxidation occurs at the anode, which is negatively charged
**B.** Oxidation occurs at the anode, which is positively charged
**C.** Oxidation occurs at the cathode, which is positively charged
**D.** Oxidation occurs at the cathode, which is negatively charged

 

**54.** Which of the statements below is true regarding the redox reaction occurring in a nonspontaneous electrochemical cell?

$$Cd\ (s) + Zn(NO_3)_2\ (aq) + \text{electricity} \rightarrow Zn\ (s) + Cd(NO_3)_2\ (aq)$$

**A.** Oxidation half-reaction: $Zn^{2+} + 2\ e^- \rightarrow Zn$
**B.** Reduction half-reaction: $Cd \rightarrow Cd^{2+} + 2\ e^-$
**C.** Cd metal dissolves at the anode
**D.** Zn metal dissolves at the cathode

**55.** Which battery system is based on half-reactions involving zinc metal and manganese dioxide?

**A.** Alkaline batteries
**B.** Fuel cells
**C.** Lead-acid storage batteries
**D.** Dry-cell batteries

**56.** Which of the following equations is a disproportionation reaction?

**A.** $2\ H_2O \rightarrow 2\ H_2 + O_2$
**B.** $H_2SO_3 \rightarrow H_2O + SO_2$
**C.** $HNO_2 \rightarrow NO + HNO_3$
**D.** $Mg + H_2SO_4 \rightarrow MgSO_4 + H_2$

**57.** How is electrolysis different from the chemical process inside a battery?

**A.** Pure compounds cannot be generated *via* electrolysis
**B.** Electrolysis only uses electrons from a cathode
**C.** Electrolysis does not use electrons
**D.** They are the same process in reverse

**58.** Which fact about fuel cells is FALSE?

**A.** Fuel cell automobiles are powered by water and only emit hydrogen
**B.** Fuel cells are based on the tendency of some elements to gain electrons from other elements
**C.** Fuel cell automobiles are quiet
**D.** Fuel cell automobiles are environmentally friendly

**59.** Which process occurs when copper is refined using the electrolysis technique?

**A.** Impure copper goes into solution at the anode, and pure copper plates out on the cathode
**B.** Impure copper goes into solution at the cathode, and pure copper plates out on the anode
**C.** Pure copper goes into solution from the anode and forms a precipitate at the bottom of the tank
**D.** Pure copper goes into solution from the cathode and forms a precipitate at the bottom of the tank

**60.** Which of the following statements describes electrolysis?

**A.** A chemical reaction which results when electrical energy is passed through a metallic liquid
**B.** A chemical reaction which results when electrical energy is passed through a liquid electrolyte
**C.** The splitting of atomic nuclei by electrical energy
**D.** The splitting of atoms by electrical energy

 

---

**Practice Set 4: Questions 61–80**

---

**61.** In a battery, which of the species in the two oxidation-reduction reactions is undergoing oxidation?

Reaction I: $Zn + 2\ OH^- \rightarrow ZnO + H_2O + 2\ e^-$

Reaction II: $2\ MnO_2 + H_2O + 2\ e^- \rightarrow Mn_2O_3 + 2\ OH^-$

**A.** $H_2O$
**B.** $MnO_2$
**C.** $ZnO$
**D.** Zn

**62.** What is the term for an electrochemical cell where the anode and cathode reactions do not occur in aqueous solutions?

**A.** Voltaic cell
**B.** Electrolytic cell
**C.** Dry cell
**D.** Galvanic cell

**63.** How many electrons are needed to balance the charge for the following half-reaction in an acidic solution?

$C_2H_6O \rightarrow HC_2H_3O_2$

**A.** 6 electrons to the right side
**B.** 4 electrons to the right side
**C.** 3 electrons to the left side
**D.** 2 electrons to the left side

**64.** Which of the statements is true regarding the following redox reaction occurring in an electrolytic cell?

$$3\ C\ (s) + 4\ AlCl_3\ (l) \xrightarrow{\textit{Electricity}} 4\ Al\ (l) + 3\ CCl_4\ (g)$$

**A.** $Al^{3+}$ is produced at the cathode
**B.** $CCl_4$ gas is produced at the anode
**C.** Reduction half-reaction: $C + 4\ Cl^- \rightarrow CCl_4 + 4\ e^-$
**D.** Oxidation half-reaction: $Al^{3+} + 3\ e^- \rightarrow Al$

**65.** What is the term for an electrochemical cell that spontaneously produces electrical energy?

I. half-cell　　II. electrolytic cell　　III. dry cell

**A.** I and II only
**B.** II only
**C.** III only
**D.** II and III only

**66.** The anode of a battery is indicated with a negative (–) sign because the anode is:

**A.** where electrons are adsorbed
**B.** positive
**C.** negative
**D.** where electrons are generated

**67.** Which is the strongest reducing agent for the following half-reaction potentials?

$Sn^{4+}$ (*aq*) + 2 $e^-$ → $Sn^{2+}$ (*aq*)　　E° = –0.13 V

$Ag^+$ (*aq*) + $e^-$ → Ag (*s*)　　E° = +0.81 V

$Cr^{3+}$ (*aq*) + 3 $e^-$ → Cr (*s*)　　E° = –0.75 V

$Fe^{2+}$ (*aq*) + 2 $e^-$ → Fe (*s*)　　E° = –0.43 V

**A.** Cr (*s*)
**B.** $Fe^{2+}$ (*aq*)
**C.** $Sn^{2+}$ (*aq*)
**D.** Ag (*s*)

**68.** Which is true for the redox reaction in a spontaneous electrochemical cell?

Zn (*s*) + $Cd^{2+}$ (*aq*) → Cd (*s*) + $Zn^{2+}$ (*aq*)

**A.** Electrons flow from the Cd electrode to the Zn electrode
**B.** Anions in the salt bridge flow from the Cd half-cell to the Zn half-cell
**C.** Zn is reduced at the anode
**D.** $Cd^{2+}$ is oxidized at the cathode

**69.** Based upon the reduction potential: $Zn^{2+}$ + 2 $e^-$ → Zn (*s*); $E°$ = –0.76 V, does a reaction take place when Zinc (*s*) is added to aqueous HCl, under standard conditions?

**A.** Yes, because the reduction potential for $H^+$ is negative
**B.** Yes, because the reduction potential for $H^+$ is zero
**C.** No, because the oxidation potential for $Cl^-$ is positive
**D.** No, because the reduction potential for $Cl^-$ is negative

 

**70.** If 1 amp of current passes a cathode for 10 minutes, how much Zn (*s*) forms in the following reaction? (Use the molecular mass of Zn = 65 g/mole)

$Zn^{2+} + 2\ e^- \rightarrow Zn\ (s)$

**A.** 0.10 g **B.** 10.0 g **C.** 0.20 g **D.** 2.20 g

**71.** In the oxidation–reduction reaction Mg (*s*) + $Cu^{2+}$ (*aq*) → $Mg^{2+}$ (*aq*) + Cu (*s*), which atom/ion is reduced, and which atom/ion is oxidized?

**A.** The $Cu^{2+}$ ion is reduced (gains electrons) to form Cu metal, while Mg metal is oxidized (loses electrons) to form $Mg^{2+}$

**B.** Since Mg is transformed from a solid to an aqueous solution and Cu is transformed from an aqueous solution to a solid, no oxidation-reduction reaction occurs

**C.** The $Cu^{2+}$ ion is oxidized (i.e., gains electrons) from Cu metal, while Mg metal is reduced (i.e., loses electrons) to form $Mg^{2+}$

**D.** The $Mg^{2+}$ ion is reduced (i.e., gains electrons) from Cu metal, while $Cu^{2+}$ is oxidized (i.e., loses electrons) to the $Mg^{2+}$

**72.** Electrolysis is an example of a(n):

**A.** acid-base reaction
**B.** exothermic reaction
**C.** physical change
**D.** chemical change

**73.** What is the term for a process characterized by the loss of electrons?

**A.** Redox
**B.** Reduction
**C.** Electrochemistry
**D.** Oxidation

**74.** Which of the following statements is true about a galvanic cell?

**A.** The standard reduction potential for the anode reaction is always positive
**B.** The standard reduction potential for the anode reaction is always negative
**C.** The standard reduction potential for the cathode reaction is always positive
**D.** E° for the cell is always positive

**75.** Which of the following statements is true?

**A.** Galvanic cells were invented by Thomas Edison
**B.** Galvanic cells generate electrical energy rather than consume it
**C.** Electrolysis cells generate alternating current when their terminals are reversed
**D.** Electrolysis was discovered by Lewis Latimer

**76.** Which statement is correct for an electrolysis cell with two electrodes?

**A.** Reduction occurs at the anode, which is positively charged
**B.** Reduction occurs at the anode, which is negatively charged
**C.** Reduction occurs at the cathode, which is positively charged
**D.** Reduction occurs at the cathode, which is negatively charged

**77.** How long would it take to deposit 4.00 grams of Cu from a $CuSO_4$ solution if a current of 2.5 A is applied? (Use the molecular mass of Cu = 63.55 g/mol and the conversion of 1 mol $e^-$ = $9.65 \times 10^4$ C and 1 C = A·s)

**A.** 1.02 hours
**B.** 1.36 hours
**C.** 4.46 hours
**D.** 6.38 hours

**78.** Which battery system is completely rechargeable?

**A.** Alkaline batteries
**B.** Fuel cells
**C.** Lead-acid storage batteries
**D.** Dry-cell batteries

**79.** In which cell type does the following reaction occur when electrons are forced into a system by applying an external voltage?

$$Fe^{2+} + 2\ e^- \rightarrow Fe\ (s) \qquad E° = -0.44\ V$$

**A.** Electrolytic cell
**B.** Battery
**C.** Electrochemical cell
**D.** Galvanic cell

**80.** What is the term for a reaction that represents separate oxidation or reduction processes?

**A.** Reduction reaction
**B.** Redox reaction
**C.** Oxidation reaction
**D.** Half-reaction

 

*Notes for active learning*

*Notes for active learning*

 

# 11 – Atomic and Electronic Structure

**Practice Set 1: Questions 1–20**

**1.** Which statement(s) about alpha particles is/are FALSE?

I. They are a harmless form of radiation
II. They have low penetrating power
III. They have high ionization power

**A.** I only
**B.** II only
**C.** III only
**D.** I and II only

**2.** What is the term for nuclear radiation that is identical to an electron?

**A.** Positron
**B.** Gamma ray
**C.** Beta minus particle
**D.** Alpha particle

**3.** Protons are being accelerated in a particle accelerator. When the speed of the protons is doubled, by what factor does their de Broglie wavelength change? Note: consider this situation non-relativistic.

**A.** Increases by $\sqrt{2}$
**B.** Decreases by $\sqrt{2}$
**C.** Increases by 2
**D.** Decreases by 2

**4.** The Bohr model of the atom was able to explain the Balmer series because:

**A.** electrons were allowed to exist only in specific orbits and nowhere else
**B.** differences between the energy levels of the orbits matched the differences between the energy levels of the line spectra
**C.** smaller orbits require electrons to have more negative energy to match the angular momentum
**D.** differences between the energy levels of the orbits were precisely half the differences between the energy levels of the line spectra

**5.** Radioactivity is the tendency for an element to:

**A.** become ionized easily
**B.** be dangerous to living things
**C.** emit radiation
**D.** emit protons

**6.** Which is the missing species in the nuclear equation: $^{100}_{44}\text{Ru} + ^{0}_{-1}\text{e}^- \rightarrow$ ___?

**A.** $^{100}_{45}\text{Ru}$
**B.** $^{100}_{43}\text{Ru}$
**C.** $^{101}_{44}\text{Ru}$
**D.** $^{100}_{43}\text{Tc}$

**7.** The term nucleon refers to:

**A.** the nucleus of a specific isotope
**B.** protons and neutrons
**C.** positrons emitted from an atom that undergoes nuclear decay
**D.** electrons emitted from a nucleus in a nuclear reaction

**8.** An isolated $^9$Be atom spontaneously decays into two alpha particles. What can be concluded about the mass of the $^9$Be atom?

**A.** The mass is less than twice the mass of the $^4$He atom, but not equal to the mass of $^4$He
**B.** No conclusions can be made about the mass
**C.** The mass is exactly twice the mass of the $^4$He atom
**D.** The mass is greater than twice the mass of the $^4$He atom

**9.** Which of the following isotopes contains the most neutrons?

**A.** $^{178}_{84}\text{Po}$
**B.** $^{178}_{87}\text{Fr}$
**C.** $^{181}_{86}\text{Rn}$
**D.** $^{170}_{83}\text{Bi}$

**10.** Which of the following correctly balances this nuclear fission reaction?

$$^{1}_{0}\text{n} + ^{235}_{92}\text{U} \rightarrow ^{131}_{53}\text{I} + \_\_\_ + 3\,^{1}_{0}\text{n}$$

**A.** $^{102}_{39}\text{Y}$
**B.** $^{102}_{36}\text{Kr}$
**C.** $^{104}_{39}\text{Y}$
**D.** $^{105}_{36}\text{Kr}$

**11.** What is the frequency of the light emitted by atomic hydrogen according to the Balmer formula where n = 12? (Use the Balmer series constant B = $3.6 \times 10^{-7}$ m and the speed of light $c = 3 \times 10^8$ m/s)

**A.** $5.3 \times 10^6$ Hz
**B.** $8.1 \times 10^{14}$ Hz
**C.** $5.9 \times 10^{13}$ Hz
**D.** $1.2 \times 10^{11}$ Hz

**12.** Gamma rays require the heaviest shielding of the common types of nuclear radiation because gamma rays have the:

**A.** heaviest particles
**B.** lowest energy
**C.** most intense color
**D.** highest energy

**13.** In making a transition from state n = 1 to state n = 2, the hydrogen atom must [ ] a photon of [ ]. (Use Planck's constant $h = 4.14 \times 10^{-15}$ eV·s, speed of light $c = 3 \times 10^8$ m/s, Rydberg constant R = $1.097 \times 10^7$ m$^{-1}$)

**A.** absorb … 10.2 eV
**B.** absorb … 8.6 eV
**C.** emit … 8.6 eV
**D.** emit … 10.2 eV

**14.** Rubidium $^{87}_{37}\text{Rb}$ is a naturally-occurring nuclide that undergoes $\beta^-$ decay. What is the resultant nuclide from this decay?

**A.** $^{86}_{36}\text{Rb}$
**B.** $^{87}_{38}\text{Kr}$
**C.** $^{87}_{38}\text{Sr}$
**D.** $^{87}_{36}\text{Kr}$

**15.** Which of the following statements best describes the role of neutrons in the nucleus?

**A.** The neutrons stabilize the nucleus by attracting protons
**B.** The neutrons stabilize the nucleus by balancing charge
**C.** The neutrons stabilize the nucleus by attracting other nucleons
**D.** The neutrons stabilize the nucleus by repelling other nucleons

**16.** A Geiger–Muller counter detects radioactivity by:

**A.** ionizing argon gas in a chamber which produces an electrical signal
**B.** analyzing the mass and velocity of each particle
**C.** developing film which is exposed by radioactive particles
**D.** slowing the neutrons using a moderator and then counting the secondary charges produced

**17.** What percentage of the radionuclides in a given sample remains after three half-lives?

**A.** 25%
**B.** 12.5%
**C.** 6.25%
**D.** 33.3%

**18.** The Lyman series is formed by electron transitions in hydrogen that:

**A.** begin on the n = 2 shell
**B.** end on the n = 2 shell
**C.** begin on the n = 1 shell
**D.** end on the n = 1 shell

**19.** Most of the volume of an atom is occupied by:

**A.** neutrons
**B.** empty space
**C.** electrons
**D.** protons

**20.** Alpha and beta minus particles are deflected in opposite directions in a magnetic field because:

I. they have opposite charges
II. alpha particles contain nucleons, and beta minus particles do not
III. they spin in opposite directions

**A.** I only
**B.** II only
**C.** III only
**D.** I and II only

---

**Practice Set 2: Questions 21–40**

---

**21.** The conversion of mass to energy is measurable only in:

**A.** chemiluminescent transformations
**B.** spontaneous chemical reactions
**C.** endothermic reactions
**D.** nuclear reactions

**22.** What is the term for a radioactive substance undergoing $3.7 \times 10^{10}$ disintegrations per sec?

**A.** Rem
**B.** Rad
**C.** Curie
**D.** Roentgen

**23.** An isolated $^{235}U$ atom spontaneously undergoes fission into two approximately equal-sized fragments. What is missing from the product side of the reaction:

$$^{235}U \rightarrow {}^{141}Ba + {}^{92}Kr + \_\_\_?$$

**A.** A neutron
**B.** Two neutrons
**C.** Two protons and two neutrons
**D.** Two protons and a neutron

**24.** How many protons and neutrons are in $^{34}_{16}S$?

**A.** 18 neutrons and 34 protons
**B.** 16 neutrons and 18 protons
**C.** 16 protons and 34 neutrons
**D.** 16 protons and 18 neutrons

**25.** The radioactive isotope Z has a half-life of 12 hours. What is the fraction of the original amount remaining after 2 days?

**A.** 1/16
**B.** 1/8
**C.** 1/4
**D.** 1/2

**26.** Which of the following nuclear equations correctly describes alpha emission?

**A.** $^{238}_{92}U \rightarrow {}^{242}_{94}Pu + {}^{4}_{2}He$

**B.** $^{238}_{92}U \rightarrow {}^{4}_{2}He$

**C.** $^{238}_{92}U \rightarrow {}^{234}_{90}Th + {}^{4}_{2}He$

**D.** $^{238}_{92}U \rightarrow {}^{235}_{90}Th + {}^{4}_{2}He$

**27.** A hydrogen atom transitions downward from the n = 20 state to the n = 5 state. Find the wavelength of the emitted photon. (Use Planck's constant $h = 4.14 \times 10^{-15}$ eV·s, the speed of light $c = 3 \times 10^8$ m/s and the Rydberg constant $R = 1.097 \times 10^7 \text{ m}^{-1}$)

**A.** 1.93 μm
**B.** 2.82 μm
**C.** 1.54 μm
**D.** 2.43 μm

**28.** A nuclear equation is balanced when the:

**A.** same elements are found on both sides of the equation
**B.** sums of the atomic numbers of the particles and atoms are equal on both sides of the equation
**C.** sum of the mass numbers of the particles and the sum of atoms are the same on both sides of the equation
**D.** sum of the mass numbers and the sum of the atomic numbers of the particles and atoms are the same on both sides of the equation

**29.** A blackbody is an ideal system that:

**A.** absorbs 50% of the light incident upon it and emits 50% of the radiation it generates
**B.** absorbs 0% of the light incident upon it and emits 100% of the radiation it generates
**C.** absorbs 100% of the light incident upon it and emits 100% of the radiation it generates
**D.** emits 100% of the light it generates and absorbs 50% of the radiation incident upon it

**30.** Recent nuclear bomb tests have created an extra-high level of atmospheric $^{14}C$. When future archaeologists date samples without knowing of these nuclear tests, will the dates they calculate be correct?

**A.** Correct, because biological materials do not gather $^{14}C$ from bomb tests
**B.** Correct, since the $^{14}C$ decays within the atmosphere at the natural rate
**C.** Incorrect, they would appear too old
**D.** Incorrect, they would appear too young

**31.** When an isotope releases gamma radiation, the atomic number:

**A.** and the mass number remain the same
**B.** and the mass number decrease by one
**C.** and the mass number increase by one
**D.** remains the same, and the mass number increases by one

 

**32.** If $^{14}$Carbon is a beta emitter, what is the likely product of radioactive decay?

**A.** $^{22}$Silicon
**B.** $^{13}$Boron
**C.** $^{14}$Nitrogen
**D.** $^{12}$Carbon

**33.** In a balanced nuclear equation, the:

I. sum of the mass numbers on both sides must be equal
II. sum of the atomic numbers on both sides must be equal
III. daughter nuclide appears on the right side of the arrow

**A.** I only
**B.** II only
**C.** III only
**D.** I, II and III

**34.** What is the term for the number that characterizes an element and indicates the number of protons found in the nucleus of the atom?

**A.** Mass number
**B.** Atomic number
**C.** Atomic mass
**D.** Neutron number

**35.** Hydrogen atoms can emit four spectral lines with visible colors from red to violet. These four visible lines emitted by hydrogen atoms are produced by electrons that:

**A.** end in the ground state
**B.** end in the n = 3 level
**C.** end in the n = 2 level
**D.** start in the ground state

**36.** The electron was discovered through experiments with:

**A.** quarks
**B.** foil
**C.** light
**D.** electricity

Questions **37-39** are based on the following:

The image shows a beam of radiation passing between two electrically-charged plates.

I. a

II. b

III. c

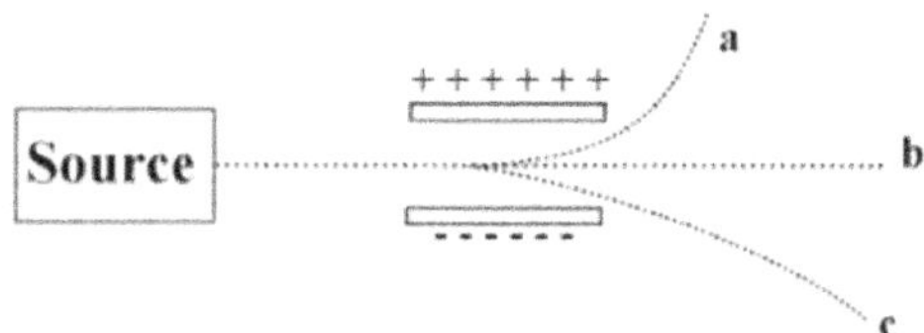


**37.** Which of the beams is due to an energetic lightwave?

**A.** I only
**B.** II only
**C.** III only
**D.** I and II only

**38.** Which of the beams is/are composed of particles?

**A.** I only
**B.** II only
**C.** I, II and III
**D.** I and III only

**39.** Which of the beams is due to a positively-charged helium nucleus?

**A.** I only
**B.** II only
**C.** III only
**D.** I, II and III

**40.** Lithium atoms can absorb photons transitioning from the ground state (at –5.37 eV) to an excited state with one electron removed from the atom, corresponding to the zero-energy state. What is the wavelength of light associated with this transition? (Use Planck's constant $h = 4.14 \times 10^{-15}$ eV·s and speed of light $c = 3 \times 10^8$ m/s)

**A.** $6.6 \times 10^{-6}$ m
**B.** $2.3 \times 10^{-7}$ m
**C.** $3.6 \times 10^{6}$ m
**D.** $4.2 \times 10^{5}$ m

---

**Practice Set 3: Questions 41–60**

---

**41.** All of the elements with atomic numbers of 84 and higher are radioactive because:

**A.** strong attractions between their nucleons make them unstable
**B.** their atomic numbers are larger than their mass numbers
**C.** strong repulsions between their electrons make them unstable
**D.** strong repulsions between their protons make their nuclei unstable

**42.** Which of the following statements about β particles is FALSE?

**A.** They have a smaller mass than α particles
**B.** They have high energy and a charge
**C.** They are created when neutrons become protons and vice versa
**D.** They are a harmless form of radioactivity

**43.** Which of the following provides the minimum amount of protection needed to block gamma radiation?

**A.** Thick leather
**B.** T-shirt
**C.** Lead suit
**D.** Suntan lotion

**44.** Which statement regarding Planck's constant is true?

**A.** It relates mass to the amount of energy that can be emitted
**B.** It sets a lower limit to the amount of energy that can be absorbed or emitted
**C.** It sets an upper limit to the amount of energy that can be absorbed
**D.** It sets an upper limit to the amount of energy that can be absorbed or emitted

**45.** The decay rate of a radioactive isotope will NOT be increased by increasing:

I. surface area II. pressure III. temperature

**A.** I only
**B.** II only
**C.** III only
**D.** I, II and III

**46.** Why are some smaller nuclei such as $^{14}$Carbon often radioactive?

I. The attractive force of the nucleons has a limited range
II. The neutron to proton ratio is too large or too small
III. Most smaller nuclei are not stable

**A.** I only
**B.** II only
**C.** III only
**D.** II and III only

**47.** Scandium $^{44}$Sc decays by emitting a positron. What is the resultant nuclide produced by this decay?

**A.** $^{43}_{21}Sc$
**B.** $^{45}_{21}Sc$
**C.** $^{44}_{20}Ca$
**D.** $^{43}_{20}Ca$

**48.** A scintillation counter detects radioactivity by:

**A.** analyzing the mass and velocity of each electron
**B.** ionizing argon gas in a chamber which produces an electrical signal
**C.** emitting light from a NaI crystal when radioactivity passes through the crystal
**D.** developing film which is exposed by radioactive particles

**49.** Which of the following is indicated by each detection sound by a Geiger counter?

**A.** One half-life
**B.** One nucleus decaying
**C.** One neutron being emitted
**D.** One positron being emitted

**50.** According to the Pauli Exclusion Principle, how many electrons in an atom may have a particular set of quantum numbers?

**A.** 1
**B.** 2
**C.** 3
**D.** 4

**51.** The atomic number of an atom identifies the number of:

**A.** excited states
**B.** electron orbits
**C.** neutrons
**D.** protons

**52.** Which of the following correctly characterizes gamma radiation?

**A.** High penetrating power; charge = –1; mass = 0 amu
**B.** Low penetrating power; charge = –1; mass = 0 amu
**C.** High penetrating power; charge = 0; mass = 0 amu
**D.** High penetrating power; charge = 0; mass = 4 amu

**53.** The rest mass of a proton is 1.0072764669 amu, and that of a neutron is 1.0086649156 amu. The $^4$He nucleus weighs 4.002602 amu. What is the total binding energy of the nucleus? (Use the speed of light $c = 3 \times 10^8$ m/s and 1 amu = $1.6606 \times 10^{-27}$ kg)

**A.** $2.7 \times 10^{-11}$ J
**B.** $4.4 \times 10^{-12}$ J
**C.** $1.6 \times 10^{-7}$ J
**D.** $2.6 \times 10^{-12}$ J

**54.** Which is the correct electron configuration for ground state boron ($Z = 5$)?

**A.** $1s^21p^22s$
**B.** $1s^22p^23s$
**C.** $1s^22p^3$
**D.** $1s^22s^22p$

**55.** The Sun produces $3.85 \times 10^{26}$ J each second. How much mass does it lose per second from nuclear processes alone? (Use the speed of light $c = 3 \times 10^8$ m/s)

**A.** $9.8 \times 10^1$ kg
**B.** $2.4 \times 10^9$ kg
**C.** $4.3 \times 10^9$ kg
**D.** $1.1 \times 10^8$ kg

**56.** The damaging effects of radiation on the body are a result of:

**A.** extensive damage to nerve cells
**B.** transmutation reactions in the body
**C.** formation of radioactive particles in the body
**D.** formation of unstable ions or radicals in the body

**57.** How does the emission of a gamma-ray affect the radioactive atom?

I. The atomic mass increases
II. The atom has a smaller amount of energy
III. The atom gains energy for further radioactive particle emission

**A.** I only
**B.** II only
**C.** III only
**D.** I and II only

 

**58.** The nuclear particle, which is described by the symbol $^{1}_{0}n$ is a(n):

**A.** neutron
**B.** gamma ray
**C.** beta particle
**D.** electron

**59.** Heisenberg's uncertainty principle states that:

**A.** at times, a photon appears to be a particle, and at other times it appears to be a wave
**B.** whether a photon is a wave, or a particle cannot be determined with certainty
**C.** the position and the momentum of a particle cannot be simultaneously known with absolute certainty
**D.** the properties of an electron cannot be known with absolute certainty

**60.** The material used in nuclear bombs is $^{239}Pu$, with a half-life of about 20,000 years. What is the approximate time needed for a buried stockpile of this substance to decay to 3% of its original $^{239}Pu$ mass?

**A.** 0.8 thousand years
**B.** 65 thousand years
**C.** 90 thousand years
**D.** 101 thousand years

===============================================================

**Practice Set 4: Questions 61–80**

===============================================================

**61.** The emitted particle with the mass of an electron but carrying a +1 charge is a:

**A.** plusion
**B.** positron
**C.** proelectron
**D.** proton

**62.** Which type of nuclear radiation results in the release of particles identical to electrons and are deflected toward the positive electrode as they pass between electrically-charged plates?

**A.** Alpha
**B.** Gamma
**C.** Beta
**D.** Nuclide

**63.** The radial distance between the nucleus and the orbital shell in a hydrogen atom:

**A.** varies randomly with increasing values of n
**B.** remains constant for all values of n
**C.** decreases with increasing values of n
**D.** increases with increasing values of n

**64.** The first part of an atom to be discovered was the:

**A.** electron
**B.** nucleus
**C.** proton
**D.** neutron

**65.** Which is the best description of an alpha particle?

**A.** Charge of –1; a mass of 0 amu; low penetrating power
**B.** Charge of –2; a mass of 4 amu; medium penetrating power
**C.** Charge of +2; a mass of 4 amu; low penetrating power
**D.** Charge of +2; a mass of 4 amu; high penetrating power

**66.** Which isotope has the maximum binding energy per nucleon and therefore is the maximum in the binding energy per nucleon curve?

**A.** $^{56}Fe$
**B.** $^{1}H$
**C.** $^{251}Cf$
**D.** $^{197}Au$

**67.** In nuclear fusion:

**A.** electrons and nuclei combine to form gamma rays
**B.** several small nuclei combine to form an atom of an element with a greater atomic number
**C.** an atomic nucleus emits alpha particles
**D.** an atomic nucleus splits into two fragments, each forming an atom of an element with a smaller atomic number than the source nuclei

**68.** If the nucleus $^{15}_{7}N$ is bombarded with a proton, one or more products are formed. Which of the following represents a set of products from this reaction?

**A.** $^{16}N + \gamma$
**B.** $^{14}B + ^{2}Li$
**C.** $^{15}O + \gamma$
**D.** $^{12}C + ^{4}He$

**69.** Which is a common consumer item that utilizes the concept of radioactive decay?

**A.** Carburetor
**B.** Smoke detector
**C.** Furnace
**D.** Gas stove

**70.** What happens to the atomic number of a nucleus that undergoes $\beta^-$ decay?

**A.** It increases by two
**B.** It decreases by one
**C.** It increases by one
**D.** It remains the same

**71.** Two radioactive nuclides I and II both decay to stable products. The half-life of I is about a day, while that of II is about a week. Suppose a radioactive sample consists of a mixture of these two nuclides. If the mixture is such that the radioactivity emitted from I and II are initially equal, then a few days later, the radioactivity of the sample will be due:

**A.** to nuclides I and II equally
**B.** entirely to nuclide II
**C.** predominantly to nuclide I
**D.** predominantly to nuclide II

**72.** A metal surface is illuminated with blue light, and electrons are ejected at a given rate, each with a certain amount of energy. If the intensity of the blue light is increased, electrons are ejected at:

**A.** an increased rate with no change in energy per electron
**B.** an increased rate with an increase in energy per electron
**C.** the same rate, but with an increase in energy per electron
**D.** the same rate, but with a decrease in energy per electron

**73.** How many 3*d* electron states can an atom have?

**A.** 0
**B.** 4
**C.** 6
**D.** 10

**74.** Beyond which element in the periodic table (based on the atomic number) are the successive elemental nuclei considered radioactive?

**A.** Barium
**B.** Bismuth
**C.** Uranium
**D.** Radium

**75.** What is the term for a powerful type of nuclear radiation that has neither mass nor charge?

**A.** Gamma ray
**B.** Positron
**C.** Alpha particle
**D.** Beta particle

**76.** To which of the following values of n does the shortest wavelength in the Balmer series correspond? (Use the Balmer series constant B = $3.645 \times 10^{-7}$ m)

**A.** 1
**B.** 5
**C.** 3
**D.** $\infty$

**77.** What is the name for the sum of the number of protons and neutrons in the nucleus of an atom?

**A.** Atomic weight
**B.** Atomic number
**C.** Mass number
**D.** Atomic mass

**78.** Which of the following provides the minimum amount of protection needed to block beta radiation?

**A.** Thin T-shirt
**B.** Thick leather
**C.** 1 m of water
**D.** Thick lead shielding

**79.** What is the mass number of an alpha particle?

**A.** 0
**B.** 1
**C.** 2
**D.** 4

**80.** What is the name of the particle having the following atomic notation: ${}^{0}_{-1}e^{-}$?

**A.** Neutron
**B.** Gamma
**C.** Beta
**D.** Alpha

 

## Practice Set 5: Questions 81–100

**81.** A common reaction in the Sun involves the encounter of two nuclei of light helium ($^{3}_{2}He$). If one $^{3}He$ nucleus encounters another, which products are possible?

**A.** $^{2}H + ^{2}H + ^{2}H$
**B.** $^{7}Li + ^{1}H$
**C.** $^{4}He + ^{1}H + ^{1}H$
**D.** $^{4}He + ^{2}H$

**82.** What is the term for the unit of biological radiation damage equivalent to 100 rems?

**A.** Gray
**B.** Becquerel
**C.** Rad
**D.** Sievert

**83.** Which statement about positron emission is FALSE?

I. It occurs with alpha and beta decay
II. The alternative symbol is $\beta^+$
III. It occurs when a proton is converted into a neutron and a positron

**A.** I only
**B.** II only
**C.** III only
**D.** I and II only

**84.** Why do heavy nuclei contain more neutrons than protons?

**A.** Neutrons are heavier than protons
**B.** Neutrons are repelled by the protons
**C.** Neutrons are radioactive, and so are heavy nuclei
**D.** Neutrons minimize the electric repulsion of the protons

**85.** Classical wave theory predicts that the photocurrent of the photoelectric effect is proportional to the:

**A.** intensity of light
**B.** magnitude of the electric field
**C.** frequency of light
**D.** wavelength of light

 

**86.** Suppose the half-life of some element is 2 days. 10 grams of this element is contained in a mixture produced by a laboratory 3 days ago. If the element is isolated from the mixture, how much of it would remain after 3 days from its separation from the mixture?

**A.** 0.75 grams
**B.** 1.25 grams
**C.** 2.5 grams
**D.** 4.5 grams

**87.** Nuclear fusion:

**A.** produces non-radioactive elements
**B.** is the energy source of the Sun and stars
**C.** releases a larger amount of heat than nuclear fission
**D.** all of the above

**88.** The type of nuclear radiation with the least penetrating ability is:

**A.** gamma radiation
**B.** neutrons
**C.** alpha radiation
**D.** beta radiation

**89.** Which of the following statements is true for the Bohr model of the atom?

**A.** As the electron shells increase, they get further apart, but the difference in the energy between them decreases
**B.** As electron shells increase, they get closer, but the difference in energy between them gets greater
**C.** The energy difference between all the electron shells is the same
**D.** The spacing between all the electron shells is the same

**90.** A measure of the radiation that considers the possible biological damage produced by different types of radiation is:

**A.** Roentgen
**B.** Curie
**C.** rem
**D.** rad

**91.** The accelerating voltage in an X-ray tube is doubled. Compared to the original value, what change occurs with respect to the minimum wavelength of the X-ray?

**A.** Remains the same
**B.** Decreases to one half
**C.** Doubles
**D.** Quadruples

 

**92.** Which nuclear process does not cause a change in the atomic number of the isotope?

**A.** Emission of a positron
**B.** Emission of a $\beta$ particle
**C.** Emission of a $\gamma$ ray
**D.** Emission of a $\alpha$ particle

**93.** Which is the product from the alpha decay of $^{235}_{92}U$?

**A.** $^{231}_{90}Th$
**B.** $^{233}_{80}Ra$
**C.** $^{235}_{93}Np$
**D.** $^{239}_{94}Pu$

**94.** The isotope $^{36}_{17}Cl$ most likely undergoes:

I. $\alpha$ decay II. $\beta^-$ decay III. $\beta^+$ decay

**A.** I only
**B.** II only
**C.** III only
**D.** I and II only

**95.** A positron is a particle emitted from the nucleus that has the same mass as a(n):

**A.** hydrogen atom
**B.** alpha particle
**C.** neutron
**D.** electron

**96.** Which of the following statements regarding radiation is FALSE?

**A.** The time for half of the original sample to spontaneously decay is the half-life
**B.** All radioactive elements are spontaneously decaying toward forming a stable element
**C.** All elements heavier than bismuth are radioactive
**D.** All the above are true

**97.** Isotopes are atoms of an element with similar chemical properties but with different:

**A.** numbers of electrons
**B.** atomic numbers
**C.** numbers of protons
**D.** masses

**98.** Which of the following statements regarding a nucleon is true?

I. Attraction between nucleons changes their mass
II. Some of the nucleon's mass can be converted into energy by breaking certain nuclei
III. The mass of a nucleon is different outside the nucleus

**A.** I only
**B.** II only
**C.** III only
**D.** I, II and III

**99.** A nuclear power plant provides $10^{12}$ J of electrical power each day by using heat from the fission of $^{235}U$ to turn turbines. The mass deficit due to the fission of one $^{235}U$ atom is $3 \times 10^{-25}$ grams. How much mass is converted to energy per day? (Use Planck's constant $h = 6.63 \times 10^{-34}$ J·s and the speed of light $c = 3 \times 10^8$ m/s)

**A.** $10^{-22}$ g
**B.** $10^{-7}$ g
**C.** $10^{-2}$ g
**D.** 18 g

**100.** Which statement about a photon is true for a photon to eject an electron from a metal's surface in the photoelectric effect?

**A.** Momentum must be zero
**B.** Speed must be greater than a specified minimum value
**C.** Polarization must have a component perpendicular to the surface
**D.** Frequency must be greater than a specified minimum value

 

---

**Practice Set 6: Questions 101–120**

---

**101.** Which of the following correctly characterizes a beta particle $\beta^-$?

**A.** Medium penetrating power; charge = –1; mass = 0 amu
**B.** High penetrating power; charge = –1; mass = 0 amu
**C.** High penetrating power; charge = +2; mass = 4 amu
**D.** Low penetrating power; charge = +2; mass = 4 amu

**102.** The control rods of a nuclear reactor regulate the chain reaction because they:

**A.** produce positrons
**B.** absorb the neutrons
**C.** absorb the $^{235}U$ atoms
**D.** contain catalysts

**103.** What is the term for nuclear radiation that is identical to a $^4He$ nucleus?

**A.** Positron
**B.** Gamma ray
**C.** Beta particle
**D.** Alpha particle

**104.** The electron spin quantum number can have values of:

**A.** –½ and ½
**B.** –½, 0, +½
**C.** –½, –1, +1, +½
**D.** –½, –1, 0, +1, +½

**105.** The nucleus was discovered through experiments with:

**A.** radio waves
**B.** cathode rays
**C.** radioactivity
**D.** light

**106.** Which of the following does NOT describe a nuclear fission reaction?

I. The energy released is proportional to the mass difference in the system
II. The number of nucleons in the products is different from the number in the starting material
III. The mass of the products is less than the mass of the reactants

**A.** I only
**B.** II only
**C.** III only
**D.** I and II only

**107.** In the nuclear decay of a beta-emitter, the new nucleus may contain:

**A.** 4 fewer protons
**B.** 2 more protons
**C.** 2 fewer protons
**D.** 1 more proton

**108.** The nuclear particle which is represented by the symbol ${}^{0}_{+1}e$ is a(n):

**A.** neutron
**B.** positron
**C.** alpha particle
**D.** electron

**109.** Which statement describes an atom that loses an alpha particle?

**A.** Atomic number remains the same, and the mass number decreases
**B.** Mass number decreases by 2, but its atomic number remains the same
**C.** Atomic number increases by 1, but its mass number remains the same
**D.** Atomic number decreases by 2, and its mass number decreases by 4

**110.** The radioactivity of a sample of ${}^{32}P$ was determined to be 400 mCi. How much time elapses before the radioactivity decreases to 25 mCi? (Use the half-life of ${}^{32}P$ = 14.3 days)

**A.** 28.6 days
**B.** 42.9 days
**C.** 57.2 days
**D.** 114.4 days

**111.** The symbol ${}^{0}_{-1}e$ is used to represent a(n):

**A.** alpha particle
**B.** beta particle
**C.** neutron
**D.** proton

**112.** For the atoms in a neon discharge tube to emit light characteristic of brilliant red color, it is necessary that:

**A.** the electrons gain energy to be promoted from their ground state to an excited state
**B.** the atoms are continually replaced with fresh atoms
**C.** each atom carries a net electric charge
**D.** there be no unoccupied energy levels in each atom

 

**113.** Increasing the brightness of a beam of light without changing its color increases the:

I. speed of the photons
II. energy of each photon
III. number of photons emitted by light every second

**A.** I only
**B.** II only
**C.** III only
**D.** I and II only

**114.** What is the term for the type of nuclear phenomenon in which a heavy nuclide attracts one of its inner core electrons into the nucleus?

**A.** Nuclide decay
**B.** Electron capture
**C.** Beta decay
**D.** Neutron reaction

**115.** The mass number of an atom is equal to the number of what particles in the atom?

**A.** Electrons
**B.** Positrons
**C.** Neutrons
**D.** Nucleons

**116.** Neon has 10 electrons. What is the value for Z of the element with the next larger Z that has chemical properties very similar to those of neon?

**A.** 10
**B.** 16
**C.** 18
**D.** 24

**117.** The binding energy per nucleon of a nucleus:

**A.** has a maximum near iron in the periodic table and then decreases for heavier elements
**B.** is approximately constant in the periodic table, except for very light nuclei
**C.** decreases steadily in the progression towards heavier elements
**D.** increases steadily in the progression towards heavier elements

**118.** The intensity of X-rays, gamma rays, or any other radiation is:

**A.** inversely proportional to the square of the distance from the source
**B.** inversely proportional to the distance from the source
**C.** directly proportional to the square of the distance from the source
**D.** directly proportional to the distance from the source

**119.** What percentage of a sample remains after 60 hours if the half-life of $^{24}Na$ is 15 hours?

**A.** 0%
**B.** 2%
**C.** 6.25%
**D.** 8%

**120.** Ionizing radiation is:

**A.** a neutron that has acquired a charge, resulting in forming an ion
**B.** high-energy radiation that removes electrons from atoms or molecules
**C.** radiation that only interacts with ions
**D.** equivalent to a proton

 

---

**Practice Set 7: Questions 121–140**

---

**121.** When a positron is emitted from the nucleus of an atom, what effect does this have on the nuclear mass?

**A.** It increases by 1
**B.** It decreases by 2
**C.** It increases by 2
**D.** It remains the same

**122.** What nucleus results when $^{55}Ni$ decays by positron emission?

**A.** $^{55}Ca$
**B.** $^{55}Ni$
**C.** $^{55}Co$
**D.** $^{55}Fe$

**123.** Which of the following types of radiation has the highest energy?

**A.** $\gamma$ rays
**B.** Visible light rays
**C.** $\alpha$ particles
**D.** $\beta$ particles

**124.** A radioactive nuclide of atomic number Z emits an alpha particle, and the daughter nucleus then emits a beta-minus particle. What is the atomic number of the resulting nuclide?

**A.** Z–1
**B.** Z+1
**C.** Z–2
**D.** Z–3

**125.** Which type of nuclear radiation is a helium nucleus deflected toward the negative electrode as it passes between electrically-charged plates?

**A.** Gamma
**B.** Beta
**C.** Alpha
**D.** Nuclide

**126.** In massive stars, three helium nuclei fuse, forming a carbon nucleus, and this reaction heats the star's core. What is the net mass of the three helium nuclei?

**A.** Same as the carbon nucleus because energy is always conserved
**B.** Same as the carbon nucleus because mass is always conserved
**C.** Less than that of the carbon nucleus
**D.** Greater than the carbon nucleus

**127.** Why is the planetary model of an atom, with the nucleus playing the role of the Sun and the electrons playing the role of planets, flawed?

**A.** The electrical attraction between a proton and an electron is too weak
**B.** An electron is accelerating and loses energy
**C.** The nuclear attraction between a proton and an electron is too strong
**D.** An electron is accelerating and gains energy

**128.** Which of the following types of radiation might have the greatest application for medical imaging?

**A.** Beta radiation would be best because it can be measured electrically
**B.** X-rays would be best because they interact with the DNA in the cells
**C.** Gamma radiation would be best because it penetrates the furthest
**D.** Alpha radiation would be best because it penetrates the least and does the least damage

**129.** The type of nuclear radiation having particles with the greatest charge consists of:

**A.** neutrons
**B.** gamma rays
**C.** beta particles
**D.** alpha particles

**130.** All of the statements about nuclear reactions are true EXCEPT:

**A.** they are not affected by the chemical state of the atoms involved
**B.** they can have their rate increased by the addition of a catalyst
**C.** they involve changes in the nucleus of an atom
**D.** they have energy changes much greater than in ordinary chemical reactions

 

**131.** In a nuclear reaction, the mass of the products is less than the mass of the reactants. Why is this not observed in a chemical reaction?

**A.** In chemical reactions, the mass is held constant by the nucleus
**B.** In chemical reactions, the mass deficit is balanced by a mass surplus
**C.** The mass deficit in chemical reactions is too small to be observed
**D.** The mass does not convert to energy in chemical reactions

**132.** The isotope $^{13}_{7}N$ decays by positron emission to what isotope?

**A.** $^{14}_{6}C$ **B.** $^{11}_{7}N$ **C.** $^{13}_{6}C$ **D.** $^{12}_{6}C$

**133.** The radioactive gas radon is:

I. more hazardous to smokers than nonsmokers
II. the single greatest source of human radiation exposure
III. a product of the radioactive decay series of uranium

**A.** I only
**B.** II only
**C.** III only
**D.** I, II and III

**134.** Which of the following statements best describes the strong nuclear force?

**A.** The strength of the force increases with distance
**B.** The force is very strong and is effective over a large range of distances
**C.** The electrical force is stronger than the nuclear force
**D.** The force is very strong but is effective only within a short range of distances

**135.** Natural line broadening can be understood in terms of the:

**A.** Schrodinger wave equation
**B.** Pauli exclusion principle
**C.** de Broglie wavelength
**D.** uncertainty principle

**136.** If a star has a peak intensity at 580 nm, what is its temperature? (Use Wien's displacement constant $b = 2.9 \times 10^{-3}$ K·m)

**A.** 5,000 °C
**B.** 2,000 °C
**C.** 5,000 °F
**D.** 5,000 K

**137.** Which type of nuclear radiation is powerful light energy that is *not* deflected as it passes between electrically-charged plates?

**A.** Gamma
**B.** Beta
**C.** Alpha
**D.** Nuclide

**138.** The main reason that there is a limit to the size of a stable nucleus is the:

**A.** weakness of the electrostatic force
**B.** weakness of the gravitational force
**C.** short-range effect of the strong nuclear force
**D.** limited range of the gravitational force

**139.** Elements combine in fixed mass ratios to form compounds. This requires the elements:

**A.** have unambiguous atomic numbers
**B.** are always chemically active
**C.** are composed of continuous matter without subunits
**D.** are composed of discrete subunits called atoms

**140.** What is rem?

**A.** A unit for measuring rapid electron motion
**B.** A unit for measuring radiation exposure
**C.** The number of radiation particles emitted per second
**D.** The maximum exposure limit for occupational safety

*Notes for active learning*

*Notes for active learning*

 

# 12 – Quantum Mechanics

## Practice Set 1: Questions 1–20

**1.** The work function of a certain metal is 1.90 eV. What is the longest wavelength of light that can cause photoelectron emission from this metal?

**A.** 64 nm
**B.** 98 nm
**C.** 247 nm
**D.** 653 nm

**2.** Which of the following statements is correct if the frequency of the light in a laser beam is doubled while the number of photons per second in the beam is fixed?

I. The power in the beam does not change
II. The intensity of the beam doubles
III. The energy of individual photons does not change

**A.** I only
**B.** II only
**C.** III only
**D.** I and II only

**3.** Upon being struck by 240 nm photons, a material ejects electrons with a maximum kinetic energy of 2.58 eV. What is the work function of this material?

**A.** 1.17 eV
**B.** 2.04 eV
**C.** 2.60 eV
**D.** 3.26 eV

**4.** A high-energy photon collides with matter and creates an electron-positron pair. What is the minimum frequency of the photon? (Use the $m_{electron} = 9.11 \times 10^{-31}$ kg, $c = 3.00 \times 10^8$ m/s, and $h = 6.626 \times 10^{-34}$ J·s)

**A.** greater than $1.24 \times 10^{12}$ Hz
**B.** greater than $2.47 \times 10^{16}$ Hz
**C.** greater than $2.47 \times 10^{20}$ Hz
**D.** greater than $2.47 \times 10^{22}$ Hz

**5.** What is the longest wavelength of light that can cause photoelectron emission from metal with a work function of 2.20 eV?

**A.** 216 nm
**B.** 372 nm
**C.** 484 nm
**D.** 564 nm

**6.** In 1932, C. D. Anderson:

**A.** set the limits on the probability of measurement accuracy
**B.** predicted the positron from relativistic quantum mechanics
**C.** discovered the positron using a cloud chamber
**D.** was the first to produce diffraction patterns of electrons in crystals

**7.** What is the energy of an optical photon of frequency $6.43 \times 10^{14}$ Hz (Use $h = 6.626 \times 10^{-34}$ J·s and 1 eV = $1.60 \times 10^{-19}$ J)

**A.** 1.04 eV
**B.** 1.86 eV
**C.** 2.66 eV
**D.** 3.43 eV

**8.** A photocathode has a work function of 2.4 eV. The photocathode is illuminated with monochromatic radiation, whose photon energy is 3.4 eV. What is the maximum kinetic energy of the photoelectrons produced? (Use 1 eV = $1.60 \times 10^{-19}$ J)

**A.** $3.4 \times 10^{-20}$ J
**B.** $1.6 \times 10^{-19}$ J
**C.** $4.6 \times 10^{-19}$ J
**D.** $5.8 \times 10^{-19}$ J

**9.** A photocathode whose work function is 2.9 eV is illuminated with white light with a continuous wavelength band from 400 nm to 700 nm. What is the range of the wavelength band in this white light illumination for which photoelectrons are NOT produced?

**A.** 360 to 440 nm
**B.** 400 to 480 nm
**C.** 430 to 500 nm
**D.** 430 to 700 nm

**10.** Photon A has twice the momentum of photon B as both travels in a vacuum. Which of the following statements about these photons is correct?

**A.** Both photons have the same speed
**B.** Both photons have the same wavelength
**C.** Photon A is traveling twice as fast as photon B
**D.** The energy of photon A is half as great as the energy of photon B

**11.** What is the energy of the photon emitted when an electron drops from the n = 20 state to the n = 7 state in a hydrogen atom?

**A.** 0.244 eV
**B.** 0.288 eV
**C.** 0.336 eV
**D.** 0.404 eV

**12.** Protons are being accelerated in a particle accelerator. What is the de Broglie wavelength when the energy of the protons is doubled if the protons are non-relativistic (i.e., their kinetic energy is much less than $mc^2$)?

**A.** increases by a factor of 2
**B.** increases by a factor of 3
**C.** decreases by a factor of 2
**D.** decreases by a factor of √2

**13.** A certain photon, after being scattered from a free electron at rest, moves at an angle of 120° with respect to the incident direction. If the wavelength of the incident photon is 0.591 nm, what is the wavelength of the scattered photon? (Use $m_{electron} = 9.11 \times 10^{-31}$ kg, $c = 3.00 \times 10^8$ m/s and $h = 6.626 \times 10^{-34}$ J·s)

**A.** 0.0 nm
**B.** 0.180 nm
**C.** 0.252 nm
**D.** 0.595 nm

**14.** Increasing the *brightness* of a beam of light without changing its color increases the:

**A.** speed of the photons
**B.** frequency of the light
**C.** number of photons per second traveling in the beam
**D.** energy of each photon

**15.** What is the frequency of the light emitted by atomic Hydrogen according to the Balmer formula with $m = 4$ and $n = 9$?

**A.** 820 Hz
**B.** $1.65 \times 10^{14}$ Hz
**C.** 1,820 Hz
**D.** 1,640 Hz

**16.** One of the emission lines described by the original version of the Balmer formula has a wavelength of 377 nm. What is the value of n in the Balmer formula that gives this emission line?

**A.** 5
**B.** 7
**C.** 9
**D.** 11

**17.** What is the wavelength of the most intense light emitted by a giant star of surface temperature 5000 K? (Use the constant in Wien's law = 0.00290 m·K)

**A.** 366 nm
**B.** 448 nm
**C.** 580 nm
**D.** 490 nm

**18.** In 1928, Paul Dirac:

**A.** set the limits on the probability of measurement accuracy
**B.** developed a wave equation for matter waves
**C.** suggested the existence of matter waves
**D.** predicted the positron from relativistic quantum mechanics

**19.** A photocathode has a work function of 2.4 eV. The photocathode is illuminated with monochromatic radiation, whose photon energy is 3.4 eV. What is the maximum kinetic energy of the photoelectrons produced?

**A.** $0.9 \times 10^{-19}$ J
**B.** $1.6 \times 10^{-19}$ J
**C.** $2.8 \times 10^{-19}$ J
**D.** $4.2 \times 10^{-19}$ J

**20.** In a Compton scattering experiment, which scattering angle produces the greatest change in wavelength?

**A.** 0°
**B.** 45°
**C.** 90°
**D.** 180°

 

## Practice Set 2: Questions 21–40

**21.** A photocathode has a work function of 2.4 eV. The photocathode is illuminated with monochromatic radiation, whose photon energy is 3.5 eV. What is the wavelength of the illuminating radiation?

**A.** 280 nm
**B.** 325 nm
**C.** 350 nm
**D.** 395 nm

**22.** If the wavelength of a photon is doubled, what happens to its energy?

**A.** It is reduced to one-half of its original value
**B.** It stays the same
**C.** It is doubled
**D.** It is increased to four times its original value

**23.** A photocathode has a work function of 2.8 eV. The photocathode is illuminated with monochromatic radiation. What is the threshold frequency for the monochromatic radiation required to produce photoelectrons?

**A.** $1.2 \times 10^{14}$ Hz
**B.** $2.8 \times 10^{14}$ Hz
**C.** $4.6 \times 10^{14}$ Hz
**D.** $6.8 \times 10^{14}$ Hz

**24.** If the de Broglie wavelength of an electron is 380 nm, what is the speed of this electron? (Use $m_{electron} = 9.11 \times 10^{-31}$ kg and $h = 6.626 \times 10^{-34}$ J·s)

**A.** 0.6 km/s
**B.** 1.9 km/s
**C.** 3.4 km/s
**D.** 4.6 km/s

**25.** What is the wavelength of the scattered photon if a photon of wavelength $1.50 \times 10^{-10}$ m is scattered at an angle of 90° in the Compton effect?

**A.** $1.29 \times 10^{-10}$ m
**B.** $1.52 \times 10^{-10}$ m
**C.** $1.84 \times 10^{-10}$ m
**D.** $2.42 \times 10^{-10}$ m

**26.** When the surface of a metal is exposed to blue light, electrons are emitted. Which of the following increases if the intensity of the blue light increases?

I. the maximum kinetic energy of the ejected electrons
II. the number of electrons ejected per second
III. the time lag between the onset of the absorption of light and the ejection of electrons

**A.** I only
**B.** II only
**C.** III only
**D.** I and II only

**27.** A proton has a speed of 7.2 x $10^4$ m/s. What is the energy of a photon with the same wavelength as the de Broglie wavelength of this proton? (Use $m_{proton} = 1.67 \times 10^{-27}$ kg and $c = 3.00 \times 10^8$ m/s)

**A.** 80 keV
**B.** 120 keV
**C.** 160 keV
**D.** 230 keV

**28.** In the Compton effect, as the scattering angle increases monotonically from 0° to 180°, the frequency of the X-rays scattered at that angle:

**A.** decreases by the √2
**B.** decreases monotonically
**C.** increases by the √2
**D.** increases monotonically

**29.** In the spectrum of Hydrogen, the lines obtained by setting m = 1 in the Rydberg formula are the Lyman series. What is the wavelength of the spectral line of the 15$^{th}$ member of the Lyman series?

**A.** 91.6 nm
**B.** 126.2 nm
**C.** 244.6 nm
**D.** 368.2 nm

**30.** If the frequency of the light in a laser beam is doubled while the number of photons per second in the beam is fixed, which of the following statements is correct?

I. The energy of individual photons doubles
II. The wavelength of the individual photons doubles
III. The intensity of the beam doubles

**A.** I only
**B.** II only
**C.** III only
**D.** I and III only

**31.** The spacing of the surface planes of a crystal is 159.0 pm. A beam directed at the surface of the crystal undergoes first-order diffraction at an angle of 58° from the normal. What is the energy of the neutrons if the diffraction is done with a beam of monenergistic neutrons? (Use $1.67 \times 10^{-27}$ kg for the mass of a neutron)

**A.** 0.0155 eV
**B.** 0.0106 eV
**C.** 0.0909 eV
**D.** 0.1450 eV

**32.** A photocathode whose work function is 2.5 eV is illuminated with white light with a continuous wavelength band from 360 nm to 700 nm. What is the stopping potential for this white light illumination?

**A.** 0.95 V
**B.** 1.45 V
**C.** 1.90 V
**D.** 2.6 V

**33.** If the momentum of an electron is $1.95 \times 10^{-27}$ kg·m/s, what is its de Broglie wavelength? (Use $h = 6.626 \times 10^{-34}$ J·s)

**A.** 86.2 nm
**B.** 130.6 nm
**C.** 240.8 nm
**D.** 340.0 nm

**34.** A Hydrogen atom is excited to the n = 9 level, and its decay to the n = 6 level is detected on a photographic plate. What is the frequency of the light photographed?

**A.** 3,810 Hz
**B.** 7,240 Hz
**C.** $5.08 \times 10^{13}$ Hz
**D.** $3.28 \times 10^{-9}$ Hz

**35.** How much energy is carried by a photon of light having a frequency of 110 GHz? (Use $h = 6.626 \times 10^{-34}$ J·s)

**A.** $7.3 \times 10^{-23}$ J
**B.** $2.9 \times 10^{-25}$ J
**C.** $1.7 \times 10^{-26}$ J
**D.** $1.1 \times 10^{-21}$ J

**36.** How many of the infinite number of Balmer spectrum lines are in the visible spectrum range?

**A.** 0
**B.** 2
**C.** 4
**D.** 6

**37.** Each photon in a beam of light has an energy of 4.20 eV. What is the wavelength of this light? ($c = 3.00 \times 10^8$ m/s, $h = 6.626 \times 10^{-34}$ J·s and 1 eV = $1.60 \times 10^{-19}$ J)

**A.** 118.0 nm
**B.** 296.0 nm
**C.** 365.0 nm
**D.** 462.0 nm

**38.** Electrons are emitted from a surface when the light of wavelength 500.0 nm is shone on the surface, but electrons are not emitted for longer wavelengths of light. What is the work function of the surface?

**A.** 0.5 eV
**B.** 1.6 eV
**C.** 2.5 eV
**D.** 3.8 eV

**39.** In the Bohr theory, the orbital radius depends upon the principal quantum number in what way?

**A.** n
**B.** 1/n
**C.** $n^2$
**D.** $n^3$

**40.** The uncertainty in the position of a proton is 0.053 nm. What is the minimum uncertainty in its speed? (Use $1.67 \times 10^{-27}$ kg as the proton mass)

**A.** $0.6 \times 10^3$ m/s
**B.** $1.2 \times 10^3$ m/s
**C.** $2.4 \times 10^3$ m/s
**D.** $3.6 \times 10^3$ m/s

 

---

**Practice Set 3: Questions 41–60**

---

**41.** What is the wavelength of the light emitted by atomic Hydrogen according to the Balmer formula with m = 9 and n = 11?

**A.** 8,500 nm
**B.** 14,700 nm
**C.** 22,300 nm
**D.** 31,900 nm

**42.** A certain particle's energy is known within $10^{-18}$ J. What is the minimum uncertainty in its arrival time at a detector?

**A.** $5.08 \times 10^{-12}$ s
**B.** $4.25 \times 10^{-13}$ s
**C.** $3.88 \times 10^{-14}$ s
**D.** $1.05 \times 10^{-16}$ s

**43.** Upon being struck by 240.0 nm photons, a material ejects electrons with a maximum kinetic energy of 2.58 eV. What is the work function of this material?

**A.** 1.20 eV
**B.** 2.82 eV
**C.** 2.60 eV
**D.** 3.46 eV

**44.** What is the de Broglie wavelength of a 1.30 kg missile moving at 28.10 m/s. (Use $h = 6.626 \times 10^{-34}$ J·s)

**A.** $1.85 \times 10^{-37}$ m
**B.** $2.40 \times 10^{-36}$ m
**C.** $1.81 \times 10^{-35}$ m
**D.** $3.37 \times 10^{-35}$ m

**45.** A beam of light falling on a metal surface causes electrons to be ejected from the surface. If the frequency of the light now doubles, which of the following statements is always true?

**A.** The number of electrons ejected per second doubles
**B.** Twice as many photons hit the metal surface as before
**C.** The kinetic energy of the ejected electrons doubles
**D.** None of the above statements is always true

**46.** What is the longest wavelength of a photon emitted by a hydrogen atom, for which the initial state is n = 3?

**A.** 486 nm
**B.** 510 nm
**C.** 540 nm
**D.** 656 nm

**47.** Which of the following always increases if the brightness of a beam of light is increased without changing its color?

I. the speed of the photons
II. the average energy of each photon
III. the number of photons

**A.** I only
**B.** II only
**C.** III only
**D.** I and II only

**48.** In a particular case of Compton scattering, a photon collides with a free electron and scatters backward. The wavelength after the collision is exactly double the wavelength before the collision. What is the wavelength of the incident photon? (Use $m_{electron} = 9.11 \times 10^{-31}$ kg, $c = 3.00 \times 10^8$ m/s and $h = 6.626 \times 10^{-34}$ J·s)

**A.** $3.4 \times 10^{-12}$ m
**B.** $4.8 \times 10^{-12}$ m
**C.** $5.6 \times 10^{-12}$ m
**D.** $6.8 \times 10^{-12}$ m

**49.** Two sources emit beams of microwaves. The microwaves from source A have a frequency of 15 GHz, and the microwaves from source B have a frequency of 30 GHz. This is all the information available for the two beams. Which of the following statements about these microwave beams must be correct?

**A.** The intensity of beam B is twice as great as the intensity of beam A
**B.** A photon in beam B has the same energy as a photon in beam A
**C.** Beam B carries twice as many photons per second as beam A
**D.** A photon in beam B has twice the energy of a photon in beam A

**50.** What is the shortest wavelength of a photon that can be emitted by a hydrogen atom, for which the initial state is n = 3?

**A.** 102.6 nm
**B.** 97.3 nm
**C.** 820.0 nm
**D.** 121.6 nm

 

**51.** The radius of a typical nucleus is about $5.0 \times 10^{-15}$ m. Assuming this to be the uncertainty in the position of a proton in the nucleus, what is the uncertainty in the proton's energy? (Use $1.67 \times 10^{-27}$ kg as the proton mass)

**A.** 0.06 MeV
**B.** 0.25 MeV
**C.** 0.4 MeV
**D.** 0.8 MeV

**52.** A photocathode has a work function of 2.4 eV. The photocathode is illuminated with monochromatic radiation, and electrons are emitted with a stopping potential of 1.1 volts. What is the wavelength of the illuminating radiation? (Use $c = 3.00 \times 10^8$ m/s, $h = 6.626 \times 10^{-34}$ J·s and 1 eV = $1.60 \times 10^{-19}$ J)

**A.** 300 nm
**B.** 350 nm
**C.** 390 nm
**D.** 420 nm

**53.** The Compton effect directly demonstrated which property of electromagnetic radiation?

I. energy content
II. particle nature
III. momenta

**A.** I only
**B.** II only
**C.** I and II only
**D.** II and III only

**54.** What is the wavelength of the matter-wave associated with an electron moving with a speed of $2.5 \times 10^7$ m/s? (Use $m_{electron} = 9.11 \times 10^{-31}$ kg and $h = 6.626 \times 10^{-34}$ J·s)

**A.** 17 pm
**B.** 29 pm
**C.** 39 pm
**D.** 51 pm

**55.** A laser produces a beam of 4000 nm light. A shutter allows a pulse of light, for 30.0 ps to pass. What is the uncertainty in the energy of a photon in the pulse? ($h = 6.626 \times 10^{-34}$ J·s and 1 eV = $1.60 \times 10^{-19}$ J)

**A.** $2.6 \times 10^{-2}$ eV
**B.** $4.2 \times 10^{-3}$ eV
**C.** $6.8 \times 10^{-4}$ eV
**D.** $2.2 \times 10^{-5}$ eV

**56.** Lithium atoms can absorb photons of frequency $2.01 \times 10^{14}$ Hz, transitioning from the ground state (at –5.37 eV) to an excited state. When one electron is completely removed from the atom, it corresponds to the zero-energy state. Which statement is true about a photon that ionizes a lithium atom? (Use Planck's constant $h = 4.14 \times 10^{-15}$ eV·s and the speed of light $c = 3 \times 10^8$ m/s)

**A.** It must have a frequency greater than $1.3 \times 10^{15}$ Hz
**B.** It must have a minimum frequency of $1.3 \times 10^{15}$ Hz
**C.** It must have a frequency less than $1.3 \times 10^{15}$ Hz
**D.** It must have a frequency greater than $2.6 \times 10^{15}$ Hz

**57.** Observing the emission spectrum of a hypothetical atom, a line corresponding to a wavelength of $1.25 \times 10^{-7}$ m is observed from the transition to the ground state. If the ground state has zero energy, what other energy level must exist in this atom? (Use Planck's constant $h = 6.63 \times 10^{-34}$ J·s and the speed of light $c = 3 \times 10^8$ m/s)

**A.** $-3.3 \times 10^{-18}$ J
**B.** $-1.6 \times 10^{-18}$ J
**C.** $1.6 \times 10^{-18}$ J
**D.** $3.3 \times 10^{-32}$ J

**58.** An electron in a hydrogen atom is in its n = 2 excited state. What is the wavelength of photon needed to ionize this electron? (Use Planck's constant $h = 4.135 \times 10^{-15}$ eV·s and the speed of light $c = 3 \times 10^8$ m/s)

**A.** 365 nm
**B.** 248 nm
**C.** 137 nm
**D.** 69 nm

**59.** The square of the wave function represents the:

**A.** inertia of the particle
**B.** probability density for finding the particle
**C.** velocity of the particle
**D.** momentum of the particle

**60.** A blue photon has a:

**A.** longer wavelength than a red photon and travels with a greater speed
**B.** shorter wavelength than a red photon and travels with the same speed
**C.** shorter wavelength than a red photon and travels with a greater speed
**D.** longer wavelength than a red photon and travels with a lower speed

 

*Notes for active learning*

*Notes for active learning*

# ANSWER KEYS
# &
# DETAILED EXPLANATIONS

# Answer Keys

## 1 – Phases and Phase Equilibria

| | | | | |
|---|---|---|---|---|
| 1: A | 21: D | 41: C | 61: D | 81: A |
| 2: C | 22: A | 42: D | 62: A | 82: C |
| 3: C | 23: C | 43: B | 63: B | 83: B |
| 4: A | 24: D | 44: D | 64: C | 84: A |
| 5: A | 25: D | 45: D | 65: D | 85: D |
| 6: B | 26: A | 46: A | 66: B | 86: D |
| 7: B | 27: D | 47: C | 67: A | 87: D |
| 8: B | 28: B | 48: C | 68: C | 88: A |
| 9: D | 29: C | 49: C | 69: D | 89: C |
| 10: D | 30: A | 50: B | 70: A | 90: B |
| 11: B | 31: B | 51: B | 71: D | 91: B |
| 12: C | 32: D | 52: C | 72: B | 92: C |
| 13: A | 33: C | 53: A | 73: A | 93: A |
| 14: D | 34: B | 54: D | 74: C | 94: D |
| 15: C | 35: C | 55: C | 75: A | 95: C |
| 16: A | 36: D | 56: B | 76: B | 96: C |
| 17: C | 37: A | 57: B | 77: C | 97: D |
| 18: C | 38: C | 58: B | 78: D | 98: D |
| 19: C | 39: B | 59: A | 79: D | 99: B |
| 20: B | 40: C | 60: C | 80: D | 100: C |

## 2 – Thermochemistry

| | | | |
|---|---|---|---|
| 1: A | 21: A | 41: A | 61: C |
| 2: B | 22: C | 42: C | 62: C |
| 3: D | 23: A | 43: B | 63: C |
| 4: C | 24: B | 44: D | 64: B |
| 5: A | 25: C | 45: A | 65: A |
| 6: D | 26: C | 46: D | 66: B |
| 7: D | 27: A | 47: C | 67: D |
| 8: A | 28: D | 48: C | 68: C |
| 9: C | 29: A | 49: B | 69: A |
| 10: D | 30: D | 50: B | 70: B |
| 11: C | 31: B | 51: C | 71: D |
| 12: A | 32: C | 52: C | 72: C |
| 13: B | 33: D | 53: C | 73: D |
| 14: D | 34: B | 54: C | 74: A |
| 15: C | 35: D | 55: D | 75: C |
| 16: B | 36: A | 56: A | 76: D |
| 17: A | 37: B | 57: C | 77: B |
| 18: B | 38: D | 58: D | 78: D |
| 19: C | 39: D | 59: B | 79: A |
| 20: C | 40: D | 60: A | 80: A |

 

**3 – Thermodynamics**

| | | | | |
|---|---|---|---|---|
| 1: B | 21: D | 41: D | 61: A | 81: B |
| 2: C | 22: A | 42: C | 62: D | 82: C |
| 3: D | 23: B | 43: B | 63: B | 83: D |
| 4: A | 24: D | 44: A | 64: C | 84: C |
| 5: A | 25: A | 45: D | 65: B | 85: D |
| 6: B | 26: A | 46: C | 66: D | 86: B |
| 7: D | 27: D | 47: B | 67: A | 87: A |
| 8: A | 28: B | 48: A | 68: C | 88: C |
| 9: C | 29: C | 49: D | 69: D | 89: D |
| 10: D | 30: D | 50: B | 70: A | 90: A |
| 11: B | 31: B | 51: C | 71: A | 91: C |
| 12: A | 32: D | 52: A | 72: D | 92: D |
| 13: B | 33: B | 53: D | 73: B | 93: C |
| 14: D | 34: B | 54: B | 74: A | 94: B |
| 15: C | 35: D | 55: D | 75: D | 95: C |
| 16: A | 36: A | 56: C | 76: D | 96: C |
| 17: B | 37: C | 57: B | 77: C | 97: D |
| 18: B | 38: A | 58: A | 78: B | 98: C |
| 19: C | 39: B | 59: C | 79: D | 99: B |
| 20: A | 40: B | 60: B | 80: C | 100: D |

**4 – Fluid Statics and Dynamics**

| | | | |
|---|---|---|---|
| 1: C | 21: A | 41: A | 61: C |
| 2: D | 22: C | 42: D | 62: A |
| 3: A | 23: D | 43: B | 63: D |
| 4: A | 24: B | 44: D | 64: B |
| 5: D | 25: D | 45: A | 65: C |
| 6: C | 26: B | 46: B | 66: D |
| 7: A | 27: C | 47: A | 67: B |
| 8: C | 28: B | 48: B | 68: A |
| 9: C | 29: C | 49: C | 69: C |
| 10: C | 30: D | 50: C | 70: A |
| 11: A | 31: C | 51: C | 71: A |
| 12: B | 32: A | 52: B | 72: A |
| 13: A | 33: B | 53: C | 73: B |
| 14: B | 34: A | 54: D | 74: C |
| 15: C | 35: C | 55: C | 75: C |
| 16: D | 36: D | 56: A | 76: A |
| 17: C | 37: D | 57: B | 77: A |
| 18: B | 38: D | 58: B | 78: C |
| 19: C | 39: B | 59: A | 79: B |
| 20: D | 40: C | 60: D | 80: B |

## 5 – Chemical Bonding

| | | | | |
|---|---|---|---|---|
| 1: D | 21: D | 41: C | 61: A | 81: C |
| 2: A | 22: C | 42: A | 62: D | 82: B |
| 3: C | 23: B | 43: C | 63: A | 83: C |
| 4: D | 24: D | 44: D | 64: B | 84: A |
| 5: B | 25: D | 45: C | 65: C | 85: C |
| 6: B | 26: B | 46: A | 66: A | 86: B |
| 7: A | 27: A | 47: B | 67: B | 87: B |
| 8: A | 28: C | 48: A | 68: D | 88: C |
| 9: D | 29: A | 49: D | 69: B | |
| 10: C | 30: D | 50: B | 70: B | |
| 11: B | 31: B | 51: D | 71: C | |
| 12: D | 32: C | 52: D | 72: B | |
| 13: C | 33: B | 53: D | 73: C | |
| 14: A | 34: A | 54: C | 74: C | |
| 15: D | 35: D | 55: D | 75: B | |
| 16: C | 36: D | 56: A | 76: C | |
| 17: C | 37: B | 57: C | 77: A | |
| 18: A | 38: A | 58: C | 78: B | |
| 19: D | 39: B | 59: B | 79: D | |
| 20: C | 40: A | 60: C | 80: D | |

**6 – Stoichiometry**

| 1: C | 21: D | 41: B | 61: B | 81: B | 101: B | 121: A |
|---|---|---|---|---|---|---|
| 2: C | 22: A | 42: D | 62: D | 82: C | 102: B | 122: A |
| 3: C | 23: B | 43: D | 63: D | 83: B | 103: A | 123: A |
| 4: D | 24: C | 44: A | 64: B | 84: A | 104: D | 124: C |
| 5: C | 25: C | 45: D | 65: C | 85: D | 105: C | 125: D |
| 6: B | 26: D | 46: B | 66: D | 86: C | 106: B | 126: D |
| 7: A | 27: A | 47: C | 67: B | 87: D | 107: D | 127: C |
| 8: C | 28: B | 48: A | 68: C | 88: C | 108: A | 128: B |
| 9: B | 29: D | 49: C | 69: C | 89: C | 109: C | 129: B |
| 10: C | 30: C | 50: B | 70: B | 90: B | 110: C | 130: C |
| 11: C | 31: A | 51: D | 71: A | 91: D | 111: B | 131: B |
| 12: D | 32: C | 52: A | 72: D | 92: B | 112: A | 132: B |
| 13: C | 33: B | 53: C | 73: B | 93: B | 113: A | 133: A |
| 14: D | 34: A | 54: B | 74: C | 94: A | 114: A | 134: C |
| 15: D | 35: A | 55: D | 75: D | 95: C | 115: C | 135: A |
| 16: A | 36: C | 56: B | 76: A | 96: A | 116: B | 136: B |
| 17: B | 37: D | 57: C | 77: D | 97: C | 117: C | 137: A |
| 18: C | 38: A | 58: D | 78: C | 98: B | 118: B | 138: A |
| 19: D | 39: B | 59: D | 79: D | 99: A | 119: D | 139: C |
| 20: D | 40: A | 60: D | 80: B | 100: D | 120: A | 140: B |

**7 – Chemical Kinetics and Equilibria**

| | | | | | |
|---|---|---|---|---|---|
| 1: A | 21: B | 41: A | 61: A | 81: B | 101: C |
| 2: B | 22: A | 42: D | 62: B | 82: A | 102: A |
| 3: A | 23: D | 43: B | 63: D | 83: C | 103: D |
| 4: A | 24: B | 44: D | 64: A | 84: C | 104: D |
| 5: C | 25: C | 45: B | 65: D | 85: A | 105: D |
| 6: D | 26: D | 46: C | 66: C | 86: B | 106: C |
| 7: B | 27: B | 47: B | 67: D | 87: C | 107: B |
| 8: A | 28: A | 48: D | 68: D | 88: C | 108: D |
| 9: D | 29: D | 49: C | 69: B | 89: A | 109: A |
| 10: D | 30: A | 50: D | 70: A | 90: D | 110: D |
| 11: D | 31: C | 51: C | 71: B | 91: D | 111: D |
| 12: B | 32: B | 52: B | 72: B | 92: B | 112: C |
| 13: D | 33: A | 53: B | 73: B | 93: C | 113: B |
| 14: B | 34: A | 54: C | 74: D | 94: A | 114: B |
| 15: B | 35: B | 55: A | 75: D | 95: D | 115: C |
| 16: D | 36: D | 56: D | 76: B | 96: D | 116: D |
| 17: B | 37: A | 57: C | 77: D | 97: A | 117: A |
| 18: D | 38: A | 58: B | 78: C | 98: D | 118: A |
| 19: A | 39: D | 59: D | 79: B | 99: C | 119: A |
| 20: C | 40: C | 60: C | 80: D | 100: D | 120: C |

**8 – Solution Chemistry**

| | | | | | |
|---|---|---|---|---|---|
| 1: B | 21: B | 41: D | 61: D | 81: D | 101: A |
| 2: C | 22: A | 42: B | 62: A | 82: A | 102: A |
| 3: A | 23: C | 43: B | 63: A | 83: B | 103: D |
| 4: A | 24: B | 44: A | 64: A | 84: D | 104: B |
| 5: D | 25: A | 45: D | 65: C | 85: B | 105: B |
| 6: A | 26: C | 46: B | 66: D | 86: D | 106: D |
| 7: B | 27: A | 47: A | 67: C | 87: A | 107: D |
| 8: D | 28: D | 48: C | 68: B | 88: B | |
| 9: A | 29: D | 49: D | 69: D | 89: D | |
| 10: C | 30: A | 50: C | 70: C | 90: C | |
| 11: C | 31: D | 51: B | 71: D | 91: C | |
| 12: A | 32: B | 52: C | 72: A | 92: D | |
| 13: C | 33: B | 53: B | 73: B | 93: A | |
| 14: C | 34: B | 54: A | 74: A | 94: D | |
| 15: B | 35: D | 55: A | 75: A | 95: D | |
| 16: D | 36: A | 56: B | 76: C | 96: C | |
| 17: B | 37: A | 57: B | 77: C | 97: A | |
| 18: A | 38: C | 58: B | 78: C | 98: D | |
| 19: A | 39: D | 59: D | 79: B | 99: C | |
| 20: D | 40: B | 60: C | 80: D | 100: D | |

 

**9 – Acids and Bases**

| | | | | | | |
|---|---|---|---|---|---|---|
| 1: A | 21: C | 41: A | 61: A | 81: B | 101: C | 121: D |
| 2: B | 22: A | 42: A | 62: C | 82: C | 102: D | 122: C |
| 3: A | 23: D | 43: D | 63: D | 83: A | 103: A | 123: C |
| 4: C | 24: D | 44: B | 64: A | 84: B | 104: C | 124: D |
| 5: C | 25: B | 45: C | 65: A | 85: B | 105: B | 125: B |
| 6: C | 26: B | 46: D | 66: D | 86: D | 106: D | 126: A |
| 7: D | 27: B | 47: A | 67: A | 87: C | 107: C | 127: B |
| 8: A | 28: D | 48: D | 68: A | 88: D | 108: A | 128: B |
| 9: B | 29: C | 49: B | 69: B | 89: D | 109: D | 129: B |
| 10: B | 30: C | 50: B | 70: D | 90: B | 110: D | 130: D |
| 11: B | 31: A | 51: C | 71: A | 91: D | 111: A | 131: D |
| 12: D | 32: C | 52: A | 72: D | 92: D | 112: B | 132: A |
| 13: C | 33: A | 53: A | 73: C | 93: C | 113: C | 133: D |
| 14: B | 34: D | 54: D | 74: D | 94: A | 114: D | 134: C |
| 15: C | 35: A | 55: A | 75: B | 95: C | 115: B | 135: D |
| 16: A | 36: C | 56: C | 76: A | 96: B | 116: A | 136: A |
| 17: D | 37: B | 57: A | 77: C | 97: D | 117: D | 137: B |
| 18: C | 38: B | 58: C | 78: A | 98: C | 118: B | 138: D |
| 19: D | 39: B | 59: B | 79: B | 99: D | 119: C | 139: A |
| 20: C | 40: D | 60: C | 80: C | 100: A | 120: C | 140: B |

**10 – Electrochemistry**

| | | | |
|---|---|---|---|
| 1: D | 21: A | 41: C | 61: D |
| 2: C | 22: D | 42: A | 62: C |
| 3: A | 23: A | 43: C | 63: B |
| 4: B | 24: A | 44: B | 64: B |
| 5: C | 25: B | 45: B | 65: C |
| 6: A | 26: C | 46: B | 66: D |
| 7: C | 27: D | 47: A | 67: A |
| 8: D | 28: A | 48: B | 68: B |
| 9: D | 29: B | 49: D | 69: B |
| 10: B | 30: C | 50: B | 70: C |
| 11: A | 31: A | 51: A | 71: A |
| 12: C | 32: B | 52: C | 72: D |
| 13: A | 33: B | 53: B | 73: D |
| 14: B | 34: B | 54: C | 74: D |
| 15: C | 35: D | 55: D | 75: B |
| 16: D | 36: A | 56: C | 76: D |
| 17: B | 37: A | 57: D | 77: B |
| 18: C | 38: B | 58: A | 78: C |
| 19: C | 39: D | 59: A | 79: A |
| 20: D | 40: B | 60: B | 80: D |

 

## 11 – Atomic and Electronic Stucture

| | | | | | | |
|---|---|---|---|---|---|---|
| 1: A | 21: D | 41: D | 61: B | 81: C | 101: A | 121: D |
| 2: C | 22: C | 42: D | 62: C | 82: D | 102: B | 122: C |
| 3: D | 23: B | 43: C | 63: D | 83: A | 103: D | 123: A |
| 4: B | 24: D | 44: B | 64: A | 84: D | 104: A | 124: A |
| 5: C | 25: A | 45: D | 65: C | 85: A | 105: C | 125: C |
| 6: D | 26: C | 46: B | 66: A | 86: B | 106: B | 126: D |
| 7: B | 27: D | 47: C | 67: B | 87: D | 107: D | 127: B |
| 8: D | 28: D | 48: C | 68: D | 88: C | 108: B | 128: C |
| 9: C | 29: C | 49: B | 69: B | 89: A | 109: D | 129: D |
| 10: A | 30: D | 50: A | 70: C | 90: C | 110: C | 130: B |
| 11: B | 31: A | 51: D | 71: D | 91: B | 111: B | 131: D |
| 12: D | 32: C | 52: C | 72: A | 92: C | 112: A | 132: C |
| 13: A | 33: D | 53: B | 73: D | 93: A | 113: C | 133: D |
| 14: C | 34: B | 54: D | 74: B | 94: B | 114: B | 134: D |
| 15: C | 35: C | 55: C | 75: A | 95: D | 115: D | 135: D |
| 16: A | 36: D | 56: D | 76: D | 96: D | 116: C | 136: D |
| 17: B | 37: B | 57: B | 77: C | 97: D | 117: A | 137: A |
| 18: D | 38: C | 58: A | 78: B | 98: D | 118: A | 138: C |
| 19: B | 39: C | 59: C | 79: D | 99: C | 119: C | 139: D |
| 20: A | 40: B | 60: D | 80: C | 100: D | 120: B | 140: B |

**12 – Quantum Mechanics**

| | | | | | |
|---|---|---|---|---|---|
| 1: D | 11: A | 21: C | 31: B | 41: C | 51: D |
| 2: B | 12: D | 22: A | 32: A | 42: D | 52: B |
| 3: C | 13: D | 23: D | 33: D | 43: C | 53: B |
| 4: C | 14: C | 24: B | 34: C | 44: C | 54: B |
| 5: D | 15: B | 25: B | 35: A | 45: D | 55: D |
| 6: C | 16: D | 26: B | 36: C | 46: D | 56: A |
| 7: C | 17: C | 27: D | 37: B | 47: C | 57: C |
| 8: B | 18: D | 28: B | 38: C | 48: B | 58: A |
| 9: D | 19: B | 29: A | 39: C | 49: D | 59: B |
| 10: A | 20: D | 30: D | 40: B | 50: A | 60: B |

 

# 1 – Phases and Phase Equilibria: Detailed Explanations

==============================================================

**Practice Set 1: Questions 1–20**

==============================================================

**1. A is correct.**

Solids, liquids, and gases have a vapor pressure that increases from solid to gas.

*Vapor pressure* is a vapor's pressure in equilibrium with its condensed phases (i.e., solid or liquid) in a closed system at a given temperature.

Vapor pressure is a colligative property of a substance and depends only on the number of solutes present, not on their identity.

**2. C is correct.**

Charles' law (i.e., the law of volumes) explains how, at constant pressure, gases behave when the temperature changes:

$$V \alpha T$$

or

$$V / T = \text{constant}$$

or

$$(V_1 / T_1) = (V_2 / T_2)$$

Volume and temperature are proportional.

Doubling the temperature at constant pressure doubles the volume.

**3. C is correct.**

*Colligative properties* include lowering vapor pressure, the elevation of boiling point, depression of freezing point, and increased osmotic pressure.

Adding solute to a pure solvent lowers the vapor pressure of the solvent; therefore, a higher temperature is required to bring the vapor pressure of the solution in an open container up to the atmospheric pressure. This increases the boiling point.

Because adding solute lowers the vapor pressure, the solution's freezing point decreases (e.g., automobile antifreeze).

**4. A is correct.**

At a pressure and temperature corresponding to a substance's triple point (point D on the graph), all three states (gas, liquid, and solid) exist in equilibrium.

*Critical point* (point E on the graph) is the endpoint of the phase equilibrium curve, where the liquid and its vapor become indistinguishable.

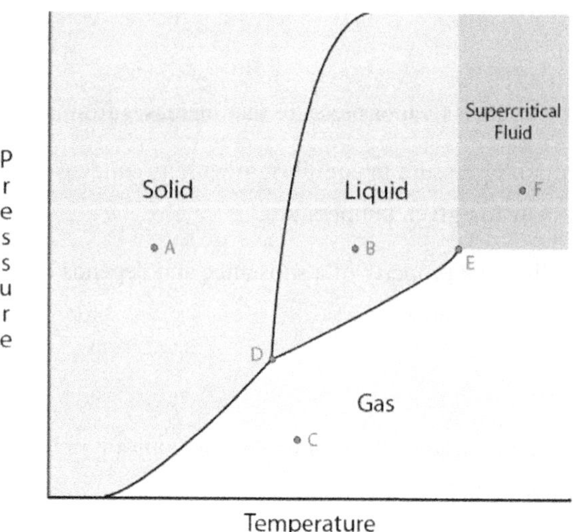


*Phase diagram of pressure vs. temperature*

**5. A is correct.**

In the van der Waals equation, *a* is the negative deviation due to attractive forces, and *b* is the positive deviation due to molecular volume.

**6. B is correct.**

R is the symbol for the ideal gas constant.

R is expressed in (L × atm) / mole × K, and the value = 0.0821.

Convert to different units (torr and mL)

$$0.0821[(\text{L} \times \text{atm}) / (\text{mole} \times \text{K})] \times (760 \text{ torr/atm}) \times (1{,}000 \text{ mL/L})$$

$$R = 62{,}396 (\text{torr} \times \text{mL}) / \text{mole} \times \text{K}$$

**7. B is correct.**

*Ideal gas law*:

$$PV = nRT$$

where P is pressure, V is volume, n is the number of molecules, R is the ideal gas constant, and T is the temperature of the gas.

R and n are constant.

From this equation, if pressure and temperature are halved, volume has no effect.

**8. B is correct.**

*Kinetic molecular theory* (KMT) of gas molecules states that the average kinetic energy per molecule in a system is proportional to the temperature of the gas.

Since containers X and Y are at the same temperature and pressure, then molecules of gases must possess the same amount of average kinetic energy.

**9. D is correct.**

*Barometers* and *manometers* are used to measure pressure. Barometers measure atmospheric pressure, while a manometer can measure lower pressure than atmospheric pressure.

A manometer has both ends of the tube open to the outside (while some may have one end closed), whereas a barometer is a type of closed-end manometer with one end of the glass tube closed and sealed with a vacuum.

Atmospheric pressure is 760 mmHg, so the barometer should be able to accommodate that.

**10. D is correct.**

Vapor pressure is a vapor's pressure in equilibrium with its condensed phases (i.e., solid or liquid) in a closed system at a given temperature.

Raoult's law states that the partial vapor pressure of each component of an ideal mixture of liquids equals the vapor pressure of the pure component multiplied by its mole fraction in the mixture.

In exothermic reactions, the vapor pressure deviates negatively from Raoult's law.

Depending on the ratios of the liquids in a solution, the vapor pressure could be lower than or just lower than X because X is a higher boiling point, thus a lower vapor pressure.

Boiling point increases from adding Y to the mixture because the vapor pressure decreases.

**11. B is correct.**

$2\ Na\ (s) + Cl_2\ (g) \rightarrow 2\ NaCl\ (s)$

In its elemental form, chlorine exists as a gas.

In its elemental form, Na exists as a solid.

**12. C is correct.**

Ideal gas molecules do not occupy significant space and exert no intermolecular forces.

In contrast, the molecules of a real gas occupy space and exert (weak attractive) intermolecular forces.

However, an ideal gas and a real gas have pressure, which is created from molecular collisions with the walls of the container.

**13. A is correct.**

*Boyle's law* (i.e., pressure-volume law) states that pressure and volume are inversely proportional:

$(P_1V_1) = (P_2V_2)$

Solve for the final pressure:

$P_2 = (P_1V_1) / V_2$

$P_2 = [(0.950\ atm) \times (2.75\ L)] / (0.450\ L)$

$P_2 = 5.80\ atm$

**14. D is correct.**

*Avogadro's law* is an experimental gas law relating the volume of a gas to the amount of substance of gas present; *equal volumes of gases have the same number of molecules at the same temperature and pressure.*

For a given mass of an ideal gas, the volume and amount (i.e., moles) of the gas are directly proportional if the temperature and pressure are constant:

$V \propto n$

Where V = volume and n = number of moles of the gas.

*continued...*

*Charles' law* (i.e., the law of volumes) explains how, at constant pressure, gases behave when the temperature changes:

$$V \alpha T$$

or

$$V / T = \text{constant}$$

or

$$(V_1 / T_1) = (V_2 / T_2)$$

Volume and temperature are proportional; an increase in one term increases the other.

*Gay-Lussac's law* (i.e., pressure-temperature law) states that pressure is proportional to temperature:

$$P \alpha T$$

or

$$(P_1 / T_1) = (P_2 / T_2)$$

or

$$(P_1T_2) = (P_2T_1)$$

or

$$P / T = \text{constant}$$

If the pressure of a gas increases, the temperature increases.

*Boyle's law* (i.e., pressure-volume law) states that pressure and volume are inversely proportional.

**15. C is correct.**

*Ideal gas law*: PV = nRT

where P is pressure, V is volume, n is the number of molecules, R is the ideal gas constant, and T is the temperature of the gas.

R and n are constant.

If T is constant, the equation becomes PV = constant. They are inversely proportional: if one of the values is reduced, the other increases.

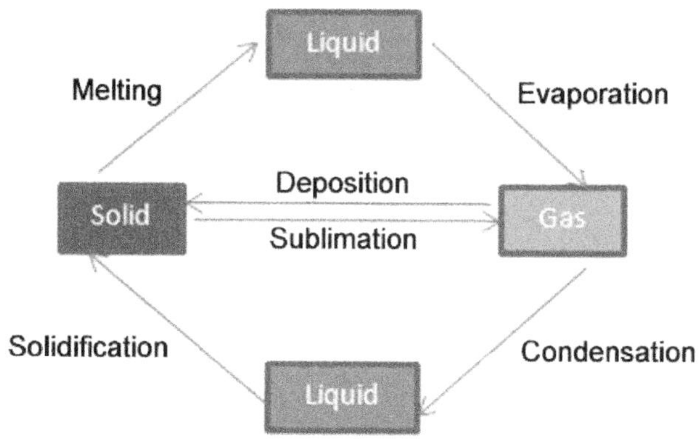


*Interconversion of states of matter*

**16. A is correct.**

*Vaporization* refers to the change of state from a liquid to a gas. Two types of vaporization: boiling and evaporation, are differentiated based on the temperature at which they occur.

Evaporation occurs at a temperature below the boiling point, while boiling occurs at or above the boiling point.

**17. C is correct.**

Ideal gas law: PV = nRT

from which simpler gas laws such as Boyle's, Charles' and Avogadro's laws are derived.

The values of n and R are constant.

The common format of the combined gas law:

$$(P_1V_1) / T_1 = (P_2V_2) / T_2$$

Try modifying the equations to recreate the formats of the equation provided by the problem.

$T_2 = T_1 \times P_1 / P_2 \times V_2 / V_1$ should be written as $T_2 = T_1 \times V_1/V_2 \times P_1/P_2$

**18. C is correct.**

*Intermolecular forces* act between neighboring molecules. Examples include hydrogen bonding, dipole-dipole, dipole-induced dipole, and van der Waals (i.e., London dispersion) forces.

*Dipole-dipole attraction* involves asymmetric, polar molecules (based on differences in electronegativity between atoms) that create a *dipole moment* (i.e., a net vector indicating the force).

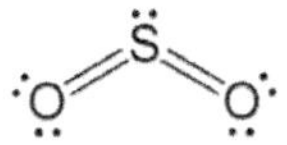

*Sulfur dioxide with an indicated bond angle*

**19. C is correct.**

Increasing the gas pressure above the liquid stresses the system's equilibrium.

Gas molecules start to collide with the liquid surface more often, which increases the rate of gas molecules entering the solution, thus increasing the solubility.

**20. B is correct.**

Avogadro's law states the correlation between volume and moles (n).

Avogadro's law is an experimental gas law relating the volume of a gas to the amount of substance of gas present. A modern statement of Avogadro's law is:

At the same pressure and temperature, equal volumes of gases have the same number of molecules.

$$V \propto n$$

or

$$V / n = k$$

where V is the volume of the gas, n is the number of moles of the gas, and $k$ is a constant equal to RT/P (where R is the universal gas constant, T is the temperature in Kelvin and P is the pressure).

For comparing the same substance under two sets of conditions, the law is expressed as:

$$V_1 / n_1 = V_2 / n_2$$

 

==========================================================================

**Practice Set 2: Questions 21–40**

==========================================================================

**21. D is correct.**

*Colligative properties* of solutions depend on the ratio of solute particles to the number of solvent molecules in a solution and not on the type of chemical species present.

Colligative properties include lowering vapor pressure, elevation of boiling point, depression of freezing point, and increased osmotic pressure.

Boiling point (BP) elevation:

$$\Delta BP = iKm$$

where i = the number of particles produced when the solute dissociates, K = boiling elevation constant, and m = molality (moles/kg solvent).

In this problem, acid molality is not known.

**22. A is correct.**

An ideal gas has no intermolecular forces, indicating that its molecules have no attraction to each other. However, the molecules of a real gas do have intermolecular forces, although these forces are extremely weak.

Therefore, the molecules of a real gas are slightly attracted to one another, although the attraction is nowhere near as strong as the attraction in liquids and solids.

**23. C is correct.**

Standard temperature and pressure (STP) have a temperature of 273.15 K (0 °C, 32 °F) and a pressure of $10^5$ Pa (100 kPa, 750.06 mmHg, 1 bar, 14.504 psi, 0.98692 atm).

mm of Hg was defined as the pressure generated by a column of mercury one millimeter high. The pressure of mercury depends on temperature and gravity.

This variation in mmHg and torr is a difference in units of about 0.000015%.

In general,

1 torr = 1 mm of Hg = 0.0013158 atm

750.06 mmHg = 0.98692 atm

 

**24. D is correct.**

At lower temperatures, the potential energy due to the intermolecular forces is more significant than the kinetic energy; this causes the pressure to be reduced because the gas molecules are attracted to each other.

**25. D is correct.**

Dalton's law (i.e., the law of partial pressures) states that *the pressure exerted by a mixture of gases equals the sum of the individual gas pressures*.

Pressure due to $N_2$ and $CO_2$:

(320 torr + 240 torr) = 560 torr

Partial pressure of $O_2$ is:

740 torr – 560 torr = 180 torr

180 torr / 740 torr = 24%

**26. A is correct.**

Ideal gas law:

PV = nRT

where P is pressure, V is volume, n is the number of molecules, R is the ideal gas constant, and T is the temperature of the gas.

If the volume is reduced by ½, the number of moles is reduced by ½.

Pressure is reduced to 90%, so the number of moles is reduced by 90%.

Therefore, the reduction in moles is (½ × 90%) = 45%.

Mass is proportional to the number of moles for a given gas so that mass reduction can be calculated directly:

New mass is 45% of 40 grams = (0.45 × 40 g) = 18 grams

**27. D is correct.**

*Evaporation* describes the phase change from liquid to gas.

The mass of the molecules and the attraction of the molecules with their neighbors (to form intermolecular attractions) determine their kinetic energy.

The increase in kinetic energy is required for individual molecules to move from the liquid to the gaseous phase.

**28. B is correct.**

The molecules of an ideal gas exert no attractive forces. Therefore, a real gas behaves most nearly like an ideal gas at high temperature and low pressure. Under these conditions, the molecules are far apart and exert little or no attractive forces on each other.

**29. C is correct.**

Hydroxyl (~OH) groups significantly increase the boiling point by forming hydrogen bonds with ~OH groups of neighboring molecules.

Hydrocarbons are nonpolar molecules, meaning London dispersion is the dominant intermolecular force. This force gets stronger as the number of atoms in each molecule increases. The stronger force increases the boiling point.

The branching of the hydrocarbon affects the boiling point. Straight molecules have slightly higher boiling points than branched molecules with the same number of atoms. The reason is that straight molecules can align parallel against each other, and atoms in the molecules are involved in the London dispersion forces.

Another factor is the presence of heteroatoms (i.e., atoms other than carbon and hydrogen). For example, the electronegative oxygen atom between carbon groups or in an ether (C–O–C) slightly increases the boiling point.

**30. A is correct.**

*Kinetic theory* explains the macroscopic properties of gases (e.g., temperature, volume, and pressure) by their molecular composition and motion.

*Gas pressure* is due to the collisions on the walls of a container from molecules moving at different velocities.

Temperature = $\frac{1}{2}mv^2$

**31. B is correct.**

Methanol ($CH_3OH$) is an alcohol that participates in hydrogen bonding.

Therefore, this gas experiences the strongest intermolecular forces.

 

**32. D is correct.**

Density = mass / volume

Gas molecules have a large amount of space between them; therefore, they can be pushed together, and thus, gases are very compressible. Because there is such a large amount of space between each molecule in a gas, the extent to which the gas molecules can be pushed together is much greater than the extent to which liquid molecules can be pushed together. Therefore, gases have greater compressibility than liquids.

Gas molecules are further apart than liquid molecules, so gases have a smaller density.

**33. C is correct.**

Vapor pressure is the pressure a vapor exerts in equilibrium with its condensed phases (i.e., solid or liquid) in a closed system at a given temperature.

Vapor pressure is inversely correlated with the strength of the intermolecular force.

The molecules are more likely to stick together in the liquid form with stronger intermolecular forces, and fewer participate in the liquid-vapor equilibrium.

Vapor pressure of a liquid decreases when a nonvolatile substance is dissolved into a liquid.

The decrease in the vapor pressure of a substance is proportional to the number of moles of the solute dissolved in a definite weight of the solvent. This is *Raoult's law*.

**34. B is correct.**

*Dalton's law* (i.e., the law of partial pressures) states that *the pressure exerted by a mixture of gases equals the sum of the individual gas pressures*. It is an empirical law observed by English chemist John Dalton and related to the ideal gas laws.

**35. C is correct.**

Solids have a definite shape and volume. For example, a granite block does not change its shape or volume, regardless of the container in which it is placed.

Molecules in a solid are very tightly packed due to the strong intermolecular attractions, which prevent the molecules from moving around.

**36. D is correct.**

Ideal gas law: PV = nRT

where P is pressure, V is volume, n is the number of molecules, R is the ideal gas constant, and T is the temperature of the gas.

At STP (standard conditions for temperature and pressure), the pressure and temperature for all three flasks are the same. It is known that the volume is the same in each case – 2.0 L.

Therefore, since R is a constant, the number of molecules "n" must be the same for the ideal gas law to hold.

**37. A is correct.**

Hydrogens, bonded directly to F, O, or N, participate in hydrogen bonds. The hydrogen is partially positive (i.e., delta plus: $\partial$+) due to the bond to these electronegative atoms. The lone pair of electrons on the F, O, or N interacts with the partial positive ($\partial$+) hydrogen to form a hydrogen bond.

D: hydrogen on the methyl carbon, and that carbon is not attached to N, O, or F. Therefore, hydrogen cannot form a hydrogen bond, even though oxygen is available on the other methanol to form a hydrogen bond.

**38. C is correct.**

Boyle's law (Mariotte's law or the Boyle-Mariotte law) is an experimental gas law that describes how the volume of a gas increases as the pressure decreases (i.e., they are inversely proportional) if the temperature is constant.

Boyle's law (i.e., pressure-volume law) states that pressure and volume are inversely proportional:

$$P_1V_1 = P_2V_2$$

or $$P \times V = \text{constant}$$

If the volume of a gas increases, its pressure decreases proportionally.

Dalton's law (i.e., the law of partial pressures) states that *the pressure exerted by a mixture of gases equals the sum of the individual gas pressures*.

Charles' law (i.e., the law of volumes) explains how, at constant pressure, gases behave when the temperature changes:

$$V \alpha T$$

*continued...*

 

or $V / T = \text{constant}$

or $(V_1 / T_1) = (V_2 / T_2)$

Volume and temperature are proportional: an increase in one results in an increase in the other.

Gay-Lussac's law (i.e., pressure-temperature law) states that pressure is proportional to temperature:

$P \alpha T$

or

$(P_1 / T_1) = (P_2 / T_2)$

or

$(P_1T_2) = (P_2T_1)$

or

$P / T = \text{constant}$

If the pressure of a gas increases, the temperature increases.

Avogadro's law is an experimental gas law relating the volume of a gas to the amount of substance of gas present. *Equal volumes of gases at the same temperature and pressure have the same number of molecules.*

**39. B is correct.**

As the automobile travels the highway, friction is generated between the road and its tires. The heat energy increases the air temperature inside the tires, causing the molecules to have more velocity. These fast-moving molecules collide with the tire walls at a higher rate, and thus, the pressure increases.

Gay-Lussac's law (i.e., pressure-temperature law) states that pressure is proportional to temperature:

$P \alpha T$

or

$(P_1 / T_1) = (P_2 / T_2)$

or

$(P_1T_2) = (P_2T_1)$

or

$P / T = \text{constant}$

If the temperature of a gas increases, the pressure increases.

**40. C is correct.**

Balanced chemical equation:

$$N_2 + 3\ H_2 \rightarrow 2\ NH_3$$

Use the balanced coefficients from the written equation and apply dimensional analysis to solve for the volume of $H_2$ needed to produce 12.5 L $NH_3$:

$$V_{H2} = V_{NH3} \times (\text{mol } H_2 / \text{mol } NH_3)$$

$$V_{H2} = (12.5\ \text{L}) \times (3\ \text{mol} / 2\ \text{mol})$$

$$V_{H2} = 18.8\ \text{L}$$

==================================================================

**Practice Set 3: Questions 41–60**

==================================================================

**41. C is correct.**

Dalton's law (i.e., the law of partial pressures) states that *the pressure exerted by a mixture of gases equals the sum of the individual gas pressures*.

Convert the masses of the gases into moles:

Moles of $O_2$:

$16 \text{ g of } O_2 \div 32 \text{ g/mole} = 0.5 \text{ mole}$

Moles of $N_2$:

$14 \text{ g of } N_2 \div 28 \text{ g/mole} = 0.5 \text{ mole}$

Mole of $CO_2$:

$88 \text{ g of } CO_2 \div 44 \text{ g/mole} = 2 \text{ moles}$

Total moles:

$(0.5 \text{ mol} + 0.5 \text{ mole} + 2 \text{ mol}) = 3 \text{ moles}$

Pressure of 1 atm (or 760 mmHg) has 38 mmHg (1 torr = 1 mmHg) contributed as $H_2O$ vapor.

Partial pressure of $CO_2$:

mole fraction × (total pressure of the gas mixture – $H_2O$ vapor)

(2 moles $CO_2$ / 3 moles total gas)] × (760 mmHg – 38 mmHg)

partial pressure of $CO_2$ = 481 mmHg

**42. D is correct.**

*Colligative properties* are solutions that depend on the ratio of solute particles to the number of solvent molecules in a solution and not on the type of chemical species present.

Colligative properties include lowering *vapor pressure*, elevation of *boiling point*, depression of *freezing point*, and increased *osmotic pressure.*

Dissolving a solute into a solvent alters the solvent's freezing point, melting point, boiling point, and vapor pressure.

**43. B is correct.**

Boyle's law (i.e., pressure-volume law) states that pressure and volume are inversely proportional:

$$(P_1V_1) = (P_2V_2)$$

or

$$P \times V = \text{constant}$$

If the volume of a gas increases, its pressure decreases proportionally.

**44. D is correct.**

Gases form homogeneous mixtures, regardless of the identities or relative proportions of the component gases. There is a relatively large distance between gas molecules (as opposed to solids or liquids where the molecules are much closer together).

When pressure is applied to gas, its volume readily decreases, and thus gases are highly compressible.

There are no attractive forces between gas molecules, so gas molecules can move about freely.

**45. D is correct.**

Solids have a definite shape and volume. For example, a granite block does not change its shape or volume, regardless of the container in which it is placed.

Molecules in a solid are very tightly packed due to the strong intermolecular attractions, which prevent the molecules from moving around.

**46. A is correct.**

Standard temperature and pressure (STP) have a temperature of 273.15 K (0 °C, 32 °F) and a pressure of $10^5$ Pa (100 kPa, 750.06 mmHg, 1 bar, 14.504 psi, 0.98692 atm).

The mm of Hg was defined as the pressure generated by a column of mercury one millimeter high. The pressure of mercury depends on temperature and gravity.

This variation in mmHg and torr is a difference in units of about 0.000015%.

In general,

1 torr = 1 mm of Hg = 0.0013158 atm.

750.06 mmHg = 0.98692 atm.

 

**47. C is correct.**

*Intermolecular forces* act between neighboring molecules. Examples include hydrogen bonding, dipole-dipole, dipole-induced dipole, and van der Waals (i.e., London dispersion) forces.

Stronger force results in a *higher boiling point.*

$CH_3COOH$ is a carboxylic acid that can form two hydrogen bonds. Therefore, it has the highest boiling point.

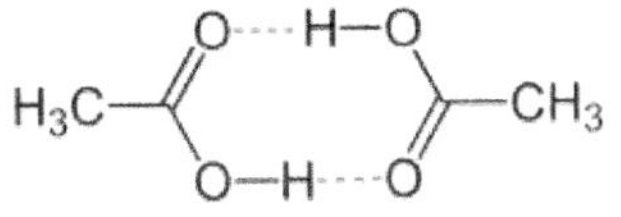

*Ethanoic acid with the two hydrogen bonds indicated on the structure*

**48. C is correct.**

Molecules in solids have the most attraction to their neighbors, followed by liquids (significant motion between the individual molecules) and then gas.

A molecule in an ideal gas has no attraction to other gas molecules. For a gas experiencing low pressure, the particles are far enough apart for no attractive forces between the individual gas molecules.

**49. C is correct.**

Ideal gas law:

$$PV = nRT$$

where P is pressure, V is volume, n is the number of molecules, R is the ideal gas constant, and T is the temperature of the gas.

Set the initial and final P, V, and T conditions equal:

$$(P_1V_1 / T_1) = (P_2V_2 / T_2)$$

Solve for the final volume of $N_2$:

$$(P_2V_2 / T_2) = (P_1V_1 / T_1)$$

$$V_2 = (T_2\ P_1V_1) / (P_2T_1)$$

$$V_2 = [(295\ K) \times (750.06\ mmHg) \times (0.190\ L\ N_2)] / [(660\ mmHg) \times (298.15\ K)]$$

$$V_2 = 0.214\ L$$

**50. B is correct.**

Volatility is the tendency of a substance to vaporize (phase change from liquid to vapor).

Volatility is related to a substance's vapor pressure. A substance with higher vapor pressure vaporizes more readily than one with lower vapor pressure at a given temperature.

Molecules with weak intermolecular attraction can increase their kinetic energy by transferring less heat due to a smaller molecular mass. The increase in kinetic energy is required for individual molecules to move from the liquid to the gaseous phase.

**51. B is correct.**

Barometers and manometers measure pressure. They are designed to measure atmospheric pressure, while a manometer can measure lower pressure than atmospheric pressure.

A manometer has both ends of the tube open to the outside (while some may have one end closed), whereas a barometer is a type of closed-end manometer with one end of the glass tube closed and sealed with a vacuum.

The difference in mercury height on both necks indicates the capacity of a manometer.

$$820 \text{ mm} - 160 \text{ mm} = 660 \text{ mm}$$

Historically, the pressure unit of torr equals 1 mmHg or the rise/dip of 1 mm of mercury in a manometer.

Because the manometer uses mercury, the height difference (660 mm) equals its measuring capacity in torr (660 torrs).

**52. C is correct.**

The conditions of the ideal gases are the same, so the number of moles (i.e., molecules) is equal. At STP, the temperature is the same, so the kinetic energy of the molecules is the same.

However, the molar mass of oxygen and nitrogen are different. Therefore, the density is different.

**53. A is correct.**

Vapor pressure is the pressure exerted by a vapor in equilibrium with condensed phases (i.e., solid or liquid) in a closed system at a given temperature.

The compound exists as a liquid if the external pressure > compound's vapor pressure.

A substance boils when vapor pressure = external pressure.

 

**54. D is correct.**

Gas molecules have a large amount of space between them; therefore, they can be pushed together, and gases are thus very compressible.

Molecules in solids and liquids are close together; therefore, they cannot get significantly closer and are thus nearly incompressible.

**55. C is correct.**

Charles' law (i.e., the law of volumes) explains how, at constant pressure, gases behave when the temperature changes:

$V \alpha T$

or $V / T = \text{constant}$

or $(V_1 / T_1) = (V_2 / T_2)$

Volume and temperature are proportional. Therefore, an increase in one results in an increase in the other.

Gay-Lussac's law (i.e., pressure-temperature law) states that pressure is proportional to temperature:

$P \alpha T$

or

$(P_1 / T_1) = (P_2 / T_2)$

or

$(P_1T_2) = (P_2T_1)$

or $P / T = \text{constant}$

If the pressure of a gas increases, the temperature increases.

Dalton's law (i.e., the law of partial pressures) states that *the pressure exerted by a mixture of gases is equal to the sum of the individual gas pressures*

Boyle's law (i.e., pressure-volume law) states that pressure and volume are inversely proportional:

$(P_1V_1) = (P_2V_2)$ or $P \times V = \text{constant}$

If the volume of a gas increases, its pressure decreases proportionally.

Avogadro's law is an experimental gas law relating the volume of a gas to the amount of substance of gas present. *Equal volumes of gases at the same temperature and pressure have the same number of molecules.*

 

**56. B is correct.**

Sublimation is the direct state change from solid to gas, skipping intermediate liquid phase.

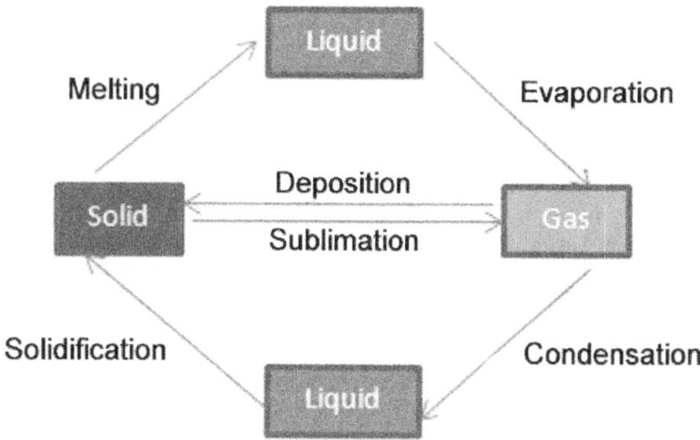


*Interconversion of states of matter*

Solid carbon dioxide (i.e., dry ice) is an example of a compound that undergoes sublimation. $CO_2$ changes phases from solid to gas (i.e., bypasses the liquid phase) and is often used as a cooling agent.

**57. B is correct.**

Graham's law of effusion states that the rate of effusion (i.e., escaping through a small hole) of a gas is inversely proportional to the square root of the molar mass of its particles.

Rate 1 / Rate 2 = √(molar mass gas 2 / molar mass gas 1)

The diffusion rate is the inverse root of the molecular weights of the gases.

Therefore, the rate of effusion is:

$O_2$ / $H_2$ = √(2 / 32)

rate of diffusion = 1:4

**58. B is correct.**

To calculate the number of molecules, calculate the moles of gas using the ideal gas law:

PV = nRT

n = PV / RT

*continued...*

Because the gas constant R is in $L \cdot atm\ K^{-1}\ mol^{-1}$, pressure must be converted into atm:

$$320\ mmHg \times (1\ /\ 760\ atm/mmHg) = 320\ mmHg\ /\ 760\ atm$$

(Leave it in this fraction form because the answer choices are in this format.)

Convert the temperature to Kelvin:

$$10\ °C + 273 = 283\ K$$

Substitute those values into the ideal gas equation:

$$n = PV\ /\ RT$$

$$n = (320\ mmHg\ /\ 760\ atm) \times 6\ L\ /\ (0.0821\ L \cdot atm\ K^{-1}\ mol^{-1} \times 283\ K)$$

This expression represents the number of gas moles present in the container.

Calculate the number of molecules:

$$\text{number of molecules} = \text{moles} \times \text{Avogadro's number}$$

$$\text{number of molecules} = (320\ mmHg\ /\ 760\ atm) \times 6\ /\ (0.0821 \times 283) \times 6.02 \times 10^{23}$$

$$\text{number of molecules} = (320\ /\ 760) \cdot (6) \cdot (6 \times 10^{23})\ /\ (0.0821) \cdot (283)$$

**59. A is correct.**

At STP (standard conditions for temperature and pressure), the pressure and temperature are the same regardless of the gas.

Ideal gas law:

$$PV = nRT$$

where P is pressure, V is volume, n is the number of molecules, R is the ideal gas constant, and T is the temperature of the gas.

Therefore, one molecule of each gas occupies the same volume at STP.

Since $CO_2$ molecules have the largest mass, $CO_2$ gas has a greater mass in the same volume, and thus it has the greatest density.

**60. C is correct.**

Hydrogens, bonded directly to F, O, or N, participate in hydrogen bonds. The hydrogen is partially positive (i.e., delta plus: $\partial+$) due to the bond to these electronegative atoms.

The lone pair of electrons on the F, O, or N interacts with the partial positive ($\partial+$) hydrogen to form a hydrogen bond.

==================================================================

**Practice Set 4: Questions 61–80**

==================================================================

**61. D is correct.**

Calculate the moles of each gas:

moles = mass / molar mass

moles $H_2$ = 9.50 g / (2 × 1.01 g/mole)

moles $H_2$ = 4.70 moles

moles Ne = 14.0 g / (20.18 g/mole)

moles Ne = 0.694 moles

Calculate the mole fraction of $H_2$:

Mole fraction of $H_2$ = moles of $H_2$ / moles in mixture

Mole fraction of $H_2$ = 4.70 moles / (4.70 moles + 0.694 moles)

Mole fraction of $H_2$ = 4.70 moles / (5.394 moles)

Mole fraction of $H_2$ = 0.87 moles

**62. A is correct.**

Vapor pressure is the pressure exerted by a vapor in equilibrium with its condensed phases (i.e., solid or liquid) in a closed system at a given temperature.

Boiling occurs when the vapor pressure of a liquid equals atmospheric pressure.

Vapor pressure increases as the temperature increases.

Liquid A boils at a lower temperature than B because the vapor pressure of Liquid A is closer to the atmospheric pressure.

**63. B is correct.**

Colligative properties of solutions depend on the ratio of solute particles to the number of solvent molecules in a solution and not on the type of chemical species present.

Colligative properties include lowering vapor pressure, the elevation of boiling point, depression of freezing point, and increased osmotic pressure.

*continued...*

 

Freezing point (FP) depression:

$$\Delta FP = -iKm$$

where i = the number of particles produced when the solute dissociates, K = freezing point depression constant, and m = molality (moles/kg solvent).

$$-iKm = -2\text{ K}$$

$$-(1)\cdot(40)\cdot(x) = -2$$

$$x = -2 / -1(40)$$

$$x = 0.05\text{ molal}$$

Assume that compound $x$ does not dissociate.

0.05 mole compound ($x$) / kg camphor = 25 g / kg camphor

Therefore, if 0.05 mole = 25 g

1 mole = 500 g

**64. C is correct.**

*Boiling* occurs when the vapor pressure of a liquid equals atmospheric pressure.

*Vapor pressure* is the pressure exerted by a vapor in equilibrium with its condensed phases (i.e., solid or liquid) in a closed system at a given temperature.

*Atmospheric pressure* is the pressure exerted by the weight of air in the atmosphere.

Vapor pressure is inversely correlated with the strength of the intermolecular force.

Molecules are more likely to stick together in liquid form due to stronger intermolecular forces. Few participate in the liquid-vapor equilibrium; therefore, the molecule would boil at a higher temperature.

**65. D is correct.**

*Van der Waals equation* describes factors that must be accounted for when the ideal gas law calculates values for nonideal gases.

Terms that affect the pressure and volume of the ideal gas law are intermolecular forces and the volume of nonideal gas molecules.

**66. B is correct.**

*Ideal gas law*:

$$PV = nRT$$

where P is pressure, V is volume, n is the number of molecules, R is the ideal gas constant, and T is the temperature of the gas.

Units of R can be calculated by rearranging the expression:

$$R = PV/nT$$

$$R = \text{atm}\cdot\text{L/mol}\cdot\text{K}$$

**67. A is correct.**

*Ideal gas law*:

$$PV = nRT$$

where P is pressure, V is volume, n is the number of molecules, R is the ideal gas constant, and T is the temperature of the gas.

Set the initial and final P/V/T conditions equal:

$$(P_1V_1 / T_1) = (P_2V_2 / T_2)$$

STP condition is the temperature of 0 °C (273 K) and a pressure of 1 atm.

Solve for the final temperature:

$$(P_2V_2 / T_2) = (P_1V_1 / T_1)$$

$$T_2 = (P_2V_2T_1) / (P_1V_1)$$

$$T_2 = [(0.80 \text{ atm}) \times (0.155 \text{ L}) \times (273 \text{ K})] / [(1.00 \text{ atm}) \times (0.120 \text{ L})]$$

$$T_2 = 282.1 \text{ K}$$

Convert temperature units to degrees Celsius:

$$T_2 = (282.1 \text{ K} - 273 \text{ K})$$

$$T_2 = 9.1 \text{ °C}$$

**68. C is correct.**

*Atmospheric pressure* is the pressure exerted by the weight of air in the atmosphere.

*Boiling* occurs when the vapor pressure of the liquid is higher than the atmospheric pressure.

At standard atmospheric pressure and 22 °C, the vapor pressure of water is less than the atmospheric pressure, and it does not boil.

However, when a vacuum pump is used, the atmospheric pressure is reduced until it has a *lower vapor pressure* than water, allowing water to boil at a much lower temperature.

**69. D is correct.**

Kinetic theory of gases describes a gas as many small particles in constant rapid motion. These particles collide with each other and with the walls of the container.

Average kinetic energy depends only on the absolute temperature of the system.

temperature = $\frac{1}{2}mv^2$

At high temperatures, the particles are moving at at greater velocity, and at a temperature of absolute zero (i.e., 0 K), gas particles do not move.

Therefore, as temperature decreases, kinetic energy decreases, and so does the velocity of the gas molecules.

**70. A is correct.**

Dalton's law (i.e., the law of partial pressures) states that *the pressure exerted by a mixture of gases equals the sum of the individual gas pressures*.

Partial pressure of molecules in a mixture is proportional to their molar ratios.

Use the coefficients of the reaction to determine the molar ratio.

Based on that information, calculate the partial pressure of $O_2$:

(coefficient $O_2$) / (sum of coefficients in a mixture) × total pressure

[1 / (2 + 1)] × 1,250 torr

417 torr = partial pressure of $O_2$

**71. D is correct.**

$2\ Na\ (s) + Cl_2\ (g) \rightarrow 2\ NaCl\ (s)$

In its elemental form, Na exists as a solid. In its elemental form, chlorine exists as a gas.

**72. B is correct.**

Charles' law (i.e., the law of volumes) explains how, at constant pressure, gases behave when the temperature changes:

$$V \alpha T$$

or

$$V / T = \text{constant}$$

or

$$(V_1 / T_1) = (V_2 / T_2)$$

Volume and temperature are proportional.

Therefore, an increase in one term increases the other.

**73. A is correct.**

Gay-Lussac's law (i.e., pressure-temperature law) states that pressure is proportional to temperature:

$$P \alpha T$$

or

$$(P_1 / T_1) = (P_2 / T_2)$$

or

$$(P_1T_2) = (P_2T_1)$$

or

$$P / T = \text{constant}$$

If the pressure of a gas increases, the temperature increases proportionally.

Quadrupling the temperature increases the pressure fourfold.

Boyle's law (i.e., pressure-volume law) states that pressure and volume are inversely proportional: $(P_1V_1) = (P_2V_2)$

or

$$P \times V = \text{constant}$$

Reducing the volume by half increases pressure twofold.

Therefore, the increase in pressure would be by a factor of $4 \times 2 = 8$.

 

**74. C is correct.**

Scientists found that the relationships between pressure, temperature, and volume of a sample of gas hold for all gases, and the gas laws were developed.

Boyle's law (i.e., pressure-volume law) states that pressure and volume are inversely proportional:

$$(P_1V_1) = (P_2V_2)$$

or

$$P \times V = \text{constant}$$

If the volume of a gas increases, its pressure decreases proportionally.

Charles' law (i.e., the law of volumes) explains how, at constant pressure, gases behave when the temperature changes:

$$(V_1 / T_1) = (V_2 / T_2)$$

Gay-Lussac's law (i.e., pressure-temperature law) states that pressure is proportional to temperature:

$$P \alpha T$$

or

$$(P_1 / T_1) = (P_2 / T_2) \text{ or } (P_1T_2) = (P_2T_1)$$

or

$$P / T = \text{constant}$$

If the pressure of a gas increases, the temperature increases.

**75. A is correct.**

Sublimation is the direct state change from solid to gas, skipping intermediate liquid phase.

Solid carbon dioxide (i.e., dry ice) is an example of a compound that undergoes sublimation. $CO_2$ changes phases from solid to gas (i.e., bypasses the liquid phase) and is often used as a cooling agent.

*continued…*

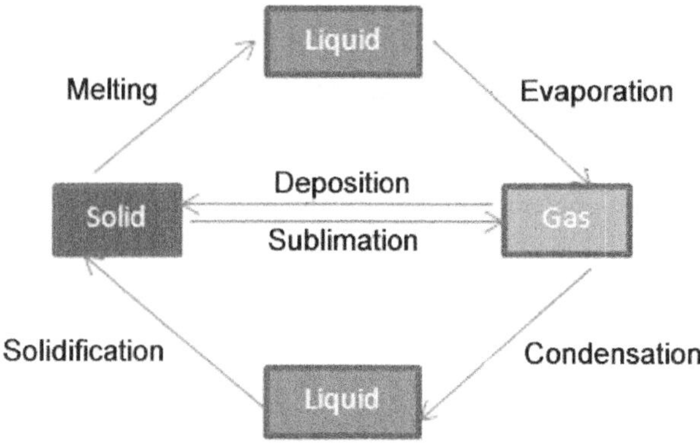


*Interconversion of states of matter*

**76. B is correct.**

Graham's law of effusion states that the rate of effusion (i.e., escaping through a small hole) of a gas is inversely proportional to the square root of the molar mass of its particles.

$$\text{Rate 1 / Rate 2} = \sqrt{(\text{molar mass gas 2 / molar mass gas 1})}$$

Set the rate of effusion of krypton over the rate of effusion of methane:

$$\text{Rate}_{Kr} / \text{Rate}_{CH4} = \sqrt{[(M_{Kr}) / (M_{CH4})]}$$

Solve for the ratio of effusion rates:

$$\text{Rate}_{Kr} / \text{Rate}_{CH4} = \sqrt{[(83.798\text{ g/mol}) / (16.04\text{ g/mol})]}$$

$$\text{Rate}_{Kr} / \text{Rate}_{CH4} = 2.29$$

Larger gas molecules must effuse at a slower rate; the effusion rate of Kr gas molecules must be slower than methane molecules:

$$\text{Rate}_{Kr} = [\text{Rate}_{CH4} / (2.29)]$$

$$\text{Rate}_{Kr} = [(631\text{ m/s}) / (2.29)]$$

$$\text{Rate}_{Kr} = 276\text{ m/s}$$

 

**77. C is correct.**

Observe information about gas C. Temperature, pressure, and volume of the gas are indicated.

Substitute these values into the ideal gas law equation to determine the number of moles:

$PV = nRT$

$n = PV / RT$

Convert units to the units indicated in the gas constant:

volume = 668.5 mL × (1 L / 1,000 mL)

volume = 0. 669 L

pressure = 745.5 torr × (1 atm / 760 torr)

pressure = 0.981 atm

temperature = 32.0 °C + 273.15 K

temperature = 305.2 K

$n = PV / RT$

$n = (0.981 \text{ atm} \times 0.669 \text{ L}) / (0.0821 \text{ L·atm K}^{-1} \text{ mol}^{-1} \times 305.2 \text{ K})$

n = 0.026 mole

*Law of mass conservation* states that the mass of products = the mass of reactants.

Based on this law, the mass of gas C can be determined:

mass product = mass reactant

mass A = mass B + mass C

5.2 g = 3.8 g + mass C

mass C = 1.4 g

Calculate the molar mass of C:

molar mass = 1.4 g / 0.026 mole

molar mass = 53.9 g / mole

 

**78. D is correct.**

Liquids take the shape of their container (i.e., they have an indefinite shape).

For example, consider a liter of water poured into a cylindrical bucket or a square box. In both cases, it takes up the shape of the container.

Liquids have a definite volume. The volume of water in the given example is 1 L, regardless of its container.

**79. D is correct.**

Intermolecular forces act between neighboring molecules. Examples include hydrogen bonding, dipole-dipole, dipole-induced dipole, and van der Waals (i.e., London dispersion) forces.

Hydrogens, bonded directly to F, O, or N, participate in hydrogen bonds. The hydrogen is partially positive (i.e., delta plus: $\partial+$) due to the bond to these electronegative atoms.

The lone pair of electrons on the F, O, or N interacts with the partial positive ($\partial+$) hydrogen to form a hydrogen bond.

**80. D is correct.**

When the balloon is placed in a freezer, the helium temperature in the balloon decreases because the surroundings are colder than the balloon, and heat is transferred from the balloon to the surrounding air until a thermal equilibrium is reached.

When the temperature of a gas decreases, the molecules move more slowly and become closer, causing the volume of the balloon to decrease.

From the ideal gas law,

$$PV = nRT$$

If the temperature decreases, the volume must decrease (if the pressure remains constant).

 

==========================================================

**Practice Set 5: Questions 81–100**

==========================================================

**81. A is correct.**

Boyle's law (i.e., pressure-volume law) states that pressure and volume are inversely proportional:

$$(P_1V_1) = (P_2V_2)$$

or

$$P \times V = \text{constant}$$

If the pressure of a gas increases, its volume decreases proportionally.

Doubling the pressure reduces the volume by half.

**82. C is correct.**

*Vapor pressure* is the pressure exerted by a vapor in equilibrium with its condensed phases (i.e., solid or liquid) in a closed system at a given temperature.

Pressure of a gas is the force that it exerts on the walls of its container. This is essentially the frequency and energy with which the gas molecules collide with the container's walls.

*Partial pressure* refers to the individual pressure of a gas in a mixture of gases.

Vapor pressure is the pressure of a vapor in thermodynamic equilibrium with its solid or liquid phases.

**83. B is correct.**

*Colligative properties* of solutions depend on the ratio of solute particles to the number of solvent molecules in a solution and not on the type of chemical species present.

Colligative properties include lowering vapor pressure, the elevation of boiling point, depression of freezing point, and increased osmotic pressure.

Freezing point (FP) depression:

$$\Delta FP = -iKm$$

where i = the number of particles produced when the solute dissociates, K = freezing point depression constant, and m = molality (moles/kg solvent).

*continued...*

Given in the question stem:

$\Delta FP = -10$ K

$-iKm = -10$ K

Toluene is the solvent and does not dissociate; the K value is not used.

FP depression from the addition of benzene:

i = 10 K / Km

$i = 1 - (1)\cdot(5.0)\cdot(x)$

K value of solvent:

$x = 2$ molal

2 moles toluene/kg benzene = $x$ moles / 0.1 kg benzene

$x$ moles / 0.1 kg benzene = 2 moles toluene/kg benzene

$x$ moles = 2 moles toluene/kg benzene × 0.1 kg benzene

toluene = 0.2 mole

**84. A is correct.**

Considering the ideal gas law, there is no difference between He and Ne since pressure, volume, temperature, and the number of moles are known quantities.

The identity of the gas would only be relevant to convert from moles to mass or vice-versa.

**85. D is correct.**

Gas molecules transfer kinetic energy between each other.

**86. D is correct.**

Assume $a$ is the equivalent temperature (i.e., has the same value in °C and °F).

Set up a formula with the equivalent temperature on one side and the conversion factor to the other unit on the other side.

For example, the left side is °C, while the right side is the conversion to °F:

$a = [(9/5) \times a] + 32$

$a - (9/5)a = 32$

$(-4/5)a = 32$

*continued...*

 

$-4a = 160$

$a = -40$

At –40, the temperature is the same in °C and °F.

**87. D is correct.**

At low elevations (i.e., sea level), the boiling point of water is 100 °C.

Atmospheric pressure is the pressure exerted by the weight of air in the atmosphere.

Boiling point decreases with increasing altitude due to the reduced atmospheric pressure above the water at high altitudes.

**88. A is correct.**

Hydrocarbons are nonpolar molecules, meaning London dispersion is the dominant intermolecular force. This force increases as the number of atoms in each molecule increases.

Stronger force results in a higher boiling point.

$CH_4$ has the least atoms and, therefore, has the lowest boiling point.

**89. C is correct.**

Melting point of a substance depends on the pressure that it is at.

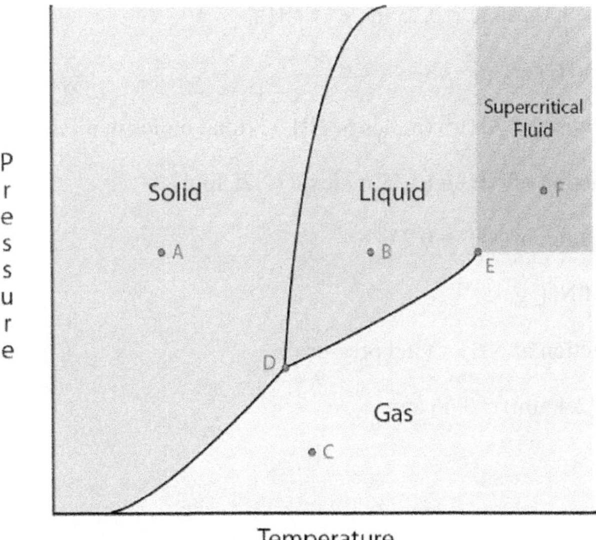


*Phase diagram of pressure vs. temperature*

*continued...*

*Triple point* (point D on the graph) is the pressure and temperature of a substance as a solid, liquid, or gas.

*Critical point* (point E on the graph) is the endpoint of the phase equilibrium curve, where the liquid and its vapor become indistinguishable.

Generally, melting points are given at standard pressure. If the substance is denser in the solid than in the liquid state (which is true of most substances), the melting point increases when pressure is increased.

This is observed with the phase diagram for $CO_2$. However, with certain substances, including water, the solid is less dense than the liquid state (e.g., ice is less dense than water), so the reverse is true, and the melting point decreases when pressure increases.

**90. B is correct.**

Dalton's law (i.e., the law of partial pressures) states that *the pressure exerted by a mixture of gases equals the sum of the individual gas pressures*.

Partial pressure of molecules in a mixture is proportional to their molar ratios.

Moles of $CH_4$:

$$32 \text{ g} \times 16 \text{ g/mol} = 2 \text{ moles of } CH_4$$

Moles of $NH_3$:

$$12.75 \text{ g} \times 17 \text{ g/mole} = 0.75 \text{ mole of } NH_3$$

Mole fraction of $NH_3$ gas:

$$\text{mole fraction of } NH_3 = (\text{moles of } NH_3) / (\text{total moles in mixture})$$

$$\text{mole fraction of } NH_3 = (0.75 \text{ moles}) / (2.75 \text{ moles})$$

$$\text{mole fraction of } NH_3 = 0.273$$

Partial pressure of $NH_3$ gas:

$$\text{mole fraction of } NH_3 \times \text{total pressure}$$

$$(0.273)\cdot(2.4 \text{ atm}) = 0.66 \text{ atm}$$

 

**91. B is correct.**

Liquids take the shape of their container (i.e., they have an indefinite shape). This can be visualized by considering a liter of water poured into a cylindrical bucket or a square box. In both cases, it takes up the shape of the container.

However, liquids have a definite volume. The volume of water in the given example is 1 L, regardless of its container.

**92. C is correct.**

Charles' law (i.e., the law of volumes) explains how, at constant pressure, gases behave when the temperature changes:

$$(V_1 / T_1) = (V_2 / T_2)$$

Solve for the final temperature:

$$T_2 = (V_1 / T_1) / V_2$$

$$T_2 = (0.050 \text{ L}) \times [(420 \text{ K}) / (0.100 \text{ L})]$$

$$T_2 = 210 \text{ K}$$

**93. A is correct.**

Gay-Lussac's law (i.e., pressure-temperature law) states that pressure is proportional to temperature:

$$P \alpha T$$

or

$$(P_1 / T_1) = (P_2 / T_2)$$

or

$$(P_1T_2) = (P_2T_1)$$

or

$$P / T = \text{constant}$$

If the pressure of a gas increases, the temperature increases.

Dalton's law (i.e., the law of partial pressures) states that *the pressure exerted by a mixture of gases equals the sum of the individual gas pressures.*

*continued...*

Charles' law (i.e., the law of volumes) explains how, at constant pressure, gases behave when the temperature changes:

$V \alpha T$ or $V / T = \text{constant}$

Volume and temperature are proportional.

Therefore, an increase in one term increases the other.

Boyle's law (i.e., pressure-volume law) states that pressure and volume are inversely proportional:

$(P_1V_1) = (P_2V_2)$ or $P \times V = \text{constant}$

If the volume of a gas increases, its pressure decreases proportionally.

**94. D is correct.**

Gay-Lussac's law (i.e., pressure-temperature law) states that pressure is proportional to temperature:

$P \alpha T$

or

$(P_1 / T_1) = (P_2 / T_2)$

or

$(P_1T_2) = (P_2T_1)$

or

$P / T = \text{constant}$

If the pressure of a gas increases, the temperature increases.

Kinetic theory of gases describes a gas as many small particles in constant rapid motion. These particles collide with each other and with the walls of the container.

Their average kinetic energy depends only on the absolute temperature of the system.

$$\text{temperature} = \tfrac{1}{2}mv^2$$

At high temperatures, the particles are moving with greater velocity.

 

**95. C is correct.**

Vaporization refers to the change of state from liquid to gas. Two types of vaporization, boiling and evaporation, are differentiated based on the temperature at which they occur.

Evaporation occurs at a temperature below the boiling point, while boiling occurs at or above the boiling point.

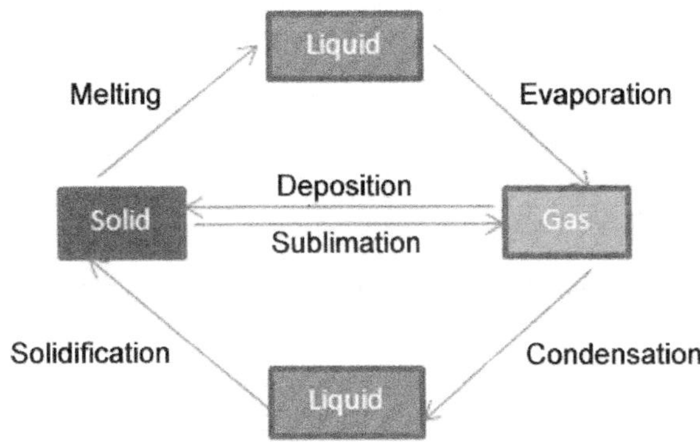


*Interconversion of states of matter*

**96. C is correct.**

*Intermolecular forces* act between neighboring molecules. Examples include hydrogen bonding, dipole-dipole, dipole-induced dipole, and van der Waals (i.e., London dispersion) forces.

Dipole-dipole attraction involves asymmetric, polar molecules (based on differences in electronegativity between atoms) that create a dipole moment (i.e., a net vector indicating the force).

$H_3C$ O $CH_3$

*Dimethyl ether with indicated bond angle*

**97. D is correct.**

Ideal gas law:

$$PV = nRT$$

where P is pressure, V is volume, n is the number of molecules, R is the ideal gas constant, and T is the temperature of the gas.

R and n are constant. Therefore, decreasing the volume, increasing temperature, or increasing the number of molecules would increase the pressure of a gas.

**98. D is correct.**

*Avogadro's law* is an experimental gas law relating the volume of gas to the amount of gas substance present. It states that equal volumes of gases have the same number of molecules at the same temperature and pressure.

For a given mass of an ideal gas, the volume and amount (moles) of the gas are directly proportional if the temperature and pressure are constant.

$$V \propto n$$

or

$$V / n = k$$

where V = volume of the gas, *n* is the amount of substance of the gas (measured in moles).

*k* is a constant equal to RT / P,

where R is the universal gas constant, T is the Kelvin temperature, and P is the pressure.

As temperature and pressure are constant, RT / P is constant and represented as *k* (derived from the ideal gas law).

At the same temperature and pressure, equal volumes of gases have the same number of molecules.

Comparing the same substance under two different sets of conditions:

$$V_1 / n_1 = V_2 / n_2$$

As the number of moles of gas increases, the volume of the gas also increases in proportion.

Similarly, if the number of gas moles is decreased, the volume also decreases.

Thus, the number of molecules or atoms in a specific volume of an ideal gas is independent of their size or the molar mass of the gas.

**99. B is correct.**

The ideal gas law:

$$PV = nRT$$

Using proportions, the volume remains the same if the temperature is increased by a factor of 1.5 and the pressure is increased by a factor of 1.5 (constant number of moles).

 

**100. C is correct.**

Since both He and Ne are at the same temperature, pressure, and volume, they have equal moles of gas.

Even though there are equal moles of each gas of He and Ne, use the molar masses given in the periodic table and notice that the samples contain different masses.

*Notes for active learning*

 

# 2 – Thermochemistry: Detailed Explanations

==============================================================

**Practice Set 1: Questions 1–20**

==============================================================

**1. A is correct.**

Reaction enthalpy is specified by ΔH (not ΔG).

ΔH determines whether a reaction is exothermic (releases heat to surroundings) or endothermic (absorbs heat from surroundings).

To predict the spontaneity of the reaction, use the Gibbs free energy equation:

$$\Delta G = \Delta H° - T\Delta S$$

Reaction is *spontaneous* (i.e., exergonic) if ΔG is *negative*.

Reaction is *nonspontaneous* (i.e., endergonic) if ΔG is *positive*.

**2. B is correct.**

Heat is energy, and the energy from the heat is transferred to the gas molecules, which increases their movement (i.e., kinetic energy).

Temperature is a measure of the *average kinetic energy* of the molecules.

**3. D is correct.**

*Heat capacity* is the amount of heat required to increase the temperature of *the whole sample* by 1 °C.

*Specific heat* is the heat required to increase the temperature of *1 gram* of sample by 1 °C.

Heat = mass × specific heat × change in temperature:

$$q = m \times c \times \Delta T$$

$$q = 21.0 \text{ g} \times 0.382 \text{ J/g·°C} \times (68.5 \text{ °C} - 21.0 \text{ °C})$$

$$q = 21.0 \text{ g} \times 0.382 \text{ J/g·°C} \times (47.5 \text{ °C})$$

$$q = 381 \text{ J}$$

**4. C is correct.**

This is a theory/memorization question. However, it can be solved by comparing the options.

Kinetic energy is correlated with temperature, so a formula that involves temperature would be a good choice.

Another approach is to analyze the units, apply them to the variables, and evaluate them. For example, choices A and B have moles, pressure (Pa, bar, or atm), and area ($m^2$), so it is not possible for them to result in energy (J) when multiplied.

Option D has molarity (moles/L) and volume (L or $m^3$); they cannot result in energy (J) when multiplied.

Option C is nRT: mol × (J/mol K) × K; units but J cancel, so this is a plausible formula for kinetic energy.

For questions using formulas, dimensional analysis limits the choices for this type of question.

**5. A is correct.**

ΔH refers to enthalpy (or heat).

Endothermic reactions have heat as a reactant.

Exothermic reactions have heat as a product.

Exothermic reactions release heat and cause the temperature of the immediate surroundings to rise (i.e., a net energy loss).

Endothermic reactions absorb heat and cool the surroundings (i.e., a net energy gain). They absorb energy to break strong bonds to form a less stable state (i.e., positive enthalpy).

Exothermic reactions release energy when forming stronger bonds to produce a more stable state (i.e., negative enthalpy).

**6. D is correct.**

Endothermic reactions absorb energy to break strong bonds to form a less stable state (i.e., positive enthalpy).

Exothermic reactions release energy when forming stronger bonds to produce a more stable state (i.e., negative enthalpy).

The spontaneity of a reaction is determined by the calculation of ΔG using:

$$\Delta G = \Delta H - T\Delta S$$

*continued...*

 

The reaction is spontaneous when the value of ΔG is negative.

In this problem, the reaction is endothermic, which means that ΔH is positive.

The reaction decreases S, which means that ΔS value is negative.

Substituting these values into the equation:

$$\Delta G = \Delta H - T(-\Delta S)$$

$\Delta G = \Delta H + T\Delta S$, with ΔH and ΔS positive values

For this problem, the value of ΔG is positive.

The reaction does not occur because ΔG needs to be negative for a spontaneous reaction to proceed; the products are more stable than the reactants.

**7. D is correct.**

Chemical energy is potential energy stored in molecular bonds.

Electrical energy is the kinetic energy associated with the motion of electrons (e.g., in wires, circuits, and lightning). Potential energy stored in a flashlight battery is electrical energy.

Heat is the spontaneous transfer of energy from a hot object to a cold one with no displacement or deformation of the objects (i.e., no work done).

**8. A is correct.**

Entropy is higher for less organized states (i.e., more random or disordered).

**9. C is correct.**

ΔH refers to enthalpy (or heat).

Endothermic reactions have heat as a reactant.

Exothermic reactions have heat as a product.

Exothermic reactions release heat and cause the temperature of the immediate surroundings to rise (i.e., a net energy loss).

Endothermic reactions absorb heat and cool the surroundings (i.e., a net energy gain).

Endothermic reactions absorb energy to break strong bonds to form a less stable state (i.e., positive enthalpy).

Exothermic reactions release energy when forming stronger bonds to produce a more stable state (i.e., negative enthalpy).

*continued...*

The reaction is nonspontaneous (i.e., endergonic) if the products are less stable than the reactants and $\Delta G$ is positive.

The reaction is spontaneous (i.e., exergonic) if the products are more stable than the reactants and $\Delta G$ is negative.

**10. D is correct.**

This cannot be determined when the given $\Delta G = \Delta H - T\Delta S$, terms cancel.

The system is at equilibrium when $\Delta G = 0$, but in this question, each side of the equation cancels:

$$\underbrace{X - RY}_{\Delta G \text{ term}} = \underbrace{X - RY}_{\substack{\Delta H - T\Delta S \\ \text{term}}}, \quad \text{so } 0 = 0$$

**11. C is correct.**

Gibbs free energy:

$$\Delta G = \Delta H - T\Delta S$$

Stable molecules are spontaneous because they have a negative (or relatively low) $\Delta G$.

$\Delta G$ is most negative (most stable) when $\Delta H$ is the smallest and $\Delta S$ is largest.

**12. A is correct.**

$\Delta H$ refers to enthalpy (or heat).

Endothermic reactions have heat as a reactant. Exothermic reactions have heat as a product.

The reaction is nonspontaneous (i.e., endergonic) if the products are less stable than the reactants and $\Delta G$ is positive.

The reaction is spontaneous (i.e., exergonic) if the products are more stable than the reactants and $\Delta G$ is negative.

Exothermic reactions release heat and cause the temperature of the immediate surroundings to rise (i.e., a net energy loss). Exothermic reactions release energy when forming stronger bonds to produce a more stable state (i.e., negative enthalpy).

Endothermic reactions absorb heat and cool the surroundings (i.e., a net energy gain).

Endothermic reactions absorb energy to break strong bonds to form a less stable state (i.e., positive enthalpy).

*continued...*

 

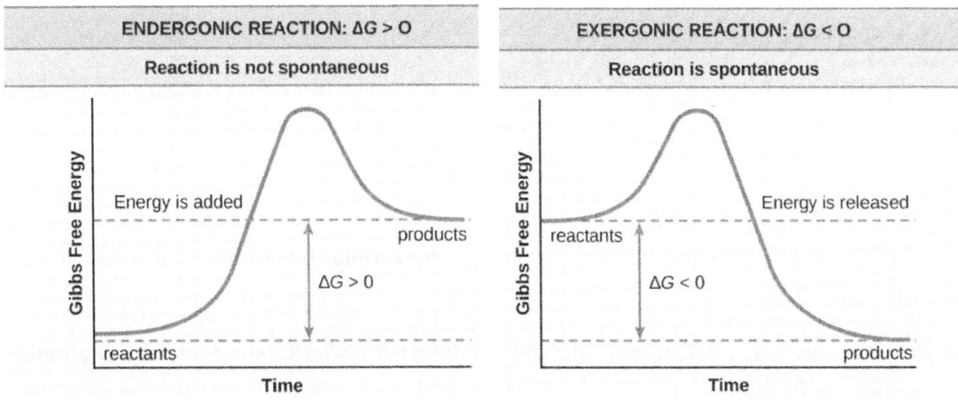


**13. B is correct.**

Gibbs free energy: $\Delta G = \Delta H - T\Delta S$

Entropy ($\Delta S$) determines the favorability of chemical reactions. If entropy is large, the reaction is more likely to proceed.

**14. D is correct.**

In a chemical reaction, bonds within reactants are broken down, and new bonds are formed to create products. Therefore, in bond dissociation problems,

$\Delta H$ reaction = sum of bond energy in reactants – sum of bond energy in products

For $H_2C{=}CH_2 + H_2 \rightarrow CH_3{-}CH_3$:

$\Delta H_{reaction}$ = sum of bond energy in reactants – sum of bond energy in products

$\Delta H_{reaction} = [(C{=}C) + 4(C{-}H) + (H{-}H)] - [(C{-}C) + 6(C{-}H)]$

$\Delta H_{reaction} = [612\ kJ + (4 \times 412\ kJ) + 436\ kJ] - [348\ kJ + (6 \times 412\ kJ)]$

$\Delta H_{reaction} = -124\ kJ$

Remember that this is the opposite of $\Delta H_f$ problems, where:

$\Delta H_{reaction}$ = (sum of $\Delta H_f$ products) – (sum of $\Delta H_f$ reactants)

**15. C is correct.**

Bond dissociation energy is required to break a bond between two gaseous atoms and is useful in estimating the enthalpy change in a reaction.

**16. B is correct.**

$\Delta H$ value is positive, which means the reaction is endothermic and absorbs energy from the surroundings.

$\Delta H$ value is expressed as the energy released/absorbed per mole (for species with a coefficient of 1). If the coefficient is 2, it is the energy transferred per 2 moles.

For example, 2 moles of NO are produced from 1 mole for each reagent for the stated reaction. Therefore, 43.2 kcal is produced for 2 moles of NO.

**17. A is correct.**

State functions depend only on the initial and final states of the system and are independent of the paths taken to reach the final state.

Common thermodynamic state functions include internal energy, enthalpy, entropy, pressure, temperature, and volume.

Work and heat relate to a system's energy change when it moves from one state to another, depending on how a system changes between states.

**18. B is correct.**

Temperature (i.e., average kinetic energy) of a substance remains constant during a phase change (e.g., solid to liquid or liquid to gas).

For example, heat (i.e., energy) breaks bonds between the ice molecules and changes phases into the liquid phase.

Since the average kinetic energy of the molecules does not change at the moment of the phase change (i.e., melting), the temperature of the molecules does not change.

**19. C is correct.**

In a chemical reaction, bonds within reactants are broken, and new bonds form to create products.

*continued...*

 

Bond dissociation:

$\Delta H_{reaction}$ = (sum of bond energy in reactants) – (sum of bond energy in products)

O=C=O + 3 $H_2$ → $CH_3$–O–H + H–O–H

$\Delta H_{reaction}$ = (sum of bond energy in reactants) – (sum of bond energy in products)

$\Delta H_{reaction}$ = [2(C=O) + 3(H–H)] – [3(C–H) + (C–O) + (O–H) + 2(O–H)]

$\Delta H_{reaction}$ = [(2 × 743 kJ) + (3 × 436 kJ)] – [(3 × 412 kJ) + 360 kJ + 463 kJ + (2 × 463 kJ)]

$\Delta H_{reaction}$ = –191 kJ

This is the reverse of $\Delta H_f$ problems, where:

$\Delta H_{reaction}$ = (sum of $\Delta H_{f\ product}$) – (sum of $\Delta H_{f\ reactant}$)

**20. C is correct.**

Entropy indicates how a system is organized; increased entropy means less order.

The entropy of the universe is increasing because a system moves towards more disorder unless energy is added to the system.

The second law of thermodynamics states that the entropy of interconnected systems, without adding energy, always increases.

---

**Practice Set 2: Questions 21–40**

---

**21. A is correct.**

Gibbs free energy:

$$\Delta G = \Delta H - T\Delta S$$

The value that is the largest positive value is the most endothermic, while the value that is the largest negative value is the most exothermic.

$\Delta H$ refers to enthalpy (or heat).

Endothermic reactions have heat as a reactant.

Exothermic reactions have heat as a product.

Exothermic reactions release heat and cause the temperature of the immediate surroundings to rise (i.e., a net energy loss).

Endothermic reactions absorb heat and cool the surroundings (i.e., a net energy gain).

Endothermic reactions absorb energy to break strong bonds to form a less stable state (i.e., positive enthalpy).

Exothermic reactions release energy when forming stronger bonds to produce a more stable state (i.e., negative enthalpy).

**22. C is correct.**

| **Potential Energy** | **Kinetic Energy** |
|---|---|
| **Stored energy and the energy of position (gravitational)** | **Energy of motion: motion of waves, electrons, atoms, molecules, and substances** |
| **Chemical Energy** | **Thermal Energy** |
| Chemical energy is the energy stored in the bonds of atoms and molecules. Examples of stored chemical energy: biomass, petroleum, natural gas, propane, and coal. | Thermal energy is internal energy in substances, the vibration, and movement of atoms and molecules within substances. Geothermal energy is an example of thermal energy. |

**23. A is correct.**

By convention, the $\Delta G$ for an element in its standard state is 0.

**24. B is correct.**

Test strategy: carefully evaluate the most descriptive answer when choosing similar explanations. However, check the statement because sometimes the most descriptive option is inaccurate.

The longest option is accurate here, so that would be the best choice.

**25. C is correct.**

ΔG is always negative for spontaneous reactions.

**26. C is correct.**

An insulator reduces conduction (i.e., thermal energy transfer through matter).

Air and vacuum are excellent insulators. Storm windows, which have air wedged between two glass panes, work by utilizing this conduction principle.

**27. A is correct.**

ΔH refers to enthalpy (or heat).

Endothermic reactions have heat as a reactant. Exothermic reactions have heat as a product.

Exothermic reactions release heat and cause the temperature of the immediate surroundings to rise (i.e., a net energy loss).

Endothermic reactions absorb heat and cool the surroundings (i.e., a net energy gain).

Reaction is nonspontaneous (i.e., endergonic) if the products are less stable than the reactants and ΔG is positive.

Reaction is spontaneous (i.e., exergonic) if the products are more stable than the reactants and ΔG is negative.

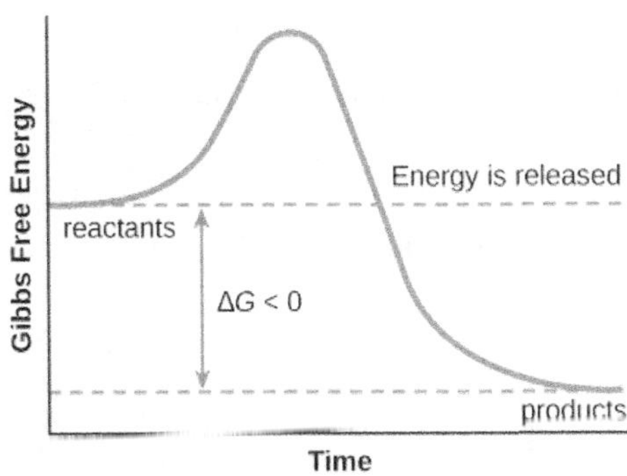


Exergonic reactions are spontaneous and have reactants with more energy than the products.

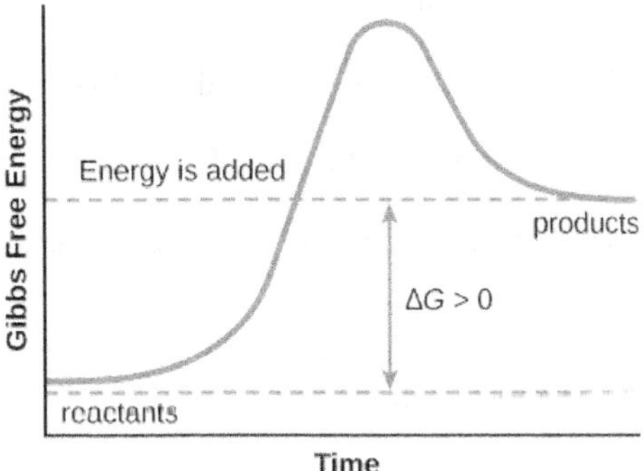


Endergonic reactions are nonspontaneous and have products with more energy than the reactants.

**28. D is correct.**

*Heat capacity* is the amount of heat required to increase temperature of *the whole sample* by 1 °C.

*Specific heat* is the heat required to increase the temperature of *1 gram* of sample by 1 °C.

$q$ = mass × heat of condensation

$q$ = 16 g × 1,380 J/g

$q$ = 22,080 J

**29. A is correct.**

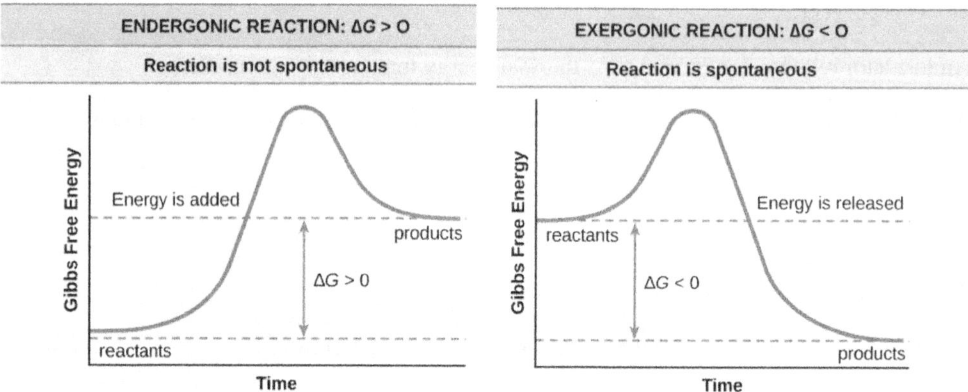


$\Delta G = \Delta H - T\Delta S$ refers to exergonic when $\Delta G$ is negative and endergonic when $\Delta G$ is positive.

If $\Delta S$ is 0 then $\Delta G = \Delta H$ because the $T\Delta S$ term cancels because it equals zero).

$\Delta H$ refers to exothermic when $\Delta H$ is negative and endothermic when $\Delta H$ is positive.

**30. D is correct.**

*Heat of formation* is defined as the heat required or released when creating one mole of the substance from its elements.

If the given reaction is reversed and divided by two, it would be the formation reaction of $NH_3$:

$\frac{1}{2} N_2 + 3/2\ H_2 \rightarrow NH_3$

$\Delta H$ of this reaction would be reversed (i.e., plus to minus sign) and divided by two:

$\Delta H = -(92.4\ \text{kJ/mol}) / 2$

$\Delta H = -46.2\ \text{kJ/mol}$

 

**31. B is correct.**

Gibbs free energy:

$$\Delta G = \Delta H - T\Delta S$$

With a positive ΔH and negative ΔS, ΔG is positive, so the reaction is nonspontaneous.

**32. C is correct.**

Entropy (ΔS) measures the degree of disorder in a system.

When a change causes the components of a system to transform into a higher energy phase (solid → liquid, liquid → gas), the system's entropy increases.

**33. D is correct.**

Only the mass and heat of specific phase changes are required during phase changes to calculate the energy released or absorbed. Because water is turning into ice, the specific heat required is the heat of solidification.

**34. B is correct.**

Nuclear energy is considered energy because splitting or combining atoms results in enormous amounts of energy used in atomic bombs and nuclear power plants.

| **Potential Energy**<br><br>**Stored energy and the energy of position (gravitational)** | **Kinetic Energy**<br><br>**Energy of motion: motion of waves, electrons, atoms, molecules, and substances** |
|---|---|
| **Nuclear Energy**<br><br>Nuclear energy is the energy stored in the nucleus of an atom. It is the energy that holds the nucleus. The nucleus of the uranium atom is an example of nuclear energy. | **Electrical Energy**<br><br>Electrical energy is the movement of electrons. Lightning and electricity are examples of electrical energy. |

**35. D is correct.**

Entropy is a measurement of disorder. Gases contain more energy than liquids of the same element and have a higher degree of randomness.

Entropy is the unavailable energy that cannot be converted into mechanical work.

**36. A is correct.**

There are two different methods available to solve this problem. Both methods are based on the definition of ΔH formation as energy consumed/released when one mole of a molecule is produced from its elemental atoms.

*Method 1.*

Rearrange and add the equations for the formation reaction of $C_2H_5OH$.

Information provided by the problem:

$C_2H_5OH + 3\ O_2 \rightarrow 2\ CO_2 + 3\ H_2O$  ΔH = 327 kcal

$H_2O \rightarrow H_2 + \frac{1}{2}\ O_2$  ΔH = 68.3 kcal

$C + O_2 \rightarrow CO_2$  ΔH = –94.1 kcal

Place $C_2H_5OH$ on the product side. For the other two reactions, arrange them so the elements are on the left and the molecules are on the right (remember that the goal is to create a formation reaction: elements forming a molecule).

When reversing the direction of a reaction, change the positive/negative sign of ΔH.

Multiplying the whole reaction by a coefficient would multiply ΔH by the same ratio.

$2\ CO_2 + 3\ H_2O \rightarrow C_2H_5OH + 3\ O_2$  ΔH = –327 kcal

$3\ H_2 + 3/2\ O_2 \rightarrow 3\ H_2O$  ΔH = –204.9 kcal

$2\ C + 2\ O_2 \rightarrow 2\ CO_2$  ΔH = –188.2 kcal

$2\ C + 3\ H_2 + 1/2\ O_2 \rightarrow C_2H_5OH$  ΔH = –720.1 kcal

ΔH formation of $C_2H_5OH$ is –720.1 kcal.

*Method 2.*

Heat of formation ($\Delta H_f$) data can be used to calculate ΔH of a reaction:

$\Delta H_{reaction}$ = the sum of $\Delta H_{f\ product}$ – sum of $\Delta H_{f\ reactant}$

Apply the formula to the first equation:

$C_2H_5OH + 3\ O_2 \rightarrow 2\ CO_2 + 3\ H_2O$  ΔH = 327 kcal

$\Delta H_{reaction}$ = the sum of $\Delta H_{f\ product}$ – sum of $\Delta H_{f\ reactant}$

$327\ \text{kcal} = [2(\Delta H_f\ CO_2) + 3(\Delta H_f\ H_2O)] - [(\Delta H_f\ C_2H_5OH) + 3(\Delta H_f\ O_2)]$

Oxygen ($O_2$) is an element, so $\Delta H_f = 0$

*continued...*

 

For $CO_2$ and $H_2O$, $\Delta H_f$ information can be obtained from the other 2 reactions:

$H_2O \rightarrow H_2 + \frac{1}{2} O_2$ $\Delta H = 68.3$ kcal

$C + O_2 \rightarrow CO_2$ $\Delta H = -94.1$ kcal

The $CO_2$ reaction is a formation reaction – creating one mole of a molecule from its elements. Therefore, $\Delta H_f$ of $CO_2 = -94.1$ kcal

If the $H_2O$ reaction is reversed, it will be a formation reaction.

$H_2 + \frac{1}{2} O_2 \rightarrow H_2O$ $\Delta H = -68.3$ kcal

Using those values, calculate $\Delta H_f$ of $C_2H_5OH$:

$$327 \text{ kcal} = [2(\Delta H_f\, CO_2) + 3(\Delta H_f\, H_2O)] - [(\Delta H_f\, C_2H_5OH) + 3(\Delta H_f\, O_2)]$$

$$327 \text{ kcal} = [2(-94.1 \text{ kcal}) + 3(-68.3 \text{ kcal})] - [(\Delta H_f\, C_2H_5OH) + 3(0 \text{ kcal})]$$

$$327 \text{ kcal} = [(-188.2 \text{ kcal}) + (-204.9 \text{ kcal})] - (\Delta H_f\, C_2H_5OH)$$

$$327 \text{ kcal} = (-393.1 \text{ kcal}) - (\Delta H_f\, C_2H_5OH)$$

$$\Delta H_f\, C_2H_5OH = -720.1 \text{ kcal}$$

**37. B is correct.**

$S + O_2 \rightarrow SO_2 + 69.8$ kcal

The reaction indicates that 69.8 kcal of energy is released (i.e., on the product side).

All species in this reaction have the coefficient of 1, which means that the $\Delta H$ value is calculated by reacting 1 mole of each reagent to create 1 mole of product.

Atomic mass of sulfur is 32.1 g, which means that this reaction uses 32.1 g of sulfur.

**38. D is correct.**

The disorder of the system is decreasing as more complex and ordered molecules are forming from the reaction of $H_2$ and $O_2$ gases.

3 moles of gas are combined to form 2 moles of gas.

Additionally, two molecules join to form one molecule.

**39. D is correct.**

In thermodynamics, an isolated system is enclosed by rigid, immovable walls through which neither matter nor energy passes.

Temperature is a measure of energy and is not a form of energy. Therefore, the temperature cannot be exchanged between the system and the surroundings.

*Closed system* can exchange energy (as heat or work) with its surroundings but not matter.

*Isolated system* cannot exchange energy (as heat or work) or matter with the surroundings.

*Open system* can exchange energy and matter with the surroundings.

**40. D is correct.**

State functions depend only on the system's initial and final states and are independent of the paths taken to reach the final state.

Common thermodynamic state functions include internal energy, enthalpy, entropy, pressure, temperature, and volume.

 

==============================================================

**Practice Set 3: Questions 41–60**

==============================================================

**41. A is correct.**

When comparing various fuels in varying forms (e.g., solid, liquid, or gas), it is easiest to compare energy released in terms of mass because matter has mass regardless of its state (as opposed to volume, which is convenient for a liquid or gas, but not for a solid).

Moles are more complicated because they must be converted to mass before a direct comparison can be made between the fuels.

**42. C is correct.**

*Convection* is heat carried by fluids (i.e., liquids and gases).

Heat is prevented from leaving the system through convection when the lid is placed on the cup.

**43. B is correct.**

For a reaction to be spontaneous, entropy ($\Delta S_{system} + \Delta S_{surrounding}$) must be positive (i.e., greater than 0).

**44. D is correct.**

The explosion of gases: chemical → heat energy

Heat converts into steam, which in turn moves a turbine: heat → mechanical energy

The generator creates electricity: mechanical → electrical energy

| **Potential Energy**<br><br>**Stored energy and the energy of position (gravitational)** | **Kinetic Energy**<br><br>**Energy of motion: motion of waves, electrons, atoms, molecules, and substances** |
|---|---|
| **Chemical Energy**<br><br>Chemical energy is the energy stored in the bonds of atoms and molecules. Examples of stored chemical energy: biomass, petroleum, natural gas, propane, and coal. | **Radiant Energy**<br><br>Radiant energy is electromagnetic energy that travels in transverse waves. Radiant energy includes visible light, x-rays, gamma rays, and radio waves. Solar energy is an example of radiant energy. |

| **Nuclear Energy** | **Thermal Energy** |
|---|---|
| Nuclear energy is the energy stored in the nucleus of an atom. It is the energy holding the nucleus. The nucleus of the uranium atom is an example of nuclear energy. | Thermal energy is internal energy in substances, the vibration, and movement of atoms and molecules within substances. Geothermal energy is an example of thermal energy. |
| **Stored Mechanical Energy** | **Motion** |
| Stored mechanical energy is energy stored in objects by applying a force. Compressed springs and stretched rubber bands are examples of stored mechanical energy. | The movement of objects or substances from one place to another is motion. Wind and hydropower are examples of motion. |
| **Gravitational Energy** | **Sound** |
| Gravitational Energy is the energy of place or position. Water in a reservoir behind a hydropower dam is an example of gravitational potential energy. When the water is released to spin the turbines, it becomes kinetic energy. | Sound is the movement of energy through substances in longitudinal (compression/rarefaction) waves. |
| | **Electrical Energy** |
| | Electrical energy is the movement of electrons. Lightning and electricity are examples of electrical energy. |

**45. A is correct.**

Relationship between enthalpy (ΔH) and internal energy (ΔE):

Enthalpy = Internal energy + work (for gases, work = PV)

$$\Delta H = \Delta E + \Delta(PV)$$

Solving for ΔE:

$$\Delta E = \Delta H - \Delta(PV)$$

According to ideal gas law:

$$PV = nRT$$

Substitute ideal gas law to the previous equation:

$$\Delta E = \Delta H - \Delta(nRT)$$

*continued...*

R and T are constant, which leaves $\Delta n$ as the variable.

Reaction is $C_2H_2\ (g) + 2H_2\ (g) \rightarrow C_2H_6\ (g)$.

There are three gas molecules on the left and one on the right, which means $\Delta n = 1 - 3 = -2$.

In this problem, the temperature is not provided. However, the presence of degree symbols ($\Delta G°$, $\Delta H°$, $\Delta S°$) indicates that those are standard values, which are measured at 25 °C or 298.15 K.

Double-check the units before calculating; $\Delta H$ is in kilojoules, while the gas constant (R) is 8.314 J/mol K. Convert $\Delta H$ to joules before calculating.

Use the $\Delta n$ and T values to calculate $\Delta E$:

$$\Delta E = \Delta H - \Delta nRT$$

$$\Delta E = -311{,}500 \text{ J/mol} - (-2 \times 8.314 \text{ J/mol K} \times 298.15 \text{ K})$$

$$\Delta E = -306{,}542 \text{ J} \approx -306.5 \text{ kJ}$$

**46. D is correct.**

Closed system can exchange energy (as heat or work) but not matter with its surroundings.

Isolated system cannot exchange energy (as heat or work) or matter with the surroundings.

Open system can exchange energy and matter with the surroundings.

**47. C is correct.**

An increase in entropy indicates an increase in disorder; find a reaction that creates more molecules on the product side compared to the reactant side.

**48. C is correct.**

When enthalpy is calculated, work done by gases is not considered; however, the energy of the reaction includes the work done by gases.

The largest difference between the energy of the reaction and enthalpy would be for reactions with the largest difference in the number of gas molecules between the product and reactants.

**49. B is correct.**

$\Delta H$ refers to enthalpy (or heat).

Endothermic reactions have heat as a reactant.

Exothermic reactions have heat as a product.

Endothermic reactions absorb energy to break strong bonds to form a less stable state (i.e., positive enthalpy).

Exothermic reactions release energy when forming stronger bonds to produce a more stable state (i.e., negative enthalpy).

The reaction is nonspontaneous (i.e., endergonic) if the products are less stable than the reactants and $\Delta G$ is positive.

The reaction is spontaneous (i.e., exergonic) if the products are more stable than the reactants and $\Delta G$ is negative.

**50. B is correct.**

All thermodynamic functions in $\Delta G = \Delta H - T\Delta S$ refer to the system.

**51. C is correct.**

To predict spontaneity of the reaction, use the Gibbs free energy equation:

$$\Delta G = \Delta H° - T\Delta S$$

Reaction is spontaneous if $\Delta G$ is negative.

Substitute the given values to the equation:

$$\Delta G = -113.4 \text{ kJ/mol} - [T \times (-145.7 \text{ J/K mol})]$$

$$\Delta G = -113.4 \text{ kJ/mol} + (T \times 145.7 \text{ J/K mol})$$

Important: note that the units aren't identical, $\Delta H°$ is in kJ and $\Delta S°$ is in J. Convert kJ to J (1 kJ = 1,000 J):

$$\Delta G = -113{,}400 \text{ J/mol} + (T \times 145.7 \text{ J/K mol})$$

It can be predicted that the value of $\Delta G$ would be negative if the value of T is small.

If T goes higher (G approaches a positive value) the reaction would be nonspontaneous.

 

**52. C is correct.**

Enthalpy:

U + PV

**53. C is correct.**

*State functions* depend only on the initial and final states of the system and are independent of the paths taken to reach the final state.

Common thermodynamic state functions include internal energy, enthalpy, entropy, pressure, temperature, and volume.

*Extensive property* is a property that changes when the size of the sample changes. Examples include mass, volume, length, and total charge.

Entropy has no absolute zero value, and a substance at zero Kelvin has zero entropy (i.e., no motion).

**54. C is correct.**

In an exothermic reaction, the bonds formed are stronger than the bonds broken.

**55. D is correct.**

Calculate value of $\Delta H_f = \Delta H_{f\ product} - \Delta H_{f\ reactant}$

$\Delta H_f = -436.8\ kJ\ mol^{-1} - (-391.2\ kJ\ mol^{-1})$

$\Delta H_f = -45.6\ kJ\ mol^{-1}$

To determine spontaneity, calculate the Gibbs free energy:

$\Delta G = \Delta H° - T\Delta S$

Reaction is spontaneous if ΔG is negative:

$\Delta G = -45.6\ kJ - T\Delta S$

Typical ΔS values are around 100–200 J.

If the ΔH value is –45.6 kJ or –45,600 J, the value of ΔG would still be negative unless TΔS is less than –45,600.

Therefore, the reaction would be spontaneous over a broad range of temperatures.

**56. A is correct.**

Enthalpy is an extensive property: it varies with the quantity of the matter.

Reaction 1 : $P_4 + 6\ Cl_2 \rightarrow 4\ PCl_3$ $\Delta H = -1{,}289$ kJ

Reaction 2 : $3\ P_4 + 18\ Cl_2 \rightarrow 12\ PCl_3$ $\Delta H = ?$

The only difference between reactions 1 and 2 is the coefficients.

In reaction 2, coefficients are three times reaction 1. Reaction 2 consumes three times as many reactants as reaction 1, and reaction 2 releases three times as much energy as reaction 1.

$\Delta H = 3 \times -1{,}289\text{ kJ} = -3{,}867\text{ kJ}$

**57. C is correct.**

ΔH refers to enthalpy (or heat).

Endothermic reactions have heat as a reactant.

Exothermic reactions have heat as a product.

Endothermic reactions absorb energy to break strong bonds to form a less stable state (i.e., positive enthalpy). Since endothermic reactions consume energy, heat should be on the reactants' side.

Exothermic reactions release energy when forming stronger bonds to produce a more stable state (i.e., negative enthalpy).

Reaction is nonspontaneous (i.e., endergonic) if the products are less stable than the reactants and ΔG is positive.

Reaction is spontaneous (i.e., exergonic) if the products are more stable than the reactants and ΔG is negative.

**58. D is correct.**

According to the second law of thermodynamics, the entropy gain of the universe must be positive (i.e., increased disorder).

Entropy of a system, however, can be negative if the surroundings experience an increase in entropy greater than the negative entropy change experienced by the system.

 

**59. B is correct.**

In thermodynamics, an isolated system is enclosed by rigid, immovable walls through which neither matter nor energy passes. Bell jars are designed not to let air or other materials in or out. Insulated means that heat cannot get in or out.

*Evaporation* indicates a phase change (vaporization) from liquid to gaseous phase; however, it does not imply that the matter has escaped its system.

*Closed system* can exchange energy (as heat or work) with its surroundings but not matter.

*Isolated system* cannot exchange energy (as heat or work) or matter with the surroundings.

*Open system* can exchange energy and matter with the surroundings.

**60. A is correct.**

*Law of Conservation of Energy* states that an isolated system's total energy (e.g., potential or kinetic) remains constant and is conserved. Energy can be neither created nor destroyed but is transformed from one form to another. For instance, chemical energy can be converted to kinetic energy in the explosion of a firecracker.

The Law of Conservation of Mass states that for a system closed to transfers of matter and energy, the mass of the system must remain constant over time.

*Law of definite proportions* (Proust's law) states that a chemical compound contains the same proportion of elements by mass. The law of definite proportions forms the basis of stoichiometry (i.e., from the known amounts of separate reactants, the amount of the product can be calculated).

*Avogadro's law* is an experimental gas law relating the volume of a gas to the amount of substance of gas present. It states that equal volumes of gases have the same number of molecules at the same temperature and pressure.

*Boyle's law* is an experimental gas law that describes how the pressure of a gas tends to increase as the volume of a gas decreases. It states that the absolute pressure exerted by a given mass of an ideal gas is inversely proportional to the volume it occupies if the temperature and amount of gas remain unchanged within a closed system.

===============================================================

## Practice Set 4: Questions 61–80

===============================================================

**61. C is correct.**

$\Delta H$ refers to enthalpy (or heat).

Endothermic reactions have heat as a *reactant.*

Exothermic reactions have heat as a *product.*

Exothermic reactions release heat and cause the temperature of the immediate surroundings to rise (i.e., a net energy loss).

Endothermic reactions absorb heat and cool the surroundings (i.e., a net energy gain).

Endothermic reactions absorb energy to break strong bonds to form a less stable state (i.e., positive enthalpy).

Exothermic reactions release energy when forming stronger bonds to produce a more stable state (i.e., negative enthalpy).

Reaction is *nonspontaneous* (i.e., endergonic) if the products are less stable than the reactants and $\Delta G$ is *positive.*

Reaction is *spontaneous* (i.e., exergonic) if the products are more stable than the reactants and $\Delta G$ is *negative.*

**62. C is correct.**

Temperature is a measure of the *average kinetic energy* of the molecules.

Kinetic energy is *proportional to temperature.*

In most thermodynamic equations, temperatures are expressed in Kelvin, so convert the Celsius temperatures to Kelvin:

$$25\ °C + 273.15 = 298.15\ K$$

$$50\ °C + 273.15 = 323.15\ K$$

Calculate the kinetic energy using simple proportions:

$$KE = (323.15\ K / 298.15\ K) \times 500\ J$$

$$KE = 540\ J$$

 

**63. C is correct.**

The equation relates the entropy (*S*) change at different temperatures.

Entropy increases at higher temperatures (i.e., increased kinetic energy).

**64. B is correct.**

A change is exothermic when energy is released to the surroundings.

Energy loss occurs when a substance changes to a more rigid phase (e.g., gas → liquid, or liquid → solid).

**65. A is correct.**

| Potential Energy | Kinetic Energy |
|---|---|
| **Stored energy and the energy of position (gravitational)** | **Energy of motion: motion of waves, electrons, atoms, molecules, and substances** |
| **Chemical Energy** | **Electrical Energy** |
| Chemical energy is the energy stored in the bonds of atoms and molecules. Examples of stored chemical energy: biomass, petroleum, natural gas, propane, and coal. | Electrical energy is the movement of electrons. Lightning and electricity are examples of electrical energy. |

**66. B is correct.**

There is no potential at equilibrium, and therefore neither reaction direction is favored.

Reaction potential (*E*) indicates the tendency of a reaction to be spontaneous in either direction.

Because the solution is in equilibrium, no further reactions occur in either direction. Therefore, $E = 0$.

**67. D is correct.**

Increased entropy (i.e., disorder) accompanies transformation from solid to gas.

Gas molecules have greater kinetic energy than the molecules of solids.

**68. C is correct.**

Reaction is nonspontaneous (i.e., endergonic) if products are less stable than reactants and $\Delta G$ is positive.

Reaction is spontaneous (i.e., exergonic) if the products are more stable than the reactants and $\Delta G$ is negative.

**69. A is correct.**

Spontaneity of a reaction is determined by evaluating the Gibbs free energy, or $\Delta G$.

$$\Delta G = \Delta H - T\Delta S$$

Reaction is spontaneous if $\Delta G < 0$ (or $\Delta G$ is negative).

For $\Delta G$ to be negative, $\Delta H$ must be less than $T\Delta S$.

Reaction is nonspontaneous if $\Delta G > 0$ (or $\Delta G$ is positive and $\Delta H$ is greater than $T\Delta S$).

**70. B is correct.**

Reaction is nonspontaneous (i.e., endergonic) if products are less stable than reactants and $\Delta G$ is positive.

Reaction is spontaneous (i.e., exergonic) if products are more stable than the reactants and $\Delta G$ is negative.

Entropy ($\Delta S$) is determined by the relative number of molecules (compounds):

Reactants have one molecule (with a coefficient total of 2). The products have two molecules (with a coefficient total of 7).

Number of molecules has increased from reactants to products. Therefore, the entropy (i.e., disorder) has increased.

**71. D is correct.**

$\Delta G$ indicates the energy available for non P–V work. This is useful for processes like batteries and living cells that do not have expanding gases.

 

**72. C is correct.**

*Internal energy* of a system is the energy contained within the system, excluding the kinetic energy of motion of the system as a whole and the potential energy of the system as a whole due to external force fields.

Internal energy of a system can be changed by transfers of matter and by work and heat transfer. When impermeable container walls prevent matter transfer, the system is said to be closed.

*First law of thermodynamics* states that the increase in internal energy equals the heat added plus the work done on the system by its surroundings. If the container walls do not pass energy or matter, the system is isolated, and its internal energy remains constant.

Bond energy is internal potential energy (PE), while thermal energy is internal kinetic energy (KE).

**73. D is correct.**

Endothermic reactions absorb energy to break strong bonds to form a less stable state (i.e., positive enthalpy).
Exothermic reactions release energy when forming stronger bonds to produce a more stable state (i.e., negative enthalpy).

To predict spontaneity of the reaction, use the Gibbs free energy equation:

$$\Delta G = \Delta H° - T\Delta S$$

Reaction is nonspontaneous (i.e., endergonic) if the products are less stable than the reactants and $\Delta G$ is positive.

Reaction is spontaneous (i.e., exergonic) if the products are more stable than the reactants and $\Delta G$ is negative.

It is the amount of energy that is determinative. Some bonds are stronger than others, so there is a net gain or net energy loss when formed.

**74. A is correct.**

Starting material is ice at −25 °C, so its temperature must be raised to 0 °C before melting.

The first term in the heat calculation is:

specific heat of ice (A) × mass × change of temperature

**75. C is correct.**

Heat of formation of water vapor is like the highly exothermic process for the heat of hydrogen combustion.

Heat is required to convert $H_2O$ (*l*) into vapor; therefore, the heat of formation of water vapor must be less exothermic than $H_2O$ (*l*).

Solid → liquid → gas: endothermic because the reaction absorbs energy to break strong bonds to form a less stable state (i.e., positive enthalpy).

Gas → liquid → solid: exothermic because the reaction releases energy when forming stronger bonds to produce a more stable state (i.e., negative enthalpy).

**76. D is correct.**

**77. B is correct.**

Einstein established the equivalence of mass and energy; the energy equivalent of mass is given by $E = mc^2$.

Energy can be converted to mass, and mass can be converted to energy if the total energy, including mass-energy, is conserved.

In nuclear fission, the mass of the parent nucleus is greater than the sum of the masses of the daughter nuclei. This deficit in mass is converted into energy.

**78. D is correct.**

Closed system can exchange energy (as heat or work) but *not matter* with its surroundings.

Isolated system *cannot* exchange energy (as heat or work) or matter with the surroundings.

Open system *can* exchange energy and matter with the surroundings.

**79. A is correct.**

Gases have the highest entropy (i.e., disorder or randomness).

Solutions have higher entropy than pure phases since the molecules are more scattered.

Liquid phase has less entropy than the aqueous phase.

**80. A is correct.**

In a chemical reaction, bonds within reactants are broken, forming new bonds.

Therefore, in bond dissociation problems:

$\Delta H_{reaction}$ = sum of bond energy in reactants – the sum of bond energy in products.

This is the opposite of $\Delta H_f$ problems, where:

$\Delta H_{reaction}$ = sum of $\Delta H_f$ product – sum of $\Delta H_f$ reactant

For $N_2 + 3\ H_2 \rightarrow 2\ NH_3$

$\Delta H_{reaction}$ = sum of bond energy in reactants – sum of bond energy in products

$\Delta H_{reaction} = [(N{\equiv}N) + 3(H{-}H)] - [2 \times 3(N{-}H)]$

$\Delta H_{reaction} = [(946\ kJ) + (3 \times 436\ kJ) - (2 \times (3 \times 389\ kJ)]$

$\Delta H_{reaction} = -80\ kJ$

*Notes for active learning*

 

# 3 – Thermodynamics: Detailed Explanations

==========================================================================

**Practice Set 1: Questions 1–20**

==========================================================================

**1. B is correct.**

Ideal gas law:

$PV = nRT$

$P_0 = nRT / V_0$

If isothermal expansion, then n, R and T are constant

$P = nRT / (1/3\ V_0)$

$P = 3(nRT / V_0)$

$P = 3P_0$

**2. C is correct.**

Area expansion equation:

$\Delta A = A_0(2\alpha\Delta T)$

$\Delta A = (\pi / 4)\cdot(1.2\text{ cm})^2\cdot(2)\cdot(19 \times 10^{-6}\text{ K}^{-1})\cdot(200\text{ °C})$

$\Delta A = 8.6 \times 10^{-3}\text{ cm}^2$

$\Delta A = A_f - A_0$

$8.6 \times 10^{-3}\text{ cm}^2 = (\pi / 4)\cdot[d_f^2 - (1.2\text{ cm})^2]$

$d_f = 1.2\text{ cm}$

**3. D is correct.**

$1\text{ Watt} = 1\text{ J/s}$

$\text{Power} \times \text{Time} = Q$

*continued...*

Heat needed to change the temperature of a mass:

$Q = mc\Delta T$

$P \times t = mc\Delta T$

$t = (mc\Delta T) / P$

$t = (90 \text{ g})\cdot(4.186 \text{ J/g}\cdot°\text{C})\cdot(30 °\text{C} - 10 °\text{C}) / (50 \text{ W})$

$t = 151 \text{ s}$

**4. A is correct.**

Convert 15 minutes to seconds:

$t = (15 \text{ min}/1)\cdot(60 \text{ s}/1 \text{ min})$

$t = 900 \text{ s}$

Find total energy generated:

$Q = P \times t$

$Q = (1{,}260 \text{ J/s})\cdot(900 \text{ s})$

$Q = 1{,}134 \text{ kJ}$

Find mass of water needed to carry away energy:

$Q = mL_v$

$m = Q / L_v$

$m = (1{,}134 \text{ kJ}) / (22.6 \times 10^2 \text{ kJ/kg})$

$m = 0.5 \text{ kg} = 500 \text{ g}$

**5. A is correct.**

Coefficient of performance assuming ideal Carnot cycle:

$C_p = Q_H / W$

$C_p = T_H / (T_H - T_C)$

$Q_H / W = T_H / (T_H - T_C)$

$W = Q_H \cdot (T_H - T_C) / T_H$

$W = (32 \times 10^3 \text{ J/s})\cdot(293 \text{ K} - 237 \text{ K}) / (293 \text{ K})$

$W = 6{,}116 \text{ J/s} \approx 6{,}100 \text{ W}$

 

**6. B is correct.**

The amount of energy needed to melt a sample of mass *m* is:

$$Q = m\,L_f$$

where $L_f$ is the latent heat of fusion.

$$Q = (55\text{ kg})\cdot(334\text{ kJ/kg})$$

$$Q = 1.8 \times 10^4\text{ kJ}$$

**7. D is correct.**

Metals are good heat and electrical conductors because of their bonding structure.

In metallic bonding, the outer electrons are held loosely and can travel freely.

Electricity and heat require high electron mobility.

Thus, the looseness of the outer electrons in the materials allows them to be excellent conductors.

**8. A is correct.**

Find the heat of phase change from steam to liquid:

$$Q_1 = mL_v$$

Find the heat of phase change from liquid to solid:

$$Q_2 = mL_f$$

Find the heat of temperature from 100 °C to 0 °C:

$$Q_3 = mc\Delta T$$

Total heat:

$$Q_{net} = Q_1 + Q_2 + Q_3$$

$$Q_{net} = mL_v + mL_f + mc\Delta T$$

To find mass:

$$Q_{net} = m(L_v + c\Delta T + L_f)$$

$$m = Q_{net} / (L_v + c\Delta T + L_f)$$

*continued...*

Solve:

$Q_{net} = 200$ kJ

$Q_{net} = 2 \times 10^5$ J

$m = (2 \times 10^5 \text{ J}) / [(22.6 \times 10^5 \text{ J/kg}) + (4{,}186 \text{ J/kg·K})·(100 \text{ °C} – 0 \text{ °C}) + (33.5 \times 10^4 \text{ J/kg})]$

$m = 0.066$ kg

**9. C is correct.**

*Fusion* is the process whereby a substance changes from a solid to liquid (i.e., melting).

*Condensation* is the process whereby a substance changes from vapor to liquid.

*Sublimation* is when a substance changes directly from a solid to the gas phase without passing through the liquid phase.

**10. D is correct.**

$Q = mc\Delta T$

$Q = (0.2 \text{ kg})·(14.3 \text{ J/g·K})·(1{,}000 \text{ g/kg})·(280 \text{ K} – 250 \text{ K})$

$Q = 86{,}000 \text{ J} = 86 \text{ kJ}$

**11. B is correct.**

The heat needed to raise the temperature of aluminum:

$Q_A = m_A c_A \Delta T$

Heat needed to raise temperature of water:

$Q_W = m_W c_W \Delta T$

Total heat to raise temperature of system:

$Q_{net} = Q_A + Q_W$

$Q_{net} = m_A c_A \Delta T + m_W c_W \Delta T$

$Q_{net} = \Delta T(m_A c_A + m_W c_W)$

$Q_{net} = (98 \text{ °C} – 18 \text{ °C})·[(0.5 \text{ kg})·(900 \text{ J/kg·K}) + (1 \text{ kg})·(4{,}186 \text{ J/kg·K})]$

$Q_{net} = 370{,}880$ J

*continued…*

Time to produce $Q_{net}$ with 500 W:

$Q_{net} = (500\text{ W})t$

$t = Q_{net} / (500\text{ W})$

$t = (370{,}880\text{ J}) / 500\text{ W}$

$t = 741.8\text{ s}$

Convert to minutes:

$t = (741.8\text{ s}/1)\cdot(1\text{ min}/60\text{ s})$

$t = 12.4\text{ min} \approx 12\text{ min}$

**12. A is correct.**

When a substance goes through a phase change, the temperature does not change.

It can be assumed that the lower plateau is $L_f$ and the upper plateau is $L_v$.

Count the columns: $L_f = 2$, $L_v = 7$

$L_v / L_f = 7 / 2$

$L_v / L_f = 3.5$

**13. B is correct.**

Specific heat is the amount of heat (i.e., energy) needed to raise the temperature of the unit mass of a substance by a given amount (usually one degree).

**14. D is correct.**

Find ½ of KE of the BB:

$\text{KE} = ½mv^2$

$½\text{KE} = ½(½mv^2)$

$½\text{KE}_{BB} = ½(½)\cdot(0.0045\text{ kg})\cdot(46\text{ m/s})^2$

$½\text{KE}_{BB} = 2.38\text{ J}$

The $½\text{KE}_{BB}$ is equal to energy taken to change temperature:

$Q = ½\text{KE}_{BB}$

$Q = mc\Delta T$

*continued...*

$mc\Delta T = \frac{1}{2}KE_{BB}$

$\Delta T = \frac{1}{2}KE_{BB} / mc$

Calculate to find $\Delta T$:

$\Delta T = (2.38 \text{ J}) / (0.0045 \text{ kg})\cdot(128 \text{ J/kg}\cdot\text{K})$

$\Delta T = 4.1 \text{ K}$

**15. C is correct.**

*Vaporization* is when a substance changes from a liquid to a gas; the process can be either boiling or evaporation.

*Sublimation* is the process whereby a substance changes from a solid to a gas.

**16. A is correct.**

Carnot efficiency:

$\eta$ = work done / total energy

$\eta = W / Q_H$

$\eta = 5 \text{ J} / 18 \text{ J}$

$\eta = 0.28$

The engine's efficiency:

$\eta = (T_H - T_C) / T_H$

$0.28 = (233 \text{ K} - T_C) / 233 \text{ K}$

$(0.28)\cdot(233 \text{ K}) = (233 \text{ K} - T_C)$

$65.2 \text{ K} = 233 \text{ K} - T_C$

$T_C = 168 \text{ K}$

 

**17. D is correct.**

The heat needed to change the temperature of a mass:

$$Q = mc\Delta T$$

Calculate to find $Q$:

$$Q = (0.92\text{ kg})\cdot(113\text{ cal/kg}\cdot°\text{C})\cdot(96\text{ °C} - 18\text{ °C})$$

$$Q = 8{,}108.9\text{ cal}$$

Convert to joules:

$$Q = (8{,}108.9\text{ cal})\cdot(4.186\text{ J/cal})$$

$$Q = 33{,}940\text{ J}$$

**18. B is correct.**

During a change of state, the addition of heat does not change the temperature (i.e., a measure of the kinetic energy).

The heat energy adds to the potential energy of the substance until the substance completely changes state.

**19. C is correct.**

The work done for a cyclic process in gas is equal to the area enclosed by the cyclic process. (Use P for pressure and V for volume on the graph).

**20. A is correct.**

Specific heat of A is larger than B:

$$c_A > c_B$$

Energy to raise the temperature:

$$Q = mc\Delta T$$

If $m$ and $\Delta T$ are equal for A and B:

$$Q_A = m_A c_A \Delta T_A$$

$$Q_B = m_B c_B \Delta T_B$$

$$Q_A > Q_B$$

This is valid because other factors are equal, and the magnitude of $Q$ only depends on $c$.

 

==================================================================

**Practice Set 2: Questions 21–40**

==================================================================

**21. D is correct.**

Find kinetic energy of meteor:

$KE = ½mv^2$

$KE = ½(0.0065\ kg)·(300\ m/s)^2$

$KE = 292.5\ J$

Find temperature rise:

$Q = KE$

$Q = mc\Delta T$

$mc\Delta T = KE$

$\Delta T = KE / mc$

Convert KE to calories:

$KE = (292.5\ J/1)·(1\ cal/4.186\ J) = 69.9\ cal$

Calculate ΔT:

$\Delta T = (69.9\ cal) / [(0.0065\ kg)·(120\ cal/kg·°C)] = 89.6\ °C \approx 90\ °C$

**22. A is correct.**

When a liquid freezes, it undergoes a phase change from liquid to solid.

For this to occur, heat energy must be dissipated (removed).

During any phase change, the temperature remains constant.

**23. B is correct.**

For an isothermal process:

$\Delta U = 0$

$\Delta U = Q - W$

$Q = W$

*continued...*

Work to expand an ideal gas in an isothermal process:

 

$W = nRT \ln(V_f / V_i)$

From the ideal gas law, $nRT = P_fV_f$, giving:

$W = P_fV_f \ln(V_f / V_i)$

$W = (130 \text{ kPa})\cdot(0.2 \text{ m}^3) \ln[(0.2 \text{ m}^3) / (0.05 \text{ m}^3)]$

$W = 36 \text{ kJ}$

$Q = W$

$Q = 36 \text{ kJ}$

Since the process is isothermal, there is no change in the internal energy. Since the surroundings are doing negative work on the system:

$\Delta E = 0 = Q + W = Q - 36 \text{ kJ}$

Therefore:

$Q = 36 \text{ kJ}$

**24. D is correct.**

Find the potential energy of 1 kg of water:

$PE = mgh$

$PE = (1 \text{ kg})\cdot(9.8 \text{ m/s}^2)\cdot(30 \text{ m})$

$PE = 294 \text{ J}$

Assume all potential energy is converted to heat for maximum temperature increase:

$PE = Q$

$Q = mc\Delta T$

$mc\Delta T = PE$

$\Delta T = PE / mc$

$\Delta T = (294 \text{ J}) / [(1 \text{ kg})\cdot(4{,}186 \text{ J/kg/K})]$

$\Delta T = 0.07 \text{ °C}$

It is not necessary to convert to Kelvin for temperature differences because a temperature change in Kelvin is equal to a temperature change in Celsius.

**25. A is correct.**

Find heat from phase change:

$Q = mL_f$

$Q = (0.75 \text{ kg})\cdot(33{,}400 \text{ J/kg})$

$Q = 25{,}050 \text{ J}$

Because the water is freezing, $Q$ should be negative due to heat being released.

$Q = -25{,}050 \text{ J}$

Find the change in entropy:

$\Delta S = Q / \text{T}$

$\Delta S = -25{,}050 \text{ J} / (0 \text{ °C} + 273 \text{ K})$

$\Delta S = -92 \text{ J} / \text{K}$

A negative change in entropy indicates that the disorder of the isolated system has decreased.

When water freezes, the entropy is negative because water is more disordered than ice.

Thus, the disorder has decreased.

**26. A is correct.**

Copper has a larger linear expansion coefficient than iron, so it expands more than iron during a given temperature change.

The bimetallic bar bends due to the difference in expansion between the copper and iron.

**27. D is correct.**

Calculate heat needed to raise temperature:

$Q = mc\Delta \text{T}$

$Q = (0.110 \text{ kg})\cdot(4{,}186 \text{ J/kg}\cdot\text{K})\cdot(30 \text{ °C} - 20 \text{ °C})$

$Q = 4{,}605 \text{ J}$

Calculate time needed to raise temperature with 60 W power source:

$Q = P \times t$

$Q = (60 \text{ W})t$

$t = \text{Q} / (60 \text{ W})$

*continued...*

 

$t = (4{,}605 \text{ J}) / (60 \text{ W})$

$t = 77 \text{ s}$

**28. B is correct.**

During a state change, the addition of heat does not change the temperature (i.e., a measure of the kinetic energy).

The heat energy only adds to the potential energy of the substance until the substance completely changes state.

**29. C is correct.**

Convert to Kelvin:

$T = -243 \text{ °C} + 273$

$T = 30 \text{ K}$

Double temperature:

$T_2 = (30 \text{ K})\cdot(2)$

$T_2 = 60 \text{ K}$

Convert back to Celsius:

$T_2 = 60 \text{ K} - 273$

$T_2 = -213 \text{ °C}$

**30. D is correct.**

Carnot efficiency engines can be written as:

$\eta = 1 - T_C / T_H$

$\eta = 1 - | Q_C / Q_H |$

Thus:

$Q_C / Q_H = T_C / T_H$

**31. B is correct.**

Convert units:

$$1.7 \times 10^5 \text{ J/kg} = 170 \text{ kJ/kg}$$

Change in internal energy = heat added ($Q$)

$$Q = mL_v$$

$$Q = (1 \text{ kg})\cdot(170 \text{ kJ/kg})$$

$$Q = 170 \text{ kJ}$$

**32. D is correct.**

$Q = mc\Delta T$

If $m$ and $c$ are constant, the relationship is directly proportional.

To double $Q$, T must be doubled:

$$5 \text{ C} + 273 = 278 \text{ K}$$

$$278 \text{ K} \times 2 = 556 \text{ K}$$

$$556 \text{ K} - 273 = 283 \text{ C}$$

**33. B is correct.**

The mass of each is required to determine which object experiences the greater temperature change when the system takes to reach thermal equilibrium.

**34. B is correct.**

Body heat gives energy to the water molecules in the sweat. This energy is transferred via collisions until some molecules have enough energy to break the hydrogen bonds and escape the liquid (i.e., evaporation).

However, if a body stayed dry, the heat would not be given to the water, and the person would stay hot because the heat is not lost due to the evaporation of the water.

 

**35. D is correct.**

Calculate heat released when 0 °C water converts to 0 °C ice:

$Q_1 = mL_f$

$Q_1 = (2{,}200 \text{ kg})\cdot(334 \times 10^3 \text{ J/kg})$

$Q_1 = 734{,}800 \text{ kJ}$

Calculate heat released for temperature drop ΔT

$Q_2 = mc\Delta T$

$Q_2 = (2{,}200 \text{ kg})\cdot(2{,}050 \text{ J/kg K})\cdot[(0\ °C - (-30\ °C)]$

ΔK = Δ°C, so units cancel:

$Q_2 = 135{,}300 \text{ kJ}$

Add heat released to get $Q_{net}$:

$Q_{net} = Q_1 + Q_2$

$Q_{net} = (734{,}800 \text{ kJ}) + (135{,}300 \text{ kJ})$

$Q_{net} = 870{,}100 \text{ kJ}$

**36. A is correct.**

Object 1 has three times the specific heat capacity and four times the mass of Object 2:

$c_1 = 3c_2;\ m_1 = 4m_2$

A single-phase substance obeys the specific heat equation:

$Q = mc\Delta T$

In this case, the same amount of heat is added to each substance, therefore:

$Q_1 = Q_2$

$m_1c_1\Delta T_1 = m_2c_2\Delta T_2$

$(4m_2)(3c_2)\Delta T_1 = m_2\,c_2\Delta T_2$

$12m_2c_2\Delta T_1 = m_2c_2\Delta T_2$

$12\ \Delta T_1 = \Delta T_2$

**37. C is correct.**

Find heat from phase change:

$$Q = mL_f$$

$$Q = (8\text{ kg})\cdot(80\text{ kcal/kg})$$

$$Q = 640\text{ kcal/kg}$$

Because the water is freezing, $Q$ should be negative due to heat being released.

$$Q = -640\text{ kcal/kg}$$

Find the change in entropy:

$$\Delta S = Q / \text{T}$$

$$\Delta S = -640\text{ kcal/kg} / (0\text{ °C} + 273\text{ K})$$

$$\Delta S = -2.3\text{ kcal/K}$$

A negative change in entropy indicates that the disorder of the isolated system has decreased.

When water freezes, the entropy is negative because water is more disordered than ice.

Thus, the disorder has decreased, and entropy is negative.

**38. A is correct.**

From the ideal gas law:

$$p_3V_3 = nRT_3$$

$$T_3 = p_3V_3 / nR$$

$$T_3 = 1.5p_1V_3 / nR$$

$$T_3 = 1.5V_3(p_1 / nR)$$

Also from the ideal gas law:

$$(p_1 / nR) = T_1 / V_1$$

$$(p_1 / nR) = (293.2\text{ K}) / (100\text{ cm}^3)$$

$$(p_1 / nR) = 2.932\text{ K/cm}^3$$

*continued...*

Calculate $T_3$:

$T_3 = 1.5V_3\ (2.932\ \text{K/cm}^3)$

$T_3 = 1.5\ (50\ \text{cm}^3)\cdot(2.932\ \text{K/cm}^3)$

$T_3 = 219.9\ \text{K}$

$T_3 = -53.3\ °\text{C} \approx -53\ °\text{C}$

Calculate $T_4$:

$T_4 = 1.5V_4\ (2.932\ \text{K/cm}^3)$

$T_4 = 1.5\ (150\ \text{cm}^3)\cdot(2.932\ \text{K/cm}^3)$

$T_4 = 659.6\ \text{K}$

$T_4 = 386.5\ °\text{C} \approx 387\ °\text{C}$

**39. B is correct.**

Steel is a very conductive material that can transfer thermal energy very well. The steel feels colder than the plastic because its higher thermal conductivity allows it to remove more heat and thus makes touching it feel colder.

**40. B is correct.**

An isobaric process involves constant pressure.

An isochoric (also isometric) process involves a closed system at constant volume.

An adiabatic process occurs without the transfer of heat or matter between a system and its surroundings.

An isothermal process involves the change of a system in which the temperature remains constant.

===============================================================

**Practice Set 3: Questions 41–60**

===============================================================

**41. D is correct.**

Carnot coefficient of performance of a refrigeration cycle:

$C_P = T_C / (T_H - T_C)$

$C_P = Q_C / W$

$Q_C / W = T_C / (T_H - T_C)$

$W = (Q_C / T_C) \cdot (T_H - T_C)$

$W = (20 \times 10^3 \text{ J} / 293 \text{ K}) \cdot (307 \text{ K} - 293 \text{ K})$

$W = 955.6 \text{ J} = 0.956 \text{ kJ}$

Power = Work / time

$P = W / t$

$P = 0.956 \text{ kJ} / 1 \text{ s}$

$P = 0.956 \text{ kW} \approx 0.96 \text{ kW}$

**42. C is correct.**

*Heat energy* is measured in units of Joules and calories.

**43. B is correct.**

*Convection* is a form of heat transfer in which mass motion of a fluid (i.e., liquids and gases) transfers energy from the heat source.

**44. A is correct.**

$Q_H = 515 \text{ J}$

$Q_C = 340 \text{ J}$

Carnot cycle efficiency:

$\eta = (Q_H - Q_C) / Q_H$

$\eta = (515 \text{ J} - 340 \text{ J}) / (515 \text{ J})$

$\eta = 0.34 = 34\%$

 

**45. D is correct.**

Convert $P_3$ to Pascals:

$P_3 = (2\text{ atm} / 1)\cdot(101{,}325\text{ Pa} / 1\text{ atm})$

$P_3 = 202{,}650\text{ Pa}$

Use the ideal gas law to find $V_3$:

$PV = nRT$

$V = (nRT) / P$

$V_3 = [(0.008\text{ mol})\cdot(8.314\text{ J/mol}\cdot\text{K})\cdot(2{,}438\text{ K})] / (202{,}650\text{ Pa})$

$V_3 = 8 \times 10^{-4}\text{ m}^3$

Convert to $cm^3$:

$V_3 = (8 \times 10^{-4}\text{ m}^3/1)\cdot(100^3\text{ cm}^3/1\text{ m}^3)$

$V_3 = 800\text{ cm}^3$

**46. C is correct.**

An *adiabatic process* involves no heat added or removed from the system.

From the First Law of Thermodynamics:

$\Delta U = Q + W$

If $Q = 0$, then:

$\Delta U = W$

Because work is being done to expand the gas, it is considered negative, and then the change in internal energy is negative (decreases).

$-\Delta U = -W$

**47. B is correct.**

Standing in a breeze while wet feels colder than when dry because of the evaporation of water off the skin.

Water requires heat to evaporate, so this is taken from the body, making a person feel colder than if they were dry and the evaporation did not occur.

**48. A is correct.**

Conduction is a form of heat transfer in which the collisions of the molecules of the material transfer energy through the material.

Higher temperature of the material causes the molecules to collide with more energy which eventually is transferred throughout the material through subsequent collisions.

**49. D is correct.**

An *isobaric process* is a constant pressure process, so the resulting pressure is always the same.

**50. B is correct.**

This question asks which type of surface has a higher emissivity than others and can radiate more energy over a set period.

A *blackbody* is an idealized radiator and has the highest emissivity.

A surface most similar to a blackbody (the black surface) is the best radiator of thermal energy.

A black surface is considered an ideal blackbody and therefore has an emissivity of 1 (perfect emissivity).

The black surface will be the best radiator compared to another surface that cannot be considered blackbodies and emissivity of $<1$.

**51. C is correct.**

Calculate the gap between the rods:

The gap in between the rods will be filled by both expanding, so total thermal expansion equals 1.1 cm.

$$\Delta L = L_0 \alpha \Delta T$$

$$\Delta L_{tot} = \Delta L_B + \Delta L_A$$

$$\Delta L_{tot} = (\alpha_B L_B + \alpha_A L_A)\ \Delta T$$

Rearrange the equation for $\Delta T$:

$$\Delta T = \Delta L_{tot} / (\alpha_B L_B + \alpha_A L_A)$$

$$\Delta T = 1.1\ \text{cm} / [(2 \times 10^{-5}\ \text{K}^{-1})\cdot(59.1\ \text{cm}) + (2.4 \times 10^{-5}\ \text{K}^{-1})\cdot(39.3\ \text{cm})]$$

$$\Delta T = 517.6\ \text{K} \approx 518\ \text{K}$$

*continued...*

Measuring the difference in temperature in K is the same as in °C, so it is not required to convert:

$\Delta T = 518\ °C$

**52. A is correct.**

Find seconds in a day:

$t = (24\ h / 1\ day)\cdot(60\ min / 1\ h)\cdot(60\ s / 1\ min)]$

$t = 86{,}400\ s$

Find energy lost in a day:

$E = \text{Power} \times \text{time}$

$E = (60\ W)\cdot(86{,}400\ s)$

$E = 5{,}184{,}000\ J$

Convert to kcal:

$E = (5{,}184{,}000\ J/1)\cdot(1\ cal/4.186\ J)\cdot(1\ kcal/10^3\ cal)$

$E = 1{,}240\ kcal$

**53. D is correct.**

Conduction is a form of heat transfer in which the collisions of the molecules of the material transfer energy through the material.

Higher temperature of the material causes the molecules to collide with more energy which eventually is transferred throughout the material through subsequent collisions.

**54. B is correct.**

Heat given off by warmer water is equal to that absorbed by the frozen cube. This heat is split into heat needed to melt the cube and bring the temperature to equilibrium.

$Q_{H2O,1} + Q_{alcohol,Temp1} + Q_{alcohol,Phase1} = 0$

$(mc\Delta T)_{H2O,1} + (mc\Delta T)_{alcohol,Temp1} + (mL_f)_{alcohol} = 0$

$Q_{H2O,2} + Q_{alcohol,Temp2} + Q_{alcohol,Phase2} = 0$

$(mc\Delta T)_{H2O,2} + (mc\Delta T)_{alcohol,Temp2} + (mL_f)_{alcohol} = 0$

*continued...*

Set equal to each other to cancel heat from phase change (since they are equal):

$$(mc\Delta T)_{H2O,1} + (mc\Delta T)_{alcohol,Temp1} = (mc\Delta T)_{H2O,2} + (mc\Delta T)_{alcohol,Temp2}$$

$$(m_{alcohol})(c_{alcohol})\cdot(\Delta T_{alcohol1} - \Delta T_{alcohol2}) = c_{H2O}(m\Delta T_{H2O,2} - m\Delta T_{H2O,1})$$

$$c_{alcohol} = (c_{H2O} / m_{alcohol})\cdot[(m\Delta T_{H2O,2} - m\Delta T_{H2O,1}) / (\Delta T_{alcohol1} - \Delta T_{alcohol2})]$$

Solving for $c_{alcohol}$:

$$c_{alc} = [(4{,}190 \text{ J/kg·K}) / (0.22 \text{ kg})]\cdot[(0.4 \text{ kg})\cdot(10 - 30 \text{ °C}) - (0.35 \text{ kg})\cdot(5 - 26 \text{ °C})]$$

$$/ [(5 \text{ °C} - (-10 \text{ °C}) - (10 \text{ °C} - (-10 \text{ °C})]$$

$$c_{alc} = (19{,}045 \text{ J/kg·K})\cdot[(-0.65 \text{ °C}) / (-5 \text{ °C})]$$

$$c_{alc} = 2{,}475 \text{ J/kg·K}$$

**55. D is correct.**

Using the calculated value for $c_{alc}$ in the problem above:

$$Q_{H2O,1} + Q_{alcohol,Temp1} + Q_{alcohol,Phase1} = 0$$

$$(mc\Delta T)_{H2O,1} + (mc\Delta T)_{alcohol} + (mL_f)_{alcohol} = 0$$

$$L_{f\,alcohol} = [-(mc\Delta T)_{H2O,1} - (mc\Delta T)_{alcohol}] / m_{alcohol}$$

Solve:

$L_{f\,alcohol} = -[(0.35 \text{ kg})\cdot(4{,}190 \text{ J/kg·K})\cdot(5 \text{ °C} - 26 \text{ °C}) - (0.22 \text{ kg})\cdot(2{,}475 \text{ J/kg·K})\cdot( 5 \text{ °C} - (-10 \text{ °C})] / (0.22 \text{ kg})$

$$L_{f\,alcohol} = (30{,}796.5 \text{ J} - 8{,}167.5 \text{ J}) / (0.22 \text{ kg})$$

$$L_{f\,alcohol} = 103 \times 10^3 \text{ J/kg} = 10.3 \times 10^4 \text{ J/kg}$$

**56. C is correct.**

The silver coating reflects thermal radiation into the bottle to reduce heat loss by radiation.

Radiation is a form of heat transfer in which electromagnetic waves carry energy from an emitting object and deposit the energy in an object absorbing the radiation.

 

**57. B is correct.**

Convert calories to Joules:

$E = (16\ \text{kcal}/1)\cdot(10^3\ \text{cal}/1\ \text{kcal})\cdot(4.186\ \text{J}/1\ \text{cal})$

$E = 66{,}976\ \text{J}$

Convert hours to seconds:

$t = (5\ \text{h})\cdot(60\ \text{min/h})\cdot(60\ \text{s/min})$

$t = 18{,}000\ \text{s}$

Find power expended:

$P = E\ /\ t$

$P = (66{,}976\ \text{J})\ /\ (18{,}000\ \text{s})$

$P = 3.7\ \text{W}$

**58. A is correct.**

Specific heat capacity:

$\Delta Q = cm\Delta \text{T}$

$\Delta \text{T} = \Delta Q\ /\ cm$

$\Delta \text{T} = (50\ \text{kcal})\ /\ [(1\ \text{kcal/kg}\cdot{}^\circ\text{C})\cdot(5\ \text{kg})]$

$\Delta \text{T} = 10\ {}^\circ\text{C}$

**59. C is correct.**

For melting, use $L_f$:

$Q = mL_f$

$Q = (30\ \text{kg})\cdot(334\ \text{kJ/kg})$

$Q = 1 \times 10^4\ \text{kJ}$

**60. B is correct.**

Find temperature change:

$Q = mc\Delta T$

$\Delta T = Q / mc$

$\Delta T = (160 \times 10^3 \text{ J}) / [(6 \text{ kg})\cdot(910 \text{ J/kg}\cdot\text{K})]$

$\Delta T = 29 \text{ °C}$

Find final temperature:

$\Delta T = T_f - T_i$

$T_f = T_i + \Delta T$

$T_f = 12 \text{ °C} + 29 \text{ °C}$

$T_f = 41 \text{ °C}$

 

===============================================================

**Practice Set 4: Questions 61–80**

===============================================================

**61. A is correct.**

After 5 minutes, the sample is at a constant temperature indicating a phase change from solid to liquid (a mixture of solid and liquid).

B: heat capacity of the liquid phase is much greater than that of the solid.

C: the sample begins to boil at the second plateau, representing the liquid to gas phase change.

D: the heat of fusion is less than the heat of vaporization.

**62. D is correct.**

Warm air is less dense than cold air. It follows that a mass of warm air is subject to buoyancy force, which is due to the density gradient in the cold air.

The direction of the buoyancy force is toward lower density, and air with lower density also has lower pressure.

**63. B is correct.**

Heat needed to raise temperature of iron:

$$Q_I = m_I c_I \Delta T$$

Heat needed to raise temperature of copper:

$$Q_C = m_C c_C \Delta T$$

Total heat needed for system:

$$Q_{net} = Q_I + Q_C$$

Calculate to find $Q_{net}$:

$$Q_{net} = m_I c_I \Delta T + m_C c_C \Delta T$$

$$Q_{net} = \Delta T(m_I c_I + m_C c_C)$$

$$Q_{net} = (58\ °C - 23\ °C)\ [(0.15\ kg)\cdot(470\ J/kg\cdot K) + (0.2\ kg)\cdot(390\ J/kg\cdot K)]$$

$$Q_{net} = 5{,}198\ J$$

**64. C is correct.**

$\Delta Q = cm\Delta T$

$\Delta Q = (0.11\ \text{kcal/kg·°C})·(0.3\ \text{kg})·(20\ °\text{C})$

$\Delta Q = 0.66\ \text{kcal} = 660\ \text{cal}$

**65. B is correct.**

Heat is the total energy of molecular motion in a substance, while temperature is the average energy of molecular motion.

The amount of heat is dependent on mass.

The larger container has more heat than the smaller one since it is larger in mass with more total thermal energy.

**66. D is correct.**

Find mass of water:

$V = (200\ \text{L})·(0.001\ \text{m}^3/\text{L})$

$V = 0.2\ \text{m}^3$

density = mass / volume

$m = V\rho$

$m = (0.2\ \text{m}^3)·(1{,}000\ \text{kg/m}^3)$

$m = 200\ \text{kg}$

Find heat added to raise temperature:

$Q = mc\Delta T$

$Q = (200\ \text{kg})·(4{,}186\ \text{J/kg·K})·(80\ °\text{C} - 28\ °\text{C})$

$Q = 43534.4\ \text{kJ}$

Find time needed to raise temperature:

$4\ \text{kW} = Q / t$

$t = Q / (4\ \text{kW})$

$t = 43{,}534.4\ \text{kJ} / 4\ \text{kW}$

$t = 10{,}884\ \text{s}$

*continued...*

 

Convert to hours:

$t = (10{,}884 \text{ s}) / (60 \text{ s}) / (60 \text{ min})$

$t = 3$ hours

**67. A is correct.**

The First Law of Thermodynamics states:

$\Delta U = Q + W$

This is equivalent to the Law of Conservation of Energy because the change in the internal energy of a system ($\Delta U$) is equal to the energy added as heat and work ($Q + W$).

**68. C is correct.**

Heat flux lost through conduction is:

$Q / t = kA(T_H - T_C) / d$

where $k$ = thermal conductivity and $t$ = time

$Q / t = [(0.8 \text{ W/m·°C})·(3.1 \text{ m} \times 1.8 \text{ m})·(22 \text{ °C} - 7 \text{ °C})] / (0.009 \text{ m})$

$Q / t = 7{,}440 \text{ J/s}$

To find heat energy lost in an hour, multiply by seconds in an hour:

$Q_{total} = (7{,}440 \text{ J/s})·(3{,}600 \text{ s/1h})$

$Q_{total} = 2.7 \times 10^7 \text{ J}$

**69. D is correct.**

A temperature difference is a difference in thermal energy. Heat always flows from warm (high energy) to colder (low energy) regions to balance the differences in thermal energy.

**70. A is correct.**

Find heat liberated from 30 °C to 0 °C:

$Q_1 = mc\Delta T$

$Q_1 = (0.435 \text{ kg})·(4{,}186 \text{ J/kg·K})·(30 \text{ °C} - 0 \text{ °C})$

$Q_1 = 54{,}600 \text{ J} = 54.6 \text{ kJ}$

*continued...*

Find heat liberated by phase change:

$Q_2 = mL_f$

$Q_2 = (0.435\ \text{kg})\cdot(33.5 \times 10^4\ \text{J/kg})$

$Q_2 = 145{,}700\ \text{J} = 145.7\ \text{kJ}$

Find heat liberated by 0 °C to –8 °C:

$Q_3 = mc\Delta T$

$Q_3 = (0.435\ \text{kg})\cdot(2{,}090\ \text{J/kg·K})\cdot[(0\ °\text{C} - (-8\ °\text{C})]$

$Q_3 = 7{,}300\ \text{J} = 7.3\ \text{kJ}$

Find total $Q$ heat liberated:

$Q_{net} = Q_1 + Q_2 + Q_3$

$Q_{net} = (54.6\ \text{kJ} + 145.7\ \text{kJ} + 7.3\ \text{kJ})$

$Q_{net} = 208\ \text{kJ}$

**71. A is correct.**

$Q = mL_f$

$Q = (0.4\ \text{kg})\cdot(80\ \text{kcal/kg})$

$Q = 32\ \text{kcal}$

**72. D is correct.**

Carnot efficiency:

$\eta = (T_H - T_C) / T_H$

$\eta = (420\ \text{K} - 270\ \text{K}) / 420\ \text{K}$

$\eta = 0.357$

$W = Q_H \times \eta$

$W = (3{,}650\ \text{J})\cdot(0.357)$

$W = 1{,}303\ \text{J}$

**73. B is correct.**

Find heat needed to raise temperature to boiling point:

$Q = mc\Delta T$

$Q = (0.8\ kg)\cdot(4{,}186\ J/kg\cdot K)\cdot(100\ °C - 70\ °C)$

$Q = 100{,}464\ J$

Subtract from 800 kJ:

$Q_{remaining} = (800 \times 10^3\ J) - 100{,}464\ J$

$Q_{remaining} = 699{,}536\ J$

Use $Q_{remaining}$ to find mass of water that has evaporated:

$Q_R = mL_v$

$m = Q_R / L_v$

$m = (699{,}536\ J) / (22.6 \times 10^5\ J/kg)$

$m = 0.310\ kg$

Subtract from original mass to find mass remaining:

$m_R = 0.800\ kg - 0.310\ kg$

$m_R = 0.490\ kg = 490\ g$

**74. A is correct.**

In an adiabatic process, heat is not added to the system thus:

$Q = 0$

**75. D is correct.**

$C_P = Q_H / W$

$C_P = 50\ kW / 7.5\ kW$

$C_P = 6.7$

**76. D is correct.**

Convection is a form of heat transfer in which mass motion of a fluid (i.e., liquids and gases) transfers energy from the heat source.

**77. C is correct.**

Find heat for temperature change from 300 K to 1,357 K:

$Q_1 = mc\Delta T$

$Q_1 = (740\text{ kg})\cdot(386\text{ J/kg}\cdot\text{K})\cdot(1{,}357\text{ K} - 250\text{ K})$

$Q_1 = 316.2\text{ MJ}$

Find heat for phase change:

$Q_2 = mL_f$

$Q_2 = (740\text{ kg})\cdot(205{,}000\text{ J/kg})$

$Q_2 = 151.7\text{ MJ}$

Find total heat:

$Q_{net} = Q_1 + Q_2$

$Q_{net} = 316.2\text{ MJ} + 151.7\text{ MJ}$

$Q_{net} = 468\text{ MJ}$

$Q_{net} = 4.70 \times 10^5\text{ kJ}$

**78. B is correct.**

Because the block can be thought of as infinitely large, equilibrium temperature = 0.

$\Delta Q = mc\Delta T$

$\Delta Q = (0.008\text{ kg})\cdot(1\text{ kcal/kg}\cdot{}^\circ\text{C})\cdot(100\text{ }^\circ\text{C} - 0\text{ }^\circ\text{C})$

$\Delta Q = 0.8\text{ kcal}$

This heat is transferred to the ice, which has a latent heat of fusion of 80 kcal/kg.

$Q / L_f = m$

(0.8 kcal) / (80 kcal/kg) = 0.01 kg = 10 g of ice is melted

 

**79. D is correct.**

$\eta = (T_H - T_C) / T_H$

$0.13T_H = T_H - 1.9\text{ K}$

$1.9\text{ K} = T_H - 0.13T_H$

$1.9\text{ K} = T_H(1 - 0.13)$

$1.9\text{ K} = T_H(0.87)$

$T_H = 1.9\text{ K} / (0.87)$

$T_H = 2.2\text{ K}$

**80. C is correct.**

Efficiency of a heat engine is given as:

$\eta = (Q_H - Q_C) / Q_H$

$\eta = (1{,}200\text{ J} - 800\text{ J}) / 1{,}200\text{ J}$

$\eta = 0.33 = 33\%$

===

**Practice Set 5: Questions 81–106**

===

**81. B is correct.**

Thermal conductivity equation:

$$Q / t = kA(T_H - T_C) / d$$

where $k$ is thermal conductivity constant

$$Q / t = (0.105 \text{ W/m·K})·(0.35 \text{ m} \times 0.55 \text{ m})·[30 - (-10 \text{ °C})] / (0.006 \text{ m})$$

$$Q / t = 135 \text{ W}$$

**82. C is correct.**

Heat to melt ice cube:

$$Q_1 = mL_f$$

Heat to raise the temperature:

$$Q_2 = mc\Delta T$$

Total heat:

$$Q_{total} = Q_1 + Q_2$$

$$Q_{total} = mL_f + mc\Delta T$$

$$Q_{total} = (0.3 \text{ kg})·(334 \text{ kJ/kg}) + (0.3 \text{ kg})·(4.186 \text{ kJ/kg·K})·(60 \text{ °C} - 0 \text{ °C})$$

$$Q_{total} = 176 \text{ kJ}$$

**83. D is correct.**

Assuming the system is isolated, no heat can enter or leave the system, and heat can only be exchanged between the lead and the water.

$$0 = Q_L + Q_W$$

$$0 = m_Lc_L(T_{Li} - T_f) + m_Wc_W(T_f - T_{Wi})$$

Solve for $T_f$:

$$T_f = (m_Lc_LT_{Li} + m_Wc_WT_{Wi}) / (m_Lc_L + c_WT_{Wi})$$

*continued...*

Note that this is the weighted average of the two initial temperatures.

$T_f$ = [(0.050 kg)·(0.11 kcal/kg/°C)·(100 °C) – (0.080 kg)·(1 kcal/kg/°C)·(0 °C)] /

[(0.050 kg)·(0.11 kcal/kg/°C) + (0.080 kg)·(1 kcal/kg/°C)]

$T_f$ = 6.4 °C

**84. C is correct.**

Work is equal to the area enclosed.

Area = ½(base × height)

Area of this triangle:

Area = ½(ΔP)·(ΔV)

Convert units:

ΔP = (3 – 1)$P_0$

ΔP = (2)·(4.8 atm)

ΔP = (9.6 atm)·[(1 × $10^5$ N/m$^2$) / (1 atm)]

ΔP = 9.6 × $10^5$ N/m$^2$

ΔV = 600 cm$^3$ – 200 cm$^3$

ΔV = 400 cm$^3$ × (1 m/100 cm)$^3$

ΔV = 4 × $10^{-4}$ m$^3$

W = ½(9.6 × $10^5$ N/m$^2$)·(4 × $10^{-4}$ m$^3$)

W = 192 N·m = 192 J

**85. D is correct.**

PE = $Q$

$mgh = mc\Delta T$

Cancel $m$ from both sides of the expression

$gh = c\Delta T$

$gh / c = \Delta T$

ΔT = [(10 m/s$^2$)·(60 m)] / 4,186 J/kg·K

ΔT = 0.14 °C

**86. B is correct.**

Heat Radiation:

$Q / t = Ae\sigma(T_H^4 - T_C^4)$

Set constants $Ae\sigma = n$:

$Q / t = n(T_H^4 - T_C^4)$

$80 \text{ J/s} = n[T_H^4 - (25 \text{ °C} + 273 \text{ K})^4]$

$0 \text{ J/s} = n[T_H^4 - (298 \text{ K})^4] - 80 \text{ J/s}$

$95 \text{ J/s} = n[T_H^4 - (20 \text{ °C} + 273 \text{ K})^4]$

$0 \text{ J/s} = n[T_H^4 - (293 \text{ K})^4] - 95 \text{ J/s}$

Set equal and solve for n:

$n[(T_H^4 - (298 \text{ K})^4] - 80 \text{ J/s} = n[(T_H^4 - (293 \text{ K})^4] - 95 \text{ J/s}$

$15 \text{ J/s} = n[T_H^4 - (293 \text{ K})^4] - n[T_H^4 - (298 \text{ K})^4]$

$15 \text{ J/s} = n[T_H^4 - (293 \text{ K})^4 - T_H^4 + (298 \text{ K})^4]$

$15 \text{ J/s} = n[(298 \text{ K})^4 - (293 \text{ K})^4]$

$15 \text{ J/s} = n(51.6 \times 10^7 \text{ K}^4)$

$n = 2.9 \times 10^{-8} \text{ W/K}^4$

Solve for $T_H$:

$0 = (2.9 \times 10^{-8} \text{ W/K}^4)\cdot[T_H^4 - (293 \text{ K})^4] - 95 \text{ J/s}$

$T_H^4 = 1.06 \times 10^{10} \text{ K}^4$

$T_H = 321 \text{ K}$

$T_H = 48 \text{ °C}$

**87. A is correct.**

Efficiency of a heat engine:

$\eta = (Q_H - Q_C) / Q_H$

$\eta = (6{,}000 \text{ J} - 4{,}000 \text{ J}) / (6{,}000 \text{ J})$

$\eta = 0.33 \times 100\% = 33\%$

 

**88. C is correct.**

$\Delta L = L_0 \alpha \Delta T$

$\Delta L = (40 \text{ m})\cdot(1.2 \times 10^{-5} \text{ K}^{-1})\Delta T$

$\Delta L = (40 \text{ m})\cdot(1.2 \times 10^{-5} \text{ K}^{-1})\cdot(160 \text{ °C} - 15 \text{ °C})$

$\Delta L = 0.07 \text{ m} = 70 \text{ mm}$

**89. D is correct.**

$\Delta S = -L_v m / T$

$\Delta S = -(22.6 \times 10^5 \text{ J/kg})\cdot(2 \text{ kg}) / (100 \text{ °C} + 273 \text{ K})$

$\Delta S = -12{,}118 \text{ J/K} = -12.1 \times 10^3 \text{ J/K}$

A negative change in entropy indicates that the disorder of the isolated system has decreased.

When steam condenses to water, the entropy is negative because steam is more disordered than water.

**90. A is correct.**

Change in internal energy is $Q$ + W.

**91. C is correct.**

Entropy increase for two temperatures:

$\Delta S = Q(1 / T_2 - 1 / T_1)$

Let $R$ be the rate of change of entropy. Then:

$R = (25 \text{ kW})\cdot[(1 / -20 \text{ °C} + 273 \text{ K}) - (1 / 20 \text{ °C} + 273 \text{ K})]$

$R = (25 \text{ kW})\cdot[(1 / 253 \text{ K}) - (1 / 293 \text{ K})]$

$R = 0.013 \text{ kW/K}$

$R = 13 \text{ W/K}$

**92. D is correct.**

$Q = mc\Delta T$

Find heat released for temperature drop from 160 °C to 150 °C:

$Q_1 = (3.4\ \text{kg})\cdot(400\ \text{J/kg}\cdot\text{K})\cdot(160\ °\text{C} - 150\ °\text{C})$

$Q_1 = 13{,}600\ \text{J}$

Find heat released due to condensation:

$Q_2 = mL_v$

$Q_2 = (3.4\ \text{kg})\cdot(7.2 \times 10^4\ \text{J/kg})$

$Q_2 = 244{,}800\ \text{J}$

Find heat released for temperature drop from 150 °C to 75 °C:

$Q_3 = mc\Delta T$

$Q_3 = (3.4\ \text{kg})\cdot(1{,}000\ \text{J/kg}\cdot\text{K})\cdot(150\ °\text{C} - 75\ °\text{C})$

$Q_3 = 255{,}000\ \text{J}$

Sum the heats:

$Q_{net} = Q_1 + Q_2 + Q_3$

$Q_{net} = (13{,}600\ \text{J} + 244{,}800\ \text{J} + 255{,}000\ \text{J})$

$Q_{net} = 513{,}400\ \text{J} = 513\ \text{kJ}$

**93. C is correct.**

Find heat from phase change:

$Q = mL_f$

$Q = (0.2\ \text{kg})\cdot(1.04 \times 10^5\ \text{J/kg})$

$Q = 20{,}800\ \text{J}$

Because the ethanol is freezing, $Q$ should be negative due to heat being released.

$Q = -20{,}800\ \text{J}$

Find change in entropy:

$\Delta S = Q / \text{T}$

$\Delta S = -20{,}800\ \text{J} / (-114.4\ °\text{C} + 273\ \text{K})$

$\Delta S = -131\ \text{J} / \text{K}$

 

**94. B is correct.**

$W = P\Delta V$

Isobaric means pressure is constant, and volume is changing.

**95. C is correct.**

Adiabatic means that no heat enters or leaves the system.

$Q = 0$ kJ

**96. C is correct.**

$Q = mc\Delta T$

Find heat added to aluminum calorimeter:

$Q_A = (0.08 \text{ kg})\cdot(910 \text{ J/kg}\cdot\text{K})\cdot(35 \text{ °C} - 20 \text{ °C})$

$Q_A = 1{,}092 \text{ J}$

Find heat added to water:

$Q_W = (0.36 \text{ kg})\cdot(4{,}190 \text{ J/kg}\cdot\text{K})\cdot(35 \text{ °C} - 20 \text{ °C})$

$Q_W = 22{,}626 \text{ J}$

Find total heat added to the system:

$Q_{total} = Q_A + Q_W$

$Q_{total} = 1{,}092 \text{ J} + 22{,}626 \text{ J}$

$Q_{total} = 23{,}718 \text{ J}$

Find specific heat of the metal:

$Q = mc\Delta T$

$c = Q / m\Delta T$

$c = (23{,}718 \text{ J}) / [(0.18 \text{ kg})\cdot(305 \text{ °C} - 35 \text{ °C})]$

$c = 488 \text{ J/kg}\cdot\text{K}$

**97. D is correct.**

Watt = 1 J/s

Thermal energy:

$Q$ = Power × time

$Q = mc\Delta T$

$P \times t = mc\Delta T$

$t = (mc\Delta T) / P$

$t = [(120\ g)\cdot(4.186\ J/g\cdot°C)\cdot(50\ °C - 20\ °C)] / (65\ W)$

$t = 232\ s$

**98. C is correct.**

The Second Law of Thermodynamics states that entropy is either constant or increasing over time.

A constant entropy process is an idealized process and does not exist.

Thus, entropy is always increasing over time.

**99. B is correct.**

$Q = (mc\Delta T)_{water} + (mc\Delta T)_{beaker}$

Change in temperature is the same for both:

$Q = \Delta T[(mc)_{water} + (mc)_{beaker}]$

$1{,}800\ cal = (20\ °C)\cdot[(65\ g)\cdot(1\ cal/g\cdot°C) + (m_{beaker})\cdot(0.18\ cal/g\cdot°C)]$

$90\ cal/°C = 65\ cal/°C + (m_{beaker})\cdot(0.18\ cal/g\cdot°C)$

$25\ cal/°C / (0.18\ cal/g\cdot°C) = (m_{beaker})$

$m_{beaker} = 139\ g$

 

**100. D is correct.**

Find heat from phase change:

$Q = mL_f$

$Q = (0.02 \text{ kg})\cdot(22.6 \times 10^5 \text{ J/kg})$

$Q = 45{,}200 \text{ J}$

Because the water is vaporizing, $Q$ should be positive due to heat being absorbed.

$Q = 45{,}200 \text{ J}$

Find the change in entropy:

$\Delta S = Q / \text{T}$

$\Delta S = 45{,}200 \text{ J} / (100 \text{ °C} + 273 \text{ K})$

$\Delta S = 121 \text{ J} / \text{K}$

A positive change in entropy indicates that the disorder of the isolated system has increased

When water evaporates into steam, the entropy is positive because the disorder of steam is higher than water.

*Notes for active learning*

# 4 – Fluid Statics and Dynamics: Detailed Explanations

===============================================================

**Practice Set 1: Questions 1–20**

===============================================================

**1. C is correct.**

Refer to the unknown liquid as "A" and the oil as "O."

$$\rho_A h_A g = \rho_O h_O g$$

Cancel $g$ from both sides of the expression

$$\rho_A h_A = \rho_O h_O$$

$$h_A = 5 \text{ cm}$$

$$h_O = 20 \text{ cm}$$

$$h_A = \tfrac{1}{4} h_O$$

$$\rho_A (\tfrac{1}{4}) h_O = \rho_O h_O$$

$$\rho_A = 4\rho_O$$

$$\rho_A = 4(850 \text{ kg/m}^3)$$

$$\rho_A = 3{,}400 \text{ kg/m}^3$$

**2. D is correct.**

$$P = \rho_{oil} \times V_{oil} \times g / (A_{tube})$$

$$P = [\rho_o \pi (r_{tube})^2 \times hg] / \pi (r_{tube})^2$$

Cancel $\pi(r_{tube})^2$ from both the numerator and denominator.

$$P = \rho_o g h$$

$$P = (850 \text{ kg/m}^3)\cdot(9.8 \text{ m/s}^2)\cdot(0.2 \text{ m})$$

$$P = 1{,}666 \text{ Pa}$$

*continued...*

**3. A is correct.**

$m_{oil} = \rho_{oil}V_{oil}$

$V = \pi r^2 h$

$m_{oil} = \rho_{oil}\pi r^2 h$

$m_{oil} = \pi(850\ kg/m^3)\cdot(0.02\ m)^2 \times (0.2\ m)$

$m_{oil} = 0.21\ kg = 210\ g$

**4. A is correct.**

Gauge pressure is the pressure experienced by an object referenced at atmospheric pressure.

When the block is lowered, its gauge pressure increases according to:

$P_G = \rho gh$

At $t = 0$, the block just enters the water and $h = 0$ so $P_G = 0$.

As time passes, the height of the block below the water increases linearly, so $P_G$ increases linearly as well.

**5. D is correct.**

Using Bernoulli's principle and assuming the opening of the tank is so large that the initial velocity is essentially zero:

$\rho gh = ½\rho v^2$

Cancel $\rho$ from both sides of the expression

$gh = ½v^2$

$v^2 = 2gh$

$v^2 = 2\cdot(9.8\ m/s^2)\cdot(0.8\ m)$

$v^2 = 15.68\ m^2/s^2$

$v = 3.96\ m/s \approx 4\ m/s$

Note: the diameter is not used to solve the problem.

*continued...*

 

**6. C is correct.**

The ideal gas law is:

$$PV = nRT$$

Keeping $n$RT constant:

If $P_{final} = ½P_{initial}$

$V_{final} = ½V_{initial}$

However, in an isothermal process, there is no change in internal energy.

Therefore, because energy must be conserved:

$$\Delta U = 0$$

**7. A is correct.**

Volumetric flow rate:

$$V_f = Av$$

$$A = \pi d^2 / 4$$

So, for the original pipe:

$$V_f = (\pi d^2 / 4) \cdot v$$

If diameter is doubled:

$$V_f = [\pi (2d)^2 / 4]\ v$$

$$V_f = (\pi / 4)\cdot(4d^2)\ v$$

$$V_f = 4\ [(\pi d^2 / 4)\ v]$$

where the term in square brackets is identical to the original flow rate.

Thus, the flow rate increases by a factor of 4.

**8. C is correct.**

The object sinks when the buoyant force is less than the weight of the object.

Since the buoyant force is equal to the weight of the displaced fluid, an object sinks precisely when the weight of the fluid it displaces is less than the weight of the object itself.

**9. C is correct.**

Both the solid circle and wire circle have the same diameters.

However, the wire has two surface areas exposed to the water (i.e., inner and outer circumference) where the difference in circumference length is assumed to be negligible.

The solid circle only has the outer edge exposed to the water's surface (i.e., circumference).

$L_{wire} = 2(2\pi r)$ = the relevant length for the wire circle

$2\pi r$ = the relevant length for the solid circle

Solid circle on water:

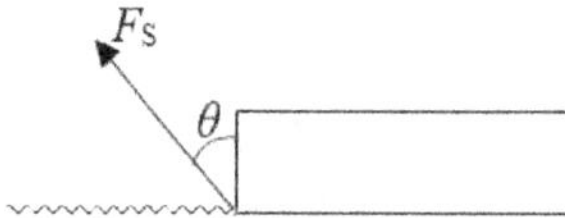


$F_{s1} = y \cos \theta L$

Wire on water:

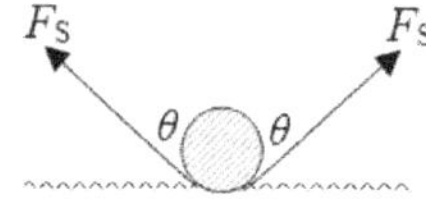


$F_{s2} = 2y \cos \theta L$

$F_{s2} = 2F_{s1}$

Balance weight vs. surface tension where $F_s$ point up and $F_w$ points down.

For solid circle:

$F_{w1} = F_{s1}$

For wire circle:

$F_{w2} = F_{s2}$

$F_{w2} = 2F_{s1}$

$F_{w2} = 2F_{w1}$

The wire circle can have double the mass without sinking.

 

**10. C is correct.**

Absolute pressure = gauge pressure + atmospheric pressure

$$P_{abs} = P_G + P_{atm}$$

$$P_{abs} = \rho gh + P_{atm}$$

Atmospheric pressure is added to the total pressure at the bottom of a volume of liquid.

Therefore, if the atmospheric pressure increases, absolute pressure increases by the same amount.

**11. A is correct.**

Surface tension increases as temperature decreases.

Generally, the cohesive forces maintaining surface tension decrease as molecular thermal activity increases.

**12. B is correct.**

By Poiseuille's Law, the volumetric flow rate of a fluid is given by:

$$V = \Delta P A r^2 / 8\eta L$$

Volumetric flow rate is the volume of fluid that passes a point per unit time:

$$V = Av$$

where $v$ is the speed of the fluid.

Therefore:

$$Av = \Delta P A r^2 / 8\eta L$$

$$v = \Delta P r^2 / 8\eta L$$

$$v = (225 \times 10^3 \text{ Pa})\cdot(0.0032 \text{ m})^2 / [8\ (0.3 \text{ Ns/m}^2)\cdot(1 \text{ m})]$$

$$v = 0.96 \text{ m/s}$$

**13. A is correct.**

The buoyant force upward must balance the weight downward.

Buoyant force = weight of the volume of water displaced

$F_B = W_{object}$

$\rho V g = W_{object}$

$W_{object} = 60 \text{ N}$

$W_{object} = (\rho_{water})\cdot(V_{water})\cdot(g)$

$60 \text{ N} = (1{,}000 \text{ kg/m}^3)\cdot(V_{water})\cdot(10 \text{ m/s}^2)$

$V_{water} = 60 \text{ N} / (1{,}000 \text{ kg/m}^3)\cdot(10 \text{ m/s}^2) = 0.006 \text{ m}^3$

**14. B is correct.**

Volume flow rate:

$Q = vA$

$Q = (2.5 \text{ m/s})\pi r^2$

$Q = (2.5 \text{ m/s})\cdot(0.015 \text{ m})^2\pi$

$Q = 1.8 \times 10^{-3} \text{ m}^3\text{/s}$

**15. C is correct.**

Force equation for the cork that is not accelerating:

$F_B - mg = 0$

Let $m$ be the mass and V be the volume of the cork.

Replace:

$m = \rho V$

$F_B = (\rho_{water})\cdot(V_{disp})\cdot(g)$

$(\rho_{water})\cdot(V_{disp})\cdot(g) = (\rho_{cork})\cdot(V)\cdot(g)$

$V_{disp} = \frac{3}{4}V$

$\rho_{water}(\frac{3}{4}Vg) = (\rho_{cork})\cdot(V)\cdot(g)\backslash$

continued...

 

Cancel $g$ and V from both sides of the expression:

$\frac{3}{4}\rho_{water} = \rho_{cork}$

$\rho_{cork} / \rho_{water} = \frac{3}{4} = 0.75$

**16. D is correct.**

volume = mass / density

$V = (600\text{ g}) / (0.93\text{ g/cm}^3)$

$V = 645\text{ cm}^3$

**17. C is correct.**

For monatomic gases:

$U = 3/2k_BT$

where U is the average KE per molecule and $k_B$ is the Boltzmann constant

**18. B is correct.**

The object weighs 150 N less while immersed because the buoyant force supports 150 N of the total weight of the object.

Since the object is totally submerged, the volume of water displaced equals the volume of the object.

$F_B = 150\text{ N}$

$F_B = \rho_{water} \times V_{water} \times g$

$V = F_B / \rho g$

$V = (150\text{ N}) / (1{,}000\text{ kg/m}^3)\cdot(10\text{ m/s}^2)$

$V = 0.015\text{ m}^3$

**19. C is correct.**

$F_B / \rho_w = (m_c g) / \rho_c$

$F_B = (\rho_w m_c g) / \rho_c$

$F_B = [(1\ g/cm^3)·(0.03\ kg)·(9.8\ m/s^2)] / (8.9\ g/cm^3)$

$F_B = 0.033\ N$

$m_{total} = m_w + (F_B / g)$

$m_{total} = (0.14\ kg) + [(0.033\ N) / (9.8\ m/s^2)]$

$m_{total} = 0.143\ kg = 143\ g$

**20. D is correct.**

$v_1 A_1 = v_2 A_2$

$v_2 = v_1 A_1 / A_2$

$v_2 = [v_1(\pi/4)d_1^2] / (\pi/4)d_2^2$

Cancel $(\pi/4)$ from both the numerator and denominator

$v_2 = v_1(d_1^2 / d_2^2)$

$v_2 = (1\ m/s)·[(6\ cm)^2 / (3\ cm)^2]$

$v_2 = 4\ m/s$

 

---

## Practice Set 2: Questions 21–40

---

**21. A is correct.**

Static fluid pressure:

$$P = \rho g h$$

Pressure is only dependent on gravity ($g$), the height ($h$) of the fluid above the object, and density ($\rho$) of the fluid.

Pressure does depend on the depth of the object but does not depend on the surface area of the object.

Both objects are submerged to the same depth, so the fluid pressure is equal. Note that the buoyant force on the blocks is NOT equal, but the pressure (force / area) is equal.

**22. C is correct.**

Surface tension force acts as the product of surface tension and the total length of contact.

$$F = AL$$

For a piece of thread, the length of contact is $l$ as shown:

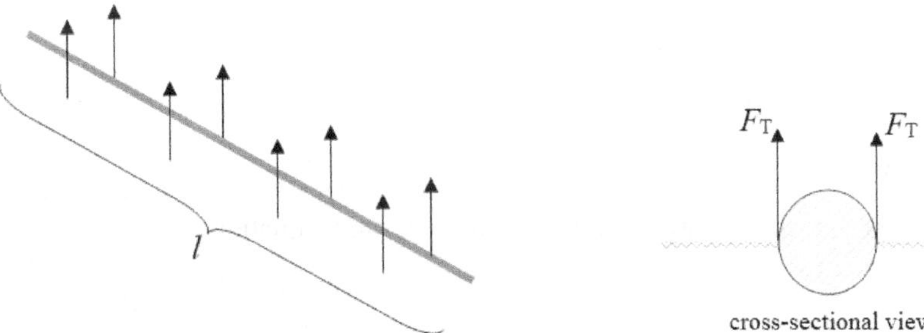


$F = 2L$ because the force acts on both sides of the thread.

Thus, for a thread rectangle, the total contact length is the total length times two.

$$L = 2(l + w + l + w)$$

$$F_{max} = 2A(l + w + l + w)$$

$$F_{max} = 2A(2l + 2w)$$

$$F_{max} = 4A(l + w)$$

**23. D is correct.**

Gauge pressure is the measure of pressure with respect to atmospheric pressure. So, if the pressure inside the tire is equal to the air pressure outside, the gauge reads zero.

**24. B is correct.**

Pressure is measured in force per unit area, which is the force divided by the area.

**25. D is correct.**

The shear stress is the force per unit area and has units of $N/m^2$.

**26. B is correct.**

Because the reservoir area is assumed to be essentially infinite, the velocity of the flow at the top of the tank is assumed to be zero.

Using Bernoulli's equation, find the speed through the 3 cm pipe:

$$(\tfrac{1}{2}\rho v^2 + \rho gh)_{out} = (\tfrac{1}{2}\rho v^2 + \rho gh)_{in}$$

$$\tfrac{1}{2}\rho v^2 = \rho gh$$

Cancel $\rho$ from both sides of the expression

$$\tfrac{1}{2}v^2 = gh$$

$$v^2 = 2gh$$

$$v = \sqrt{(2gh)}$$

$$v = \sqrt{[2(9.8\ m/s^2)\cdot(4\ m)]}$$

$$v_{3cm} = 8.9\ m/s$$

To find the speed through the 5 cm pipe use the continuity equation:

$$A_{3cm}v_{3cm} = A_{5cm}v_{5cm}$$

$$(\pi / 4)\cdot(3\ cm)^2\cdot(8.9\ m/s) = (\pi / 4)\cdot(5\ cm)^2\cdot(v_{5cm})$$

$$v_{5cm} = 3.2\ m/s$$

**27. C is correct.**

The ideal gas law is:

$PV = nRT$ where $n$, R and T are constants.

P is pressure, V is volume, n is the number of particles, R is the ideal gas law constant, and T is temperature. If $P \rightarrow 3P$, then $V \rightarrow (1/3)V$ to maintain a constant temperature.

 

**28. D is correct.**

$F = PA + F_{cover}$

$P = \rho gh$

$F = \rho ghA + F_{cover}$

$F = [(1,000\ kg/m^3)\cdot(10\ m/s^2)\cdot(1\ m)\cdot(1\ m^2)] + 1,500\ N$

$F = 11,500\ N$

**29. B is correct.**

The buoyant force on a totally submerged object is independent of its depth below the surface (since any increase in the water's density is ignored).

The buoyant force on the ball at a depth of 4 m is 20 N, the same as the buoyant force at 1 m.

When it sits at the bottom of the pool, the two upward forces (i.e., the buoyant force $F_B$ and the normal force $F_N$), must balance the downward force of gravity.

$F_B + F_N = F_g$

$(20\ N) + F_N = 80\ N$

$F_N = 80\ N - 20\ N$

$F_N = 60\ N$

**30. B is correct.**

$V_{Fluid\ Displaced} = ½V_{block}$

Buoyant force:

$F_B = \rho gV$

$\rho_F gV_F = \rho_B gV_B$

Cancel $g$ from both sides of the expression:

$\rho_F V_F = \rho_B V_B$

$\rho_F(½V_B) = \rho_B V_B$

Cancel $V_B$ from both sides of the expression

$½\rho_F = \rho_B$

$\rho_F = (1.6)\rho_{water}$

$\rho_B = (1.6)\cdot(10^3\ kg/m^3)\cdot(½)$

$\rho_B = 800\ kg/m^3$

 

**31. C is correct.**

$F_{net} = F_{object} - F_B$

$F_{net} = mg - \rho_{fluid}Vg$

$\rho_{fluid} = -(F_{net} - mg) / V_{sphere}g$

$\rho_{fluid} = -[42\text{ N} - (9.2\text{ kg})\cdot(9.8\text{ m/s}^2)] / [(9.2\text{ kg} / 3{,}650\text{ kg/m}^3)\cdot(9.8\text{ m/s}^2)]$

$\rho_{fluid} = 1{,}950\text{ kg/m}^3$

**32. A is correct.**

Pressure due to the density of a fluid surrounding an object submerged at depth $d$ below the surface is given by:

$P = \rho gd$

Since the distance that the objects are below the surface is not specified, the only conclusion that can be drawn is that object B experiences less fluid pressure than object A.

This difference is because object B is higher off the floor of the container, and thus its depth is less than object A.

**33. B is correct.**

$A_1v_1 = A_2v_2$

$A = \pi r^2$

$A_2 = \pi(2r)^2$

$A_2 = 4\pi r^2$

$A_2 = 4A_1$

If r is doubled, then area is increased by 4 times.

$A_1(14\text{ m/s}) = (4A_1)v_2$

$v_2 = (A_1 \times 14\text{ m/s}) / (4 \times A_1)$

$v_2 = (14\text{ m/s}) / (4)$

$v_2 = 3.5\text{ m/s}$

*continued...*

 

Use Bernoulli's equation to find resulting pressure:

$P_1 + ½\rho v_1^2 = P_2 + ½\rho v_2^2$

$(3.5 \times 10^4 \text{ Pa}) + ½(1{,}000 \text{ kg/m}^3)\cdot(14 \text{ m/s})^2 = P_2 + ½(1{,}000 \text{ kg/m}^3)\cdot(3.5 \text{ m/s})^2$

$P_2 = (13.3 \times 10^4 \text{ Pa}) - (6.1 \times 10^3 \text{ Pa})$

$P_2 = 12.7 \times 10^4 \text{ Pa}$

**34. A is correct.**

Bernoulli's Equation:

$P_1 + ½\rho v_1^2 + \rho g h_1 = P_2 + ½\rho v_2^2 + \rho g h_2$

When simplifying Bernoulli's Equation, the top opening and bottom opening experience atmospheric pressure, so $P_1$ and $P_2$ cancel.

Likewise, $v_1 << v_2$, so it can be assumed to be negligible and go to zero.

Finally,

$h_2 = 0$

because it is at the bottom of the tank and is the reference for further heights in the problem

This results in the simplified expression:

$\rho g h_1 = ½\rho v_2^2$

Cancel $\rho$ from both sides of the expression

$g h_1 = ½v_2^2$

$v_2 = \sqrt{(2gh_1)}$

$v_2 = \sqrt{[(2)\cdot(9.8 \text{ m/s}^2)\cdot(0.5 \text{ m})]}$

$v_2 = 3.1 \text{ m/s}$

**35. C is correct.**

The buoyant force:

$F_B = \rho_{air} V_{disp} g$, and $V_{disp}$ is the volume of the man $m / \rho_{man}$

$F_B = (\rho_{air} / \rho_{man}) mg$

$F_B = [(1.2 \times 10^{-3} \text{ g/cm}^3) / (1 \text{ g/cm}^3)]\cdot(80 \text{ kg})\cdot(9.8 \text{ m/s}^2)$

$F_B = 0.94 \text{ N}$

 

**36. D is correct.**

Graham's law states that the rate at which gas diffuses is inversely proportional to the square root of the density of the gas.

**37. D is correct.**

$F_B = \rho V g$

$F_{net} = F_{object} - F_B$

$F_{net} = (41{,}800\ N) - (1{,}000\ kg/m^3)(4.2\ m^3)(9.8\ m/s^2)$

$F_{net} = (41{,}800\ N) - (41{,}160\ N)$

$F_{net} = 640\ N$

**38. D is correct.**

$A = \pi r^2$

$A_T v_T = A_P v_P$

$v_T = v_P A_P / A_T$

The ratio of the square of the diameter is equal to the ratio of area:

$v_T = v_P(d_1 / d_2)^2$

$v_T = (0.03\ m/s)\cdot[(0.12\ m) / (0.002\ m)]^2$

$v_T = 108\ m/s$

**39. B is correct.**

Specific gravity:

$\rho_{object} / \rho_{water}$

Archimedes' principle:

$F = \rho g V$

$\rho_{water} = F / g(0.9V)$

$\rho_{object} = F / gV$

$\rho_{object} / \rho_{water} = (F / gV) / [F / g(0.9V)]$

$\rho_{object} / \rho_{water} = 0.9$

$V_{water}$ is 0.9V because only 90% of the object is in water, so 90% of the object's volume equals water displaced.

 

**40. C is correct.**

The factors considered are length, density, radius, pressure difference, and viscosity.

The continuity equation does not apply here because it can only relate velocity and radius to the flow rate.

Bernoulli's equation does not apply because it only relates density and velocity.

The Hagen-Poiseuille equation is needed because it includes all the terms except for density and therefore is the most applicable to this question.

Volumetric flow rate (Q) is:

$$Q = \Delta P\pi r^4 / 8\eta L$$

The radius is raised to the fourth power.

A 15% change to $r$ results in the greatest change.

 

## Practice Set 3: Questions 41–60

**41. A is correct.**

A force meter provides a force, and the reading indicates what the force is.

Since the hammer is not accelerating, the force equation is:

$$F_{meter} + F_B - m_h g = 0$$

$$m_h g = (0.68 \text{ kg})\cdot(10 \text{ m/s}^2)$$

$$m_h g = 6.8 \text{ N}$$

The displaced volume is the volume of the hammer:

$$V_{disp} = m_h / \rho_{steel}$$

$$V_{disp} = (680 \text{ g}) / (7.9 \text{ g/cm}^3)$$

$$V_{disp} = 86 \text{ cm}^3$$

$$F_B = \rho_{water} \times V_{disp} \times g$$

$$F_B = (1 \times 10^{-3} \text{ kg/cm}^3)\cdot(86 \text{ cm}^3)\cdot(10 \text{ m/s}^2)$$

$$F_B = 0.86 \text{ N}$$

$$F_{meter} = m_h g - F_B$$

$$F_{meter} = 6.8 \text{ N} - 0.86 \text{ N}$$

$$F_{meter} = 5.9 \text{ N}$$

**42. D is correct.**

Pascal's Principle states that pressure is transmitted undiminished in an enclosed static fluid.

 

**43. B is correct.**

The flow rate must be constant along the flow because water is incompressible.

Volumetric Flow Rate is:

$$\upsilon = Av$$

Volumetric Flow Rate at different points in a pipe with different areas:

$$A_1v_1 = A_2v_2$$

Thus, the Volumetric Flow Rate at one point must match that of the other.

$$A_2v_2 = 0.09\ \text{m}^3/\text{s}$$

**44. D is correct.**

$$P = \rho gh$$

Because the bottom of the brick is at a lower depth than the rest of the brick, it experiences the highest pressure.

**45. A is correct.**

Mass flow rate:

$$\dot{m} = \text{cross-sectional area} \times \text{density} \times \text{velocity}$$

$$\dot{m} = A_C\rho v$$

$$\dot{m} = (7\ \text{m})\cdot(14\ \text{m})\cdot(10^3\ \text{kg/m}^3)\cdot(3\ \text{m/s})$$

$$\dot{m} = 2.9 \times 10^5\ \text{kg/s}$$

**46. B is correct.**

Archimedes Principle:

$$\rho_{\text{object}} / \rho_{\text{fluid}} = W_{\text{object}} / W_{\text{fluid}}$$

$$\rho_{\text{object}} / \rho_{\text{fluid}} = (11.3\ \text{g/cm}^3) / (13.6\ \text{g/cm}^3)$$

$$\rho_{\text{object}} / \rho_{\text{fluid}} = 0.83$$

83% of the lead ball is below the surface by weight. The ball has 17% above the surface because the density is assumed to be consistent throughout the sphere.

The weight is directly correlated to volume.

 

**47. A is correct.**

$P = P_{atm} + \rho gh$

$P = (1.01 \times 10^5\ Pa) + (10^3\ kg/m^3)\cdot(10\ m/s^2)\cdot(6\ m)$

$P = (1.01 \times 10^5\ Pa) + (0.6 \times 10^5\ Pa)$

$P = 1.6 \times 10^5\ Pa$

**48. B is correct.**

density = mass / volume

**49. C is correct.**

The bulk modulus is defined as how much material is compressed under a given external pressure:

$B = \Delta P / (\Delta V / V)$

Most solids and liquids compress slightly under external pressure.

However, gases have the greatest change in volume and thus the lowest value of B.

**50. C is correct.**

$P_1 = P_2 + \rho gh$

$F_1 / A_1 = F_2 / A_2 + \rho gh$

$F_1 = A_1(F_2 / A_2 + \rho gh)$

$F_1 = \pi(0.06\ m)^2\cdot[(14{,}000\ N) / \pi(0.16\ m)^2 + (750\ kg/m^3)\cdot(9.8\ m/s^2)\cdot(1.5\ m)]$

$F_1 = \pi(0.0036\ m^2)\cdot[(14{,}000\ N) / \pi(0.0256\ m^2) + (750\ kg/m^3)\cdot(9.8\ m/s^2)\cdot(1.5\ m)]$

$F_1 = (0.0036\ m^2)\cdot[(14{,}000\ N) / (0.0256\ m^2) + (11{,}025\ N)\pi]$

$F_1 = 1{,}969\ N + 125\ N$

$F_1 = 2{,}094\ N$

**51. C is correct.**

Solve for density of ice and saltwater:

$$SG = \rho_{substance} / \rho_{water}$$

$$SG_{ice} = 0.98 = \rho_{ice} / 10^3\ kg/m^3$$

$$\rho_{ice} = 980\ kg/m^3$$

$$SG_{saltwater} = 1.03 = \rho_{saltwater} / 10^3\ kg/m^3$$

$$\rho_{saltwater} = 1{,}030\ kg/m^3$$

Solve for volume of ice:

$$F_B = F_{bear} + F_{ice}$$

$$\rho_{saltwater} V_{ice} g = m_{bear} g + \rho_{ice} V_{ice} g$$

Cancel $g$ from all terms

$$V_{ice} = m_{bear} / (\rho_{saltwater} - \rho_{ice})$$

$$V_{ice} = (240\ kg) / (1{,}030\ kg/m^3 - 980\ kg/m^3)$$

$$V_{ice} = 4.8\ m^3$$

Solve for area of ice:

$$A = V / h$$

$$A = (4.8\ m^3) / (1\ m)$$

$$A = 4.8\ m^2$$

**52. B is correct.**

Absolute pressure is measured relative to absolute zero pressure (a perfect vacuum), and gauge pressure is measured relative to atmospheric pressure.

If the atmospheric pressure increases by ΔP, the absolute pressure increases by ΔP, but the gauge pressure does not change.

Absolute pressure at an arbitrary depth $h$ in the lake:

$$P_{abs} = P_{atm} + \rho_{water} g h$$

Gauge pressure at an arbitrary depth $h$ in the lake:

$$P_{gauge} = \rho_{water} g h$$

 

**53. C is correct.**

$P_{top} = \rho gh$

$P_{top} = 108 \times 10^3$ Pa

$h = (1 / \rho)\cdot(108 \times 10^3 \text{ Pa} / 9.8 \text{ m/s}^2)$

$h = (1 / \rho)\cdot(11{,}020 \text{ kg/m}^2)$

$P_{bottom} = \rho g(h + 25 \text{ cm})$

$\rho g(h + 25 \text{ cm}) = 114 \times 10^3$ Pa

$h = (1 / \rho)\cdot(114 \times 10^3 \text{ Pa} / 9.8 \text{ m/s}^2) - 0.25$ m

$h = (1 / \rho)\cdot(11{,}633) - 0.25$ m

Set equal and solve for $\rho$:

$(1 / \rho)\cdot(11{,}020 \text{ kg/m}^2) = (1 / \rho)\cdot(11{,}633 \text{ kg/m}^2) - 0.25$ m

$(11{,}020 \text{ kg/m}^2) = (11{,}633 \text{ kg/m}^2) - 0.25 \text{ m}(\rho)$

$0.25 \text{ m}(\rho) = 613 \text{ kg/m}^2$

$\rho = (613 \text{ kg/m}^2) / (0.25 \text{ m})$

$\rho = 2{,}452 \text{ kg/m}^3$

**54. D is correct.**

By Poiseuille's Law:

$v = \Delta Pr^2 / 8\eta L$

$\eta = \Delta Pr^2 / 8Lv_{effective}$

$\eta = (970 \text{ Pa})\cdot(0.0021 \text{ m})^2 / 8\cdot(1.8 \text{ m/s})\cdot(0.19 \text{ m})$

$\eta = 0.0016 \text{ N}\cdot\text{s/m}^2$

**55. C is correct.**

*Bernoulli's equation*:

$P_1 + ½\rho_1 v_1^2 + \rho_1 gh_1 = P_2 + ½\rho_2 v_2^2 + \rho_2 gh_2$

There is no height difference, so the equation reduces to:

$P_1 + ½\rho_1 v_1^2 = P_2 + ½\rho_2 v_2^2$

*continued…*

 

If the flow of air across the wing tip is $v_1$, then:

$v_1 > v_2$

Since the air that flows across the top has a higher velocity, as it travels a larger distance (curved surface of the top) over the same period, then:

$\frac{1}{2}\rho_1 v_1^2 > \frac{1}{2}\rho_2 v_2^2$

To keep both sides equal:

$P_1 < P_2$

$\Delta P = (P_2 - P_1)$

Thus, the lower portion of the wing experiences greater pressure and therefore lifts the wing.

**56. A is correct.**

As the air bubble rises toward the surface, the volume of the bubble increases.

**57. B is correct.**

The *Bulk Modulus* is expressed as:

$B = \Delta P / (\Delta V / V)$

$B = (\Delta PV) / \Delta V$

Solve for $\Delta V$:

$\Delta V = (\Delta PV) / B$

$\Delta V = (10^7 \text{ N/m}^2)\cdot(1 \text{ m}^3) / (2.3 \times 10^9 \text{ N/m}^2)$

$\Delta V = 0.0043 \text{ m}^3$

**58. B is correct.**

*Pascal's Principle* states that pressure is transmitted undiminished in an enclosed static fluid. This principle makes hydraulic lift possible because, for equal pressure across a fluid, the force can be multiplied through an area difference:

$P_1 = P_2$

$F_1 / A_1 = F_2 / A_2$

$F_1 = (A_1 / A_2)F_2$

**59. A is correct.**

Gauge pressure is referenced relative to atmospheric pressure and is calculated by:

$P_{gauge} = \rho g h$

$P_{gauge} = (1{,}025 \text{ kg/m}^3)\cdot(9.8 \text{ m/s}^2)\cdot(11{,}030 \text{ m})$

$P_{gauge} = 1.1 \times 10^8 \text{ Pa}$

**60. D is correct.**

$P = \rho g h$

$P = (10^3 \text{ kg/m}^3)\cdot(9.8 \text{ m/s}^2)\cdot(1 \text{ m})$

$P = 9{,}800 \text{ Pa} \approx 1 \times 10^4 \text{ Pa}$

 

==========================================================

**Practice Set 4: Questions 61–80**

==========================================================

**61. C is correct.**

Assuming there is a vacuum on the inside of the sphere, and using Archimedes' principle:

$(\rho g V)_{sphere} = (\rho g V)_{disp\text{-}water}$

$(\rho V)_{sphere} = (\rho V)_{disp\text{-}water}$

$V_{sphere} = 4/3\pi r^3$

Define $r_o$ as outer radius and $r_i$ as inner radius:

$\rho_{steel}(4/3\pi)\cdot(r_o^3 - r_i^3) = \rho_{water}(4/3\pi r_o^3)$

$\rho_{steel}(r_o^3 - r_i^3) = \rho_{water} r_o^3$

$r_i^3 = [(-\rho_{water} r_o^3) / (\rho_{steel})] + r_o^3$

$r_i^3 = -[(10^3\ kg/m^3)\cdot(1.5\ m)^3 / (7{,}870\ kg/m^3)] + (1.5\ m)^3$

$r_i^3 = 2.95\ m^3$

$r_i = 1.43\ m$

Thickness = $r_o - r_i$

Thickness = 1.5 m – 1.43 m = 0.07 m (or 7 cm)

**62. A is correct.**

According to the *Bernoulli effect*, air moving with a higher velocity exerts less pressure against a surface than air with a lower velocity.

Thus, to produce lift, the pressure on the underside of a wing should be higher, and thus the air should be slower on the bottom surface of the wing compared to air on the top surface.

 

**63. D is correct.**

*Poiseuille's law*:

$Q = \pi \Delta P r^4 / 8\eta L$

$D_B = 2D_A$

$r_B = 2r_A$

$Q_A = \pi \Delta P r_A{}^4 / 8\eta L$

$Q_B = \pi \Delta P(2r_A)^4 / 8\eta L$

$Q_B = 16(\pi \Delta P r_A{}^4 / 8\eta L)$

$Q_B = 16Q_A$

**64. B is correct.**

Mass flow rate $\dot{m}$ is:

$\dot{m} = \rho v A_C$

where $\rho$ = density of the fluid, $v$ = velocity of flow, $A_C$ = cross-sectional area

The dimensions of the tank are irrelevant to this answer, so $\dot{m}$ is calculated as:

$\dot{m} = (1{,}000\ kg/m^3)\cdot(13\ m/s)\cdot(0.04\ m^2)$

$\dot{m} = 520\ kg/s$

**65. C is correct.**

Relationship between area and velocity of both exists:

$A_1 v_1 = A_2\ v_2$

$v_1 = A_2\ v_2 / A_1$

$v_1 = [(0.04\ m^2)\cdot(13\ m/s)] / [(\pi / 4)\cdot(5\ m/s)]$

$v_1 = 0.26\ m/s$

$v_1 = 26\ mm/s$

**66. D is correct.**

Gauge pressure is the pressure reference against the surrounding air pressure.

Therefore, it is valid to ignore the pressure in the tank and atmospheric pressure.

$$P = \rho gh$$

$$P = (10^3\ kg/m)\cdot(9.8\ m/s^2)\cdot(10\ m)$$

$$P = 98{,}000\ Pa = 98\ kPa$$

$$P = 98\ kPa$$

**67. B is correct.**

According to Pascal's Law, the pressure is transmitted undiminished in an enclosed static fluid.

Thus, the pressure increases by ΔP everywhere in the oil.

If the chamber is cubic, the top and bottom sides have the same area and experience the same increase in force:

$$A_{top} = A_{bottom}$$

$$\Delta P_{top} = \Delta P_{bottom}$$

$$P = F / A$$

Thus, force is directly proportional to pressure, and if the area of the top is equal to the area of the bottom:

$$\Delta F_{top} = \Delta F_{bottom}$$

**68. A is correct.**

Pressure is given as:

$$P = F / A$$

$$F = PA$$

As the tire leaks, the pressure decreases, resulting in an increase in surface area with the road.

The force remains constant if pressure decreases and the area increases at a constant rate.

**69. C is correct.**

Under one meter of water the gauge pressure is:

$$P = \rho g h$$

Jack's lungs are open to atmospheric pressure by the snorkel, so he only needs to overcome the gauge pressure.

The force needed is:

$$P = F / A$$

$$F = PA$$

$$F = (\rho g h)A$$

The area is the area of his chest, as this is what the pressure acts against.

$$F = (\text{gauge pressure})\cdot(\text{chest area})$$

**70. A is correct.**

$$F_B = mg$$

$$F_B = \rho V g$$

$$\rho V g = mg$$

Cancel $g$ from both sides of the expression

$$m = \rho V$$

$$m = (\pi r^2 h)\cdot(1\ \text{g/cm}^3)$$

$$m = \pi(1\ \text{cm})^2\cdot(14\ \text{cm})\cdot(1\ \text{g/cm}^3)$$

$$m = 14\pi\ \text{g}$$

$$m = 44\ \text{g}$$

**71. A is correct.**

The reading on the meter is the net force between the necklace weight and buoyant force.

$$F_{\text{total}} = F_O - F_B$$

$$F_{\text{total}} = mg - \rho V g$$

$$F_{\text{total}} = g(m - \rho V)$$

$$F_{\text{total}} = (9.8\ \text{m/s}^2)\cdot[(0.06\ \text{kg}) - (1\ \text{g/cm}^3)\cdot(5.7\ \text{cm}^3)\cdot(1\ \text{kg/1,000 g})]$$

$$F_{\text{total}} = 0.53\ \text{N}$$

**72. A is correct.**

The gauge pressure is referenced at ambient air pressure thus:

$$P = \rho g h$$

$$P = (1{,}000\ kg/m^3)\cdot(9.8\ m/s^2)\cdot(6\ m + 22\ m)$$

$$P = 2.7 \times 10^5\ N/m^2$$

**73. B is correct.**

Atmospheric pressure is not taken into account because it acts at the surface of the water and all around Mike's finger, so it cancels.

$$P_{water} = \rho g h$$

$$P_{water} = (10^3\ kg/m^3)\cdot(10\ m/s^2)\cdot(1\ m)$$

$$P_{water} = 10^4\ N/m^2$$

Area of hole:

$$A = (0.01\ m)\cdot(0.01\ m\ )$$

$$A = 10^{-4}\ m^2$$

$$F = PA$$

$$F = (10^4\ N/m^2)\cdot(10^{-4}\ m^2)$$

$$F = 1\ N$$

**74. C is correct.**

To calculate the buoyant force due to the water:

$$F_B = 7.86\ N - 6.92\ N$$

$$F_B = 0.94\ N$$

Volume of displaced water is equal to volume of object:

$$F_B = \rho g V$$

$$V = F_B / \rho g$$

$$V = 0.94\ N / (1{,}000\ kg/m^3)\cdot(9.8\ m/s^2)$$

$$V = 9.6 \times 10^{-5}\ m^3$$

*continued…*

Mass of the object:

$m = W / g$

$m = 7.86 \text{ N} / 9.8 \text{ m/s}^2$

$m = 0.8 \text{ kg}$

$\rho = \text{mass / volume}$

$\rho = 0.8 \text{ kg} / 9.6 \times 10^{-5} \text{ m}^3$

$\rho = 8{,}333 \text{ kg/m}^3$

**75. C is correct.**

As the block just enters the water, the total pressure will be the sum of atmospheric pressure and the pressure produced by submersion.

$P_{total} = P_{atmosphere} + \rho gh$

When the block just enters the water:

$P_{total} = P_{atmosphere} + 0$

Graph C depicts the scenario as the block will experience an initial pressure of $P_{atmosphere}$, which will linearly increase (from $\rho gh$) as the block is lowered further into the water.

**76. A is correct.**

$F_B = \rho g V$

*Buoyant force* is directly proportional to volume.

Volume of a sphere is given as:

$V = 4 / 3\pi r^3$

If radius doubles:

$V = 4 / 3\pi(2r)^3$

$V = (8)\cdot(4 / 3\pi r^3)$

Therefore, when $r$ doubles, $F_B$ increase by a factor of 8 times.

 

**77. A is correct.**

Young's Modulus ($E$) is given as:

$$E = FL / A\Delta L$$

where $F = mg$

$$E = mgL / A\Delta L$$

Rearranging for $m$:

$$m = EA\Delta L / gL_i$$

$$A = \pi r^2$$

$$m = E\pi r^2\Delta L / gL_i$$

$$m = [(2 \times 10^{11}\ N/m^2)\cdot\pi(0.9 \times 10^{-3}\ m)^2\cdot(1.8 \times 10^{-3}\ m)] / (9.8\ m/s^2)\cdot(4.2\ m)$$

$$m = 916.8\ kg\cdot m^2 / 41.2\ m^2/s^2$$

$$m = 22\ kg$$

**78. C is correct.**

$$P = F / A$$

$$F = PA$$

$$F = (3\ atm)\cdot(1.01 \times 10^5\ Pa/1\ atm)\cdot(0.2\ m)^2$$

$$F = 12{,}120\ N = 1.2 \times 10^4\ N$$

**79. B is correct.**

The equation of continuity:

$f = Av$ remains constant.

$$A_2V_2 = A_1V_1$$

$$V_2 = A_1V_1 / A_2$$

$$V_2 = \pi(16\ cm)^2\cdot(V_1) / \pi(4\ cm)^2$$

$$V_2 = 16V_1$$

If the radius decreases by a factor of 4 then the velocity increases by a factor of 16.

**80. B is correct.**

For most substances, the solid form is denser than the liquid phase, and therefore, a block of most solids sinks in the liquid.

Regarding pure water, though, a block of ice (solid phase) floats in liquid water because ice is less dense.

Like other substances, when liquid water is cooled from room temperature, it becomes increasingly dense.

However, at ~4 °C (39 °F), water reaches its maximum density, and as it is cooled further, it expands and becomes less dense. This phenomenon is *negative thermal expansion* and is attributed to strong intermolecular interactions that are orientation-dependent.

The density of water is about 1 $g/cm^3$ and depends on the temperature.

When frozen, the density of water is decreased by about 9%. This is due to the decrease in intermolecular vibrations, which allows water molecules to form stable hydrogen bonds with other water molecules around.

As these hydrogen bonds form, molecules are locking into positions similar to forming hexagonal structures.

Even though hydrogen bonds are shorter in the crystal than in the liquid, this position locking decreases the average coordination number of water molecules as the liquid reaches the solid phase.

 

*Notes for active learning*

*Notes for active learning*

# 5 – Chemical Bonding: Detailed Explanations

==========================================================

**Practice Set 1: Questions 1–20**

==========================================================

**1. D is correct.**

Valence shell is an atom's outermost shell (i.e., highest principal quantum number *n*). Valence electrons are in the outermost electron shell that can participate in a chemical bond.

The number of valence electrons for an element can be determined by its group (i.e., vertical column) on the periodic table. Except for the transition metals (i.e., groups 3-12), the group number identifies how many valence electrons are associated with a particular element: elements of the same group have the same number of valence electrons.

**2. A is correct.**

Three degenerate *p* orbitals exist for an atom with an electron configuration in the second shell or higher. The first shell only has access to *s* orbitals.

The *d* orbitals become available from n = 3 (third shell).

**3. C is correct.**

Valence shell is an atom's outermost shell (i.e., highest principal quantum number *n*).

Valence electrons are in the outermost electron shell that can participate in a chemical bond.

The number of valence electrons for an element can be determined by its group (i.e., vertical column) on the periodic table. Except for the transition metals (i.e., groups 3-12), the group number identifies how many valence electrons are associated with a particular element: elements of the same group have the same number of valence electrons.

To find the number of valence electrons in a sulfite ion, $SO_3^{2-}$, add the valence electrons of each atom:

Sulfur = 6; Oxygen = (6 × 3) = 18

Total = 24

This ion has a net charge of –2, which indicates that it has 2 extra electrons. Therefore, the valence electrons would be 24 + 2 = 26.

**4. D is correct.**

*London dispersion forces* result from the momentary flux of valence electrons and are present in all compounds; they are the attractive forces that hold molecules.

They are the weakest intermolecular forces, and their strength increases with increasing size (i.e., surface area contact) and polarity of the molecules involved.

**5. B is correct.**

Hydroxyl group (alcohol or ~OH) has oxygen with 2 lone pairs and one attached hydrogen.

Therefore, each hydroxyl group can participate in 3 hydrogen bonds:

5 hydroxyl groups × 3 bonds = 15 hydrogen bonds.

Oxygen of the ether group (C–O–C) in the ring has 2 lone pairs for an additional 2 H–bonds.

**6. B is correct.**

Valence shell is an atom's outermost shell (i.e., highest principal quantum number $n$). Valence electrons are in the outermost electron shell that can participate in a chemical bond.

The number of valence electrons for an element can be determined by its group (i.e., vertical column) on the periodic table.

Except for the transition metals (i.e., groups 3-12), the group number identifies how many valence electrons are associated with a particular element: elements of the same group have the same number of valence electrons.

```
   H H
H:C:C:H
   H H
```

*Lewis dot structure for ethane*

**7. A is correct.**

In covalent bonds, the electrons can be shared equally or unequally.

Polar covalent bonded atoms are covalently bonded compounds that involve *unequal sharing* of electrons due to large electronegativity differences (Pauling units of 0.4 to 1.7) between the atoms.

An example is water, where there is a polar covalent bond between oxygen and hydrogen. Water is a polar molecule with partially negative oxygen, while the hydrogens are partially positive.

 

**8. A is correct.**

*Van der Waals forces* involve nonpolar (hydrophobic) molecules like hydrocarbons. It is the total of attractive or repulsive forces between molecules and, therefore, can be attractive or repulsive. It can include the force between two permanent dipoles, between a permanent dipole and a temporary dipole, or between two temporary dipoles.

Hydrogens, bonded directly to F, O, or N, participate in hydrogen bonds. Hydrogen is partially positive (i.e., delta plus or $\partial+$) due to the bond to these electronegative atoms. The lone pair of electrons on the F, O, or N interacts with the $\partial+$ hydrogen to form a hydrogen bond.

Hydrogen bonds are a type of dipole-dipole and are the strongest intermolecular forces (i.e., between molecules), followed by other dipole-dipole, dipole-induced dipole, and van der Waals forces (i.e., London dispersion).

**9. D is correct.**

Representative structures include:

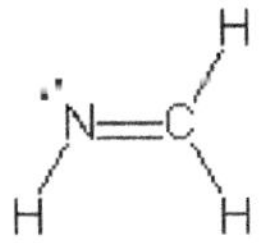

$HNCH_2$ : one single and one double bond

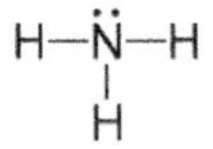

$NH_3$ : three single bonds

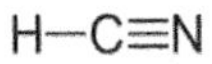

H—C≡N

HCN : one triple bond and one single bond

**10. C is correct.**

The valence shell is an atom's outermost shell (i.e., highest principal quantum number $n$).

Valence electrons are in the outermost electron shell that can participate in a chemical bond.

Number of valence electrons for an element can be determined by its group (i.e., vertical column) on the periodic table.

Except for the transition metals (i.e., groups 3-12), the group number identifies how many valence electrons are associated with a particular element: elements of the same group have the same number of valence electrons.

*continued...*

 

| H· | | | | | | | He: |
|---|---|---|---|---|---|---|---|
| Li· | ·Be· | ·B· | ·C· | ·N· | :O· | :F· | :Ne: |
| Na· | ·Mg· | ·Al· | ·Si· | ·P· | :S· | :Cl· | :Ar: |
| K· | ·Ca· | ·Ga· | ·Ge· | ·As· | :Se· | :Br· | :Kr: |
| Rb· | ·Sr· | ·In· | ·Sn· | ·Sb· | :Te· | :I· | :Xe: |

*Sample Lewis dot structures for some elements*

**11. B is correct.**

Nitrite ion has the chemical formula $NO_2^-$ with the negative charge distributed between the two oxygen atoms.

*Two resonance structures of the nitrite ion*

**12. D is correct**

Nitrogen has 5 valence electrons. Nitrogen has 3 bonds to hydrogen in ammonia, and a lone pair remains on the central nitrogen atom.

Ammonium ion ($NH_4^+$) has 4 hydrogens, and the lone pair of nitrogen has been used to bond to the $H^+$ added to ammonia.

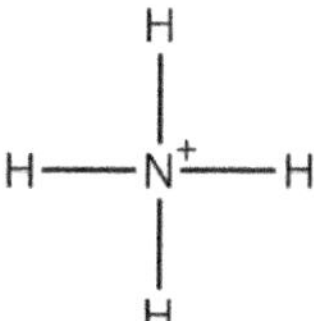

*Formal charge shown on the hydronium ion*

**13. C is correct.**

In atoms or molecules that are ions, the number of electrons is not equal to the number of protons, which gives the molecule a charge (either positive or negative).

Ionic bonds form when the difference in electronegativity of atoms in a compound is greater than 1.7 Pauling units.

*continued...*

 

*Ionic bonds* transfer an electron from the electropositive element (along the left-hand column/group) to the electronegative element (along right-hand column/group) on periodic table.

Oppositely charged ions are attracted to each other, resulting in ionic bonds.

In an ionic bond, electron(s) are transferred from a metal to a nonmetal, giving both molecules a full valence shell and causing the molecules to associate with each other.

**14. A is correct.**

*Electronegativity* is a chemical property describing an atom's tendency to attract electrons to itself.

The most electronegative atom is F, while the least electronegative atom is Fr. The trend for increasing electronegativity within the periodic table is up and toward the right (i.e., fluorine).

**15. D is correct.**

Hydrogen bonds are the strongest intermolecular forces (i.e., between molecules), followed by dipole-dipole, dipole-induced dipole, and van der Waals forces (i.e., London dispersion).

*London dispersion forces* are present in compounds, the attractive forces holding molecules. They are the weakest of the intermolecular forces, and their strength increases with the increasing size and polarity of the molecules involved.

**16. C is correct.**

For asymmetrical molecules (e.g., water is bent), use the atoms' geometry and difference of electronegativity values. For water, H (2.1 Pauling units) is more electropositive than O (3.5 Pauling units).

For sulfur dioxide, S (2.5 Pauling units) is more electropositive than O (3.5 Pauling units). The reported dipole moment for water is 1.8 D compared to 1.6 D for sulfur dioxide.

$CO_2$ does not have nonbonding electrons on the central carbon atom, and it is symmetrical, so it is non-polar and has zero dipole moment.

$CCl_4$ is symmetrical as a tetrahedron and, therefore, does not exhibit a net dipole.

**17. C is correct.**

*Cohesion* is the property of like molecules sticking together. Hydrogen bonds join water molecules.

*Adhesion* is the attraction between unlike molecules (e.g., the meniscus observed from water molecules adhering to the graduated cylinder).

*Polarity* is the differences in electronegativity between bonded molecules. It gives rise to the delta plus (on H) and the delta minus (on O), permitting hydrogen bonds to form between water molecules.

**18. A is correct.**

With little or no difference in electronegativity between the elements (i.e., Pauling units < 0.4), it is a nonpolar covalent bond, whereby the electrons are shared between the two bonding atoms.

Among the answer choices, H, C, and O atoms are closest in magnitude for Pauling units for electronegativity.

**19. D is correct.**

Positively charged nuclei repel each other while each attracts the bonding electrons. These opposing forces reach equilibrium at the bond length.

**20. C is correct.**

In a carbonate ion ($CO_3^{2-}$), the carbon atom is bonded to 3 oxygen atoms. Two of those bonds are single covalent bonds, and the oxygen atoms each have an extra (third) lone pair of electrons, which imparts a negative formal charge. The remaining oxygen has a double bond with carbon.

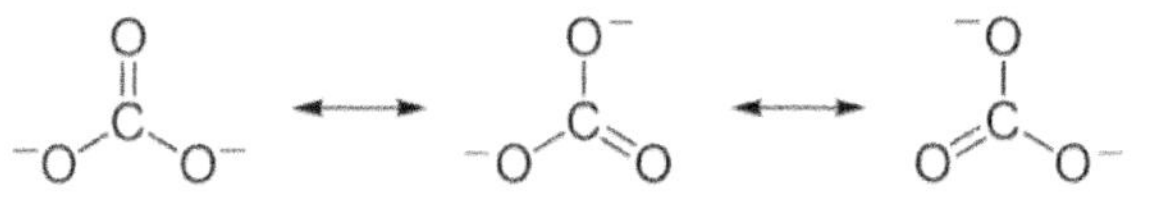

*Three resonance structures for the carbonate ion $CO_3^{2-}$*

 

---

**Practice Set 2: Questions 21–40**

---

**21. D is correct.**

*Dipole* is a separation of full (or partial) positive and negative charges due to differences in the electronegativity of atoms.

**22. C is correct.**

*Ionic bonds* transfer an electron from the electropositive element (along the left-hand column/group) to the electronegative atom (along the right-hand column/group) on the periodic table.

Ca is a group II element with 2 electrons in its valence shell.

I is a group VII element with 7 electrons in its valence shell.

Ca becomes $Ca^{2+}$, and each of the two electrons is joined to I, which becomes $I^-$.

**23. B is correct.**

An atom with 4 valence electrons can make a maximum of 4 bonds. 1 double and 1 triple bond equals 5 bonds, which exceeds the maximum allowable bonds.

**24. D is correct.**

Water molecules stick to each other (i.e., *cohesion*) due to the collective action of hydrogen bonds between individual water molecules. These hydrogen bonds are constantly breaking and reforming; these bonds hold many molecules together.

Water also sticks to surfaces (i.e., *adhesion*) because of water's polarity. The water may form a thin film on a smooth surface (e.g., glass) because the molecular forces between glass and water molecules (adhesive forces) are stronger than the cohesive forces between the water molecules.

**25. D is correct.**

*Hydrogen bonds* are the strongest intermolecular forces (i.e., between molecules), followed by dipole-dipole, dipole-induced dipole, and van der Waals forces (i.e., London dispersion).

When ionic compounds are dissolved, each ion is surrounded by more than one water molecule. Several water molecules' combined force ion-dipole interactions are stronger than a single ionic bond.

 

**26. B is correct.**

Hydrogens, bonded directly to F, O, or N, participate in hydrogen bonds. The hydrogen is partially positive (i.e., delta plus or $\partial+$) due to the bond to these electronegative atoms. The lone pair of electrons on the F, O, or N interacts with the $\partial+$ hydrogen to form a hydrogen bond.

The two lone pairs of electrons on the oxygen atom can each participate as a hydrogen bond acceptor. The molecule does not have hydrogen bonded directly to an electronegative atom (F, O, or N) and cannot be a hydrogen bond donor.

**27. A is correct.**

Water is a bent molecule with a partial negative charge on the oxygen and a partial positive charge on each hydrogen. Therefore, the water molecule exists as a dipole. The $Na^+$ is an ion attracted to the partial negative charge on the oxygen in the water molecule.

**28. C is correct.**

$H_2CO$ (formaldehyde) molecule is shown below and has one C=O bond and two C–H bonds.

Electronegative oxygen pulls electron density away from the carbon atom and creates a net dipole towards the oxygen, resulting in a polar covalent bond.

Electronegativity values of carbon and hydrogen are similar and result in a covalent bond (i.e., about equal sharing of bonded electrons).

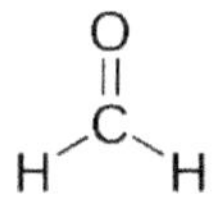

O
||
C
H H

**29. A is correct.**

Valence shell is the outermost shell (i.e., the highest principal quantum number $n$) of an atom.

Valence electrons are in the outermost electron shell that can participate in a chemical bond.

The number of valence electrons for an element can be determined by its group (i.e., vertical column) on the periodic table.

Except for the transition metals (i.e., groups 3-12), the group number identifies how many valence electrons are associated with a particular element: elements of the same group have the same number of valence electrons.

 

**30. D is correct.**

Group IA elements (e.g., Li, Na, and K) tend to lose 1 electron to achieve a complete octet to be cations with a +1 charge.

Group IIA elements (e.g., Mg and Ca) tend to lose 2 electrons to achieve a complete octet to be cations with a +2 charge.

Group VIIA elements (halogens such as F, Cl, Br, and I) tend to gain 1 electron to achieve a complete octet of anions with a –1 charge.

**31. B is correct.**

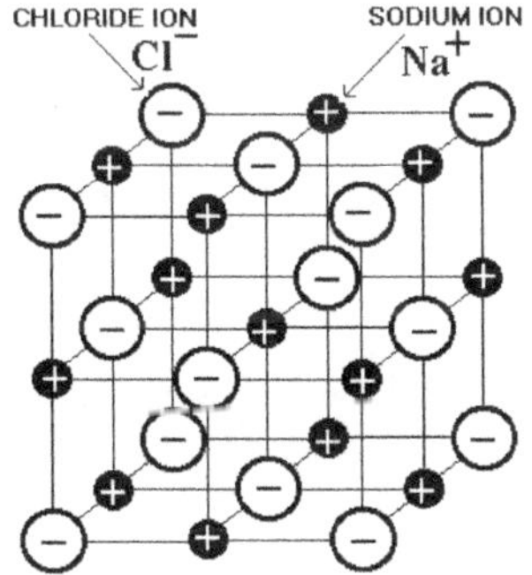


*Lattice structure of sodium chloride*

**32. C is correct.**

Each positively charged nuclei attracts the bonding electrons.

**33. B is correct.**

*Electronegativity* measures how strongly an element attracts electrons within a bond.

Electronegativity is the relative attraction of the nucleus for bonding electrons. It increases from left to right (i.e., periods) and from bottom to top along a group (like the trend for ionization energy).

Most electronegative atom is F, while the least is Fr.

The greater the difference in electronegativity between two atoms in a compound, the more polar of a bond these atoms form. The atom with the higher electronegativity is the partial (delta) negative end of the dipole.

**34. A is correct.**

*Hydrogen bonds* are the strongest intermolecular forces (i.e., between molecules), followed by dipole-dipole, dipole-induced dipole, and van der Waals forces (i.e., London dispersion).

Hydrogens, bonded directly to F, O, or N, participate in hydrogen bonds. H–bonding is a polar interaction involving hydrogen forming bonds to the electronegative atoms F, O, or N, accounting for water's high boiling points.

The hydrogen is partially positive (i.e., delta plus or $\partial+$) due to the bond to these electronegative atoms. The lone pair of electrons on the F, O, or N interacts with the $\partial+$ hydrogen to form a hydrogen bond.

S
H H

*Molecular geometry of $H_2S$*

Polar molecules have high boiling points because of polar interaction. $H_2S$ is a polar molecule that forms dipole-dipole interactions and does not form hydrogen bonds.

**35. D is correct.**

With 4 electron pairs, the starting shape of this element is tetrahedral.

O
H H

After removing 2 of the 4 groups surrounding the central atom, the central atom has two groups bound (e.g., $H_2O$). The molecule has the hybridization (and original angles) from a tetrahedral shape, bent rather than linear.

**36. D is correct.**

*Ionic bonds* hold salt crystals. Salts are composed of cations and anions and are electrically neutral. When salts are dissolved in solution, they separate into their constituent ions by breaking of non-covalent interactions.

Water: the hydrogen atoms in the molecule are covalently bonded to the oxygen atom.

Hydrogen peroxide: the molecule has one more oxygen atom than a water molecule and is also held together by covalent bonds.

Ester (RCOOR'): undergoes hydrolysis and breaks covalent bonds to separate into its constituent carboxylic acid (RCOOH) and alcohol (ROH).

 

**37. B is correct.**

An ionic compound consists of a metal ion and a nonmetal ion. Ionic bonds are formed between elements with an electronegativity difference greater than 1.7 Pauling units (e.g., a metal atom and a non-metal atom).

(K and I) form ionic bonds because K is a metal and I is a nonmetal.

(C and Cl) only have nonmetals, while (Fe and Mg) have only metals.

(Ga and Si) has a metal (Ga) and a metalloid (Si); they *might* form weak ionic bonds.

**38. A is correct.**

*Coordinate bond* is a covalent bond (i.e., a shared pair of electrons) in which both electrons come from the same atom.

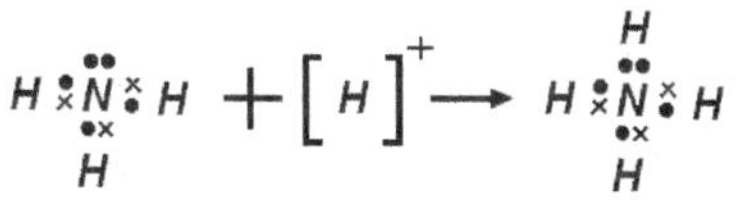

*The lone pair of the nitrogen is donated to form the fourth N–H bond*

**39. B is correct.**

*Dipole moment* depends on the overall shape of the molecule, the length of the bond, and whether the electrons are pulled to one side of the bond (or the molecule overall).

For a large dipole moment, one element pulls electrons more strongly (i.e., differences in electronegativity).

*Electronegativity* measures how strongly an element attracts electrons within a bond.

Electronegativity is the relative attraction of the nucleus for bonding electrons. It increases from left to right (i.e., periods) and from bottom to top along a group (like the trend for ionization energy). The most electronegative atom is F, while the least electronegative is Fr.

The greater the difference in electronegativity between two atoms in a compound, the more polar a bond these atoms form. The atom with higher electronegativity is the dipole's partial (delta) negative end.

**40. A is correct.**

Valence shell is an atom's outermost shell (i.e., the highest principal quantum number n). Noble gas configuration refers to eight electrons in the atom's outermost valence shell, a complete octet.

Depending on how many electrons it starts with, an atom may have to lose, gain, or share an electron to obtain the noble gas configuration.

*Notes for active learning*

 

---

**Practice Set 3: Questions 41–60**

---

**41. C is correct.**

The bond between the oxygens is nonpolar, while the bonds between the oxygens and hydrogens are polar (due to the differences in electronegativity).

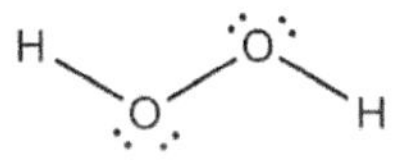

*Line bond structure of $H_2O_2$ with lone pairs shown*

**42. A is correct.**

*Octet rule* states that atoms of main-group elements tend to combine, so each atom has eight electrons in its valence shell. This occurs because electron arrangements involving eight valence electrons are highly stable, as with noble gases.

Selenium (Se) is in group VI and has 6 valence electrons. It has a complete octet (i.e., stable) by gaining two electrons.

**43. C is correct.**

When new compounds form by rearranging atoms, it is an example of a chemical reaction.

**44. D is correct.**

HBr experiences dipole-dipole interactions due to the electronegativity difference, resulting in a partial negative charge on bromine and a partial positive charge on hydrogen.

**45. C is correct.**

An ion is an atom (or a molecule) in which the number of electrons is not equal to the number of protons. Therefore, the atom (or molecule) has a net positive or negative electrical charge.

If a neutral atom loses one or more electrons, it has a net positive charge (i.e., cation). If a neutral atom gains one or more electrons, it has a net negative charge (i.e., anion).

Aluminum is a group III atom with proportionally more protons per electron once the cation forms.

All the other elements listed are from groups I or II.

 

**46. A is correct.**

*Dipole-dipole* (e.g., $CH_3Cl...CH_3Cl$) attraction occurs between neutral molecules, while ion-dipole interaction involves dipole interactions with charged ions (e.g., $CH_3Cl...^{-}OOCCH_3$).

Hydrogen bonds are the strongest intermolecular forces (i.e., between molecules), followed by dipole-dipole, dipole-induced dipole, and van der Waals forces (i.e., London dispersion).

**47. B is correct.**

When more than two $H_2O$ molecules are present (e.g., liquid water), more bonds (between 2 and 4) are possible because the oxygen of one water molecule has two lone pairs of electrons, each of which can form a hydrogen bond with hydrogen on another water molecule.

This bonding can repeat such that every water molecule is H–bonded with up to four other molecules (two through its two lone pairs of O and two through its two hydrogen atoms).

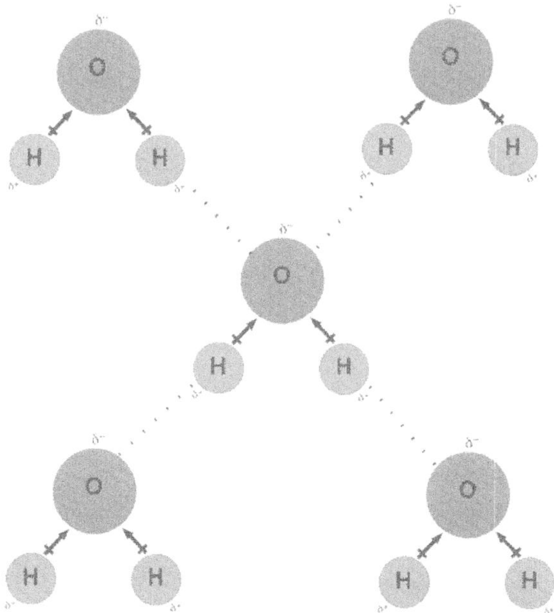


*Ice with 4 hydrogen bonds between the water molecules*

Hydrogen bonding affects the crystal structure of ice (hexagonal lattice). The density of ice is less than the density of water at the same temperature.

Thus, the solid phase of water (ice) floats on the liquid. The creation of the additional H–bonds (up to 4 for ice) forces the individual molecules further, giving them less density, unlike most other substances in the solid phase.

*continued...*

 

For most substances, the solid form is denser than the liquid phase. Therefore, a block of most solids sinks in the liquid. Regarding pure water, a block of ice (solid phase) floats in liquid water because ice is less dense.

Like other substances, when liquid water is cooled from room temperature, it becomes increasingly dense. However, at approximately 4 °C (39 °F), water reaches its maximum density, and as it is cooled further, it expands and becomes less dense. This phenomenon is *negative thermal expansion* and is attributed to strong intermolecular interactions that are orientation-dependent.

Density of water is about 1 g/cm$^3$ and depends on the temperature. When it freezes, the density of water is decreased by about 9%. This is due to the decrease in intermolecular vibrations, allowing water molecules to form stable hydrogen bonds with other water molecules around. As these hydrogen bonds form, molecules are locked into positions like the hexagonal structure.

Even though hydrogen bonds are shorter in the crystal than in the liquid, this position locking decreases the average coordination number of water molecules as the liquid reaches the solid phase.

**48. A is correct.**

The phrase "from its elements" in the question stem implies the need to create the formation reaction of the compound from its elements.

Chemical equation:

$$3\ Na^+ + N^{3-} \rightarrow Na_3N$$

Each sodium loses one electron to form $Na^+$ ions.

Nitrogen gains 3 electrons to form the $N^{3-}$ ion.

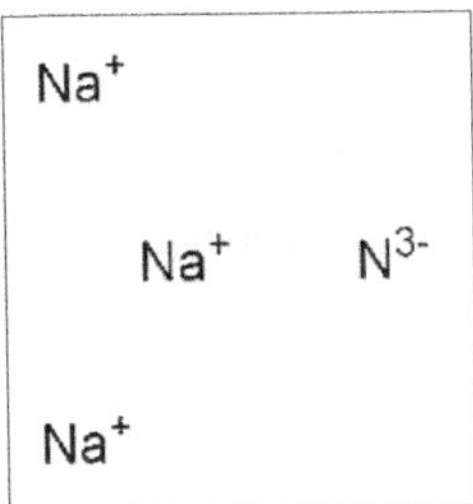


*Schematic of an ionic compound*

**49. D is correct.**

Valence shell is an atom's outermost shell (i.e., the highest principal quantum number n).

Valence electrons are in the outermost electron shell that can participate in a chemical bond.

The number of valence electrons for a neutral element can be determined by its group (i.e., vertical column) on the periodic table. Except for the transition metals (i.e., groups 3-12), the group number identifies how many valence electrons are associated with a particular element: elements of the same group have the same number of valence electrons.

Lewis dot structures show the valence electrons of an individual atom as dots around the element's symbol. Non-valence electrons (i.e., inner shell electrons) are not represented in Lewis structures.

| | | | | | | | |
|---|---|---|---|---|---|---|---|
| H· | | | | | | | He: |
| Li· | ·Be· | ·B· | ·C· | ·N· | :O· | :F· | :Ne: |
| Na· | ·Mg· | ·Al· | ·Si· | ·P· | :S· | :Cl· | :Ar: |
| K· | ·Ca· | ·Ga· | ·Ge· | ·As· | :Se· | :Br· | :Kr: |

*Sample Lewis dot structures for some neutral elements*

$Mg^{2+}$ has ten electrons but zero valance (outer shell) electrons. The valence shell (i.e., 3s subshell) has 0 electrons, and therefore, there are no dots shown in Lewis dot structure: $Mg^{2+}$

$S^{2-}$ has eight valence electrons (same electronic configuration as Ar).

$Ga^{+}$ has two valence electrons (same electronic configuration as Ca).

$Ar^{+}$ has seven valence electrons (same electronic configuration as Cl).

**50. B is correct.**

*Octet rule* states that atoms of main-group elements tend to combine, so each atom has eight electrons in its valence shell.

**51. D is correct.**

In a chemical reaction, the bonds formed differ from those broken.

**52. D is correct.**

Hydrogen sulfide (i.e., sewer gas) is $H_2S$.

Hydrosulfide (bisulfide) ion is $HS^-$. Sulfide ion is $S^{2-}$

**53. D is correct.**

An ionic compound consists of a metal ion and a nonmetal ion.

Ionic bonds are formed between elements with an electronegativity difference greater than 1.7 Pauling units (e.g., a metal atom and a non-metal atom). The charge on bicarbonate ion ($HCO_3$) is −1, while the charge on Mg is +2.

The proper formula is $Mg(HCO_3)_2$

**54. C is correct.**

Lattice: crystalline structure – consists of unit cells. Unit cell: the smallest unit of solid crystalline pattern

*Covalent unit* is not a valid term.

**55. D is correct.**

Formula to calculate dipole moment:

$$\mu = qr$$

where μ is dipole moment (coulomb-meter or C·m), $q$ is the charge (coulomb or C), and $r$ is the radius (meter or m)

Convert the unit of dipole moment from Debye to C·m:

$$0.16 \times 3.34 \times 10^{-30} = 5.34 \times 10^{-31}\ \text{C·m}$$

Convert the unit of radius to meter:

$$115\ \text{pm} \times 1 \times 10^{-12}\ \text{m/pm} = 115 \times 10^{-12}\ \text{m}$$

Rearrange the dipole moment equation to solve for $q$:

$$q = \mu / r$$

$$q = (5.34 \times 10^{-31}\ \text{C·m}) / (115 \times 10^{-12}\ \text{m})$$

$$q = 4.64 \times 10^{-21}\ \text{C}$$

*continued...*

Express the charge in terms of electron charge (e).

$$(4.64 \times 10^{-21}\text{ C}) / (1.602 \times 10^{-19}\text{ C/e}) = 0.029\text{ e}$$

In this NO molecule, oxygen is the more electronegative atom. Therefore, the charge experienced by the oxygen atom is negative: –0.029e.

**56. A is correct.**

The greater the difference in electronegativity between two atoms in a compound, the more polar of a bond these atoms form.

Atom with higher electronegativity is the dipole's partial (delta) negative end.

**57. C is correct.**

*Hybridization* of the central atom and the corresponding molecular geometry:

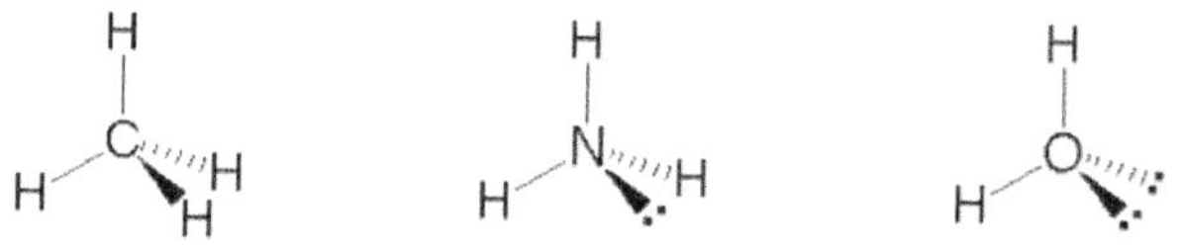

*Examples of $sp^3$ hybridized atoms with 4 substituents ($CH_4$), 3 substituents ($NH_3$) and two substituents ($H_2O$)*

Hybridization - Shape

*sp* – linear

$sp^2$ – trigonal planar

$sp^3$ – tetrahedral

$sp^3d$ – trigonal bipyramid

**58. C is correct.**

*Hydrogen bonds* are the strongest intermolecular forces (i.e., between molecules), followed by dipole-dipole, dipole-induced dipole, and van der Waals forces (i.e., London dispersion).

Hydrogen is a very electropositive element, and O is a very electronegative element. Therefore, these elements will be attracted to each other, within the same water molecule and between water molecules.

*Ionic and covalent bonding* only form within a molecule and not between adjacent molecules.

 

**59. B is correct.**

In a compound, the sum of ionic charges must equal zero.

To balance the charges of $K^+$ and $CO_3^{2-}$, there must be 2 $K^+$ ions for every $CO_3^{2-}$ ion.

The formula for this balanced molecule is $K_2CO_3$.

**60. C is correct.**

The substantial difference between the two forms of carbon (i.e., graphite and diamonds) is mainly due to their crystal structure, hexagonal for graphite and cubic for diamond.

The conditions for converting graphite into diamond are high pressure and high temperature. That is why creating synthetic diamonds is time-consuming, energy-intensive, and expensive since carbon is forced to change its bonding structure.

==================================================================

**Practice Set 4: Questions 61–88**

==================================================================

**61. A is correct.**

*Lewis acids* are electron-pair acceptors, whereas *Lewis bases* are electron-pair donors.

Boron has an atomic number of five and has a vacant $2p$ orbital to accept electrons.

**62. D is correct.**

$CH_3SH$ experiences dipole-dipole interactions due to the electronegativity difference, resulting in a partial negative charge on sulfur and a partial positive charge on hydrogen.

$CH_3CH_2OH$ experiences dipole-dipole interactions (due to the electronegativity difference, resulting in a partial negative charge on oxygen and a partial positive charge on hydrogen. This example is a type of dipole-dipole known as hydrogen bonding (when hydrogen is bonded directly to fluorine, oxygen, or nitrogen).

**63. A is correct.**

There is an overlap of $sp^2$ orbitals and a $p$ orbital on the adjacent carbon atoms in carbon-carbon double bonds.

$sp^2$ orbitals overlap head-to-head as a *sigma* ($\sigma$) bond, whereas the $p$ orbitals overlap sideways as a *pi* ($\pi$) bond.

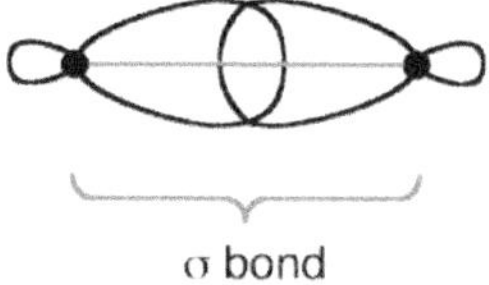


*Sigma bond formation showing electron density along the internucleus axis*

*Bond length* and *bond strength* ($\sigma$ or $\pi$) depend on the size and shape of the atomic orbitals and the density to overlap effectively.

$\sigma$ bonds are stronger than $\pi$ bonds because head-to-head orbital overlaps involve more shared electron density than sideways overlaps.

$\sigma$ bonds formed from two $2s$ orbitals are shorter than those formed from two $2p$ or two $3s$ orbitals.

*continued...*

 

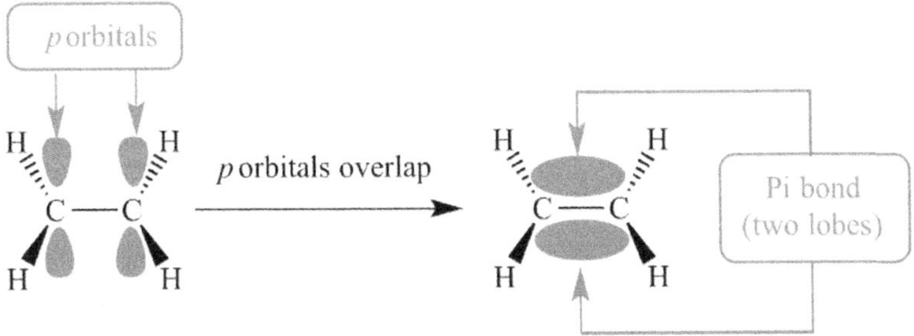


*Two pi orbitals showing the pi bond formation during sideways overlap – note the absence of electron density (i.e., node) along the internucleus axis.*

Carbon, oxygen, and nitrogen are in the second period ($n = 2$), while sulfur (S), phosphorus (P), and silicon (Si) are in the third period.

Therefore, S, P, and Si use $3p$ orbitals to form $\pi$ bonds, while C, N, and O use $2p$ orbitals. The $3p$ orbitals are much larger than the $2p$ orbitals, and therefore, there is a reduced probability for an overlap of the $2p$ orbital of C and the $3p$ orbital of S, P, and Si.

B: S, P, and Si can hybridize, but these elements can combine *s* and *p* orbitals and (unlike C, O, and N) have *d* orbitals.

C: S, P, and Si (in their ground state electron configurations) have partially occupied *p* orbitals which form bonds.

D: carbon combines with elements below the second row of the periodic table. For example, carbon forms bonds with higher principal quantum numbers ($n > 2$) halogens (e.g., F, Cl, Br, and I).

**64. B is correct.**

Magnesium atom has a +2 charge, and the sulfur atom has a –2 charge.

**65. C is correct.**

Hydrogen bonds are the strongest intermolecular forces (i.e., between molecules), followed by dipole-dipole, dipole-induced dipole, and van der Waals forces (i.e., London dispersion).

Larger atoms have more electrons; they can exert stronger induced dipole forces than smaller atoms.

Bromine is under chlorine on the periodic table, which means it has more electrons and stronger induced dipole forces.

**66. A is correct.**

Hydrogen bonds are the strongest intermolecular forces (i.e., between molecules), followed by dipole-dipole, dipole-induced dipole, and van der Waals forces (i.e., London dispersion).

Hydrogens, bonded directly to F, O, or N, participate in hydrogen bonds. The hydrogen is partially positive (i.e., delta plus: $\partial+$) due to the bond to these electronegative atoms. The lone pair of electrons on the F, O, or N interacts with the partial positive ($\partial+$) hydrogen to form a hydrogen bond.

When ~NH or ~OH groups are present in a molecule, they form hydrogen bonds with other molecules. Hydrogen bonding is the strongest and major intermolecular force in this substance.

**67. B is correct**

Potassium oxide ($K_2O$) contains metal and nonmetal, which form a salt.

Salts contain ionic bonds and dissociate in aqueous solutions and, therefore, are strong electrolytes.

Polyatomic (i.e., molecular) ion is a charged chemical species composed of two or more atoms covalently bonded or composed of a metal complex acting as a single unit. An example is the hydroxide ion with one oxygen atom and one hydrogen atom; hydroxide has a charge of −1.

Polyatomic ions bond with other ions and has various charges.

A polyatomic ion might contain only metals or nonmetals.

A polyatomic ion is not neutral.

| oxidation state | −1 | +1 | +3 | +5 | +7 |
|---|---|---|---|---|---|
| anion name | chloride | hypochlorite | chlorite | chlorate | perchlorate |
| formula | $Cl^-$ | $ClO^-$ | $ClO_2^-$ | $ClO_3^-$ | $ClO_4^-$ |

**68. D is correct.**

Valence shell is an atom's outermost shell (i.e., highest principal quantum number *n*).

Valence electrons are in the outermost electron shell that can participate in a chemical bond.

 

**69. B is correct.**

Nitrogen ($NH_3$) is neutral with three bonds, negative with two bonds ($^-NH_2$), and positive with 4 bonds ($^+NH_4$)

Line bond structure of methylamine with the lone pair on nitrogen shown:

```
    H    H
    |    |
H———C————N :
    |    |
    H    H
```

**70. B is correct.**

*Sulfide ion* is $S^{2-}$.

*Hydrogen sulfide* (i.e., sewer gas) is $H_2S$.

*Hydrosulfide* (bisulfide) ion is $HS^-$.

*Sulfite ion* is the conjugate base of bisulfite and has the molecular formula of $SO_3^{2-}$

*Sulfur* (S) is a chemical element with an atomic number of 16.

*Sulfate ion is* a polyatomic anion (i.e., two or more atoms covalently bonded or metal complex) with the molecular formula of $SO_4^{2-}$

**71. C is correct.**

*Ionic bonds* involve electrostatic interactions between oppositely charged atoms.

Electrons from the metallic atom are transferred to the nonmetallic atom, giving both atoms a full valence shell.

**72. B is correct.**

If the charge of O is −1, the proper formula of its compound with K (+1) should be KO.

**73. C is correct.**

*Lattice energy* of a crystalline solid is usually defined as the energy of the formation of the crystal from infinitely separated ions (i.e., an exothermic process and hence the negative sign).

Some older textbooks define lattice energy with the opposite sign. The older notation refers to the energy required to convert the crystal into infinitely separated gaseous ions in a vacuum (i.e., an endothermic process and the positive sign).

*continued...*

Lattice energy decreases downward for a group.

Lattice energy increases with charge.

Because NaCl and LiF contain charges of +1 and −1, they would have smaller lattice energy than the higher-charged ions on other options.

Na is lower than Li, and Cl is lower than F in their respective groups, so NaCl would have lower lattice energy than LiF.

The bond between ions of opposite charge is strongest when the ions are small.

The force of attraction between oppositely charged particles is directly proportional to the product of the charges on the two objects ($q_1$ and $q_2$) and inversely proportional to the square of the distance between the objects ($r^2$):

$$F = q_1 \times q_2 / r^2$$

Therefore, the strength of the bond between the ions of opposite charge in an ionic compound depends on the charges on the ions and the distance between the centers of the ions when they pack to form a crystal.

For NaCl, the lattice energy is the energy released by the reaction:

$Na^+$ (g) + $Cl^-$ (g) → NaCl (s), lattice energy = –786 kJ/mol

Other examples:

$Al_2O_3$, lattice energy = –15,916 kJ/mol

KF, lattice energy = –821 kJ/mol

LiF, lattice energy = –1,036 kJ/mol

**74. C is correct.**

Formula to calculate dipole moment:

$$\mu = qr$$

where μ is dipole moment (coulomb per meter or C/m), $q$ is charge (coulomb or C), and $r$ is radius (meter or m)

Convert the unit of radius to meters:

$$154 \text{ pm} \times 1 \times 10^{-12} \text{ m/pm} = 154 \times 10^{-12} \text{ m}$$

*continued…*

 

Convert the charge to coulombs:

$$0.167\ e \times 1.602 \times 10^{-19}\ C/e^- = 2.68 \times 10^{-20}\ C$$

Use these values to calculate the dipole moment:

$$\mu = qr$$

$$\mu = 2.68 \times 10^{-20}\ C \times 154 \times 10^{-12}\ m$$

$$\mu = 4.13 \times 10^{-30}\ C{\cdot}m$$

Finally, convert the dipole moment to Debye:

$$\mu = 4.13 \times 10^{-30}\ C{\cdot}m\ /\ 3.34 \times 10^{-30}\ C{\cdot}m{\cdot}D^{-1}$$

$$\mu = 1.24\ D$$

**75. B is correct.**

The greater the difference in electronegativity between two atoms in a compound, the more polar a bond these atoms form. The atom with higher electronegativity is the dipole's partial (delta) negative end.

Although each C–Cl bond is very polar, the dipole moments of the four bonds in $CCl_4$ (carbon tetrachloride) cancel because the molecule is a symmetric tetrahedron.

**76. C is correct.**

p$K_a$ of carboxylic acid is about 5 and is deprotonated (exists as an anion) at a pH of 10.

p$K_a$ of the alcohol is about 15 and is protonated (neutral) at a pH of 10.

**77. A is correct.**

Hybridization of the central atom and the corresponding electron geometry.

| Hybridization | Electron groups | Bonding groups | Lone pairs | Approx. bond angles | Electron geometry | Molecular geometry |
|---|---|---|---|---|---|---|
| *sp* | 2 | 2 | 0 | 180° | Linear | Linear |
| $sp^2$ | 3 | 3 | 0 | 120° | Trigonal planar | Trigonal planar |
| | 3 | 2 | 1 | <120° | Trigonal planar | Bent |
| $sp^3$ | 4 | 4 | 0 | 109.5° | Tetrahedral | Tetrahedral |
| | 4 | 3 | 1 | <109.5° | Tetrahedral | Trigonal pyramidal |
| | 4 | 2 | 2 | <<109.5° | Tetrahedral | Bent |
| $sp^3d$ | 5 | 5 | 0 | 120° (equatorial) 90° (axial) | Trigonal bipyramidal | Trigonal bipyramidal |
| | 5 | 4 | 1 | <120° (equatorial) <90° (axial) | Trigonal bipyramidal | Seesaw |
| | 5 | 3 | 2 | <90° | Trigonal bipyramidal | T-shaped |
| | 5 | 2 | 3 | 180° | Trigonal bipyramidal | Linear |
| $sp^3d^2$ | 6 | 6 | 0 | 90° | Octahedral | Octahedral |
| | 6 | 5 | 1 | <90° | Octahedral | Square pyramidal |
| | 6 | 4 | 2 | 90° | Octahedral | Square planar |

 

**78. B is correct.**

*Valence shell* is an atom's outermost shell (i.e., highest principal quantum number $n$).

*Octet rule* states that atoms of main-group elements tend to combine, so each atom has eight electrons in its valence shell. This occurs because electron arrangements involving eight valence electrons are highly stable, as with noble gases.

**79. D is correct.**

*Solubility* depends on the solvent's ability to overcome the intermolecular forces in a solute.

**80. D is correct.**

*Electronegativity* is defined as the ability of an atom to attract electrons when it bonds with another atom. The most electronegative atom is F, while the least is Fr. The trend for increasing electronegativity within the periodic table is up and toward the right (i.e., fluorine).

Chlorine has the highest electronegativity of the elements listed. Electronegativity is a characteristic of non-metals.

**81. C is correct.**

*Hydrogen bonds* are the strongest intermolecular forces (i.e., between molecules), followed by dipole-dipole, dipole-induced dipole, and van der Waals forces (i.e., London dispersion).

Hydrogen, bonded directly to N, F, and O, participates in hydrogen bonds. It is partially positive (i.e., delta plus or $\partial+$) due to the bond to electronegative atoms (i.e., F, O or N). The lone pair of electrons on the F, O, or N interacts with the $\partial+$ hydrogen to form a hydrogen bond.

*Polar compounds* have stronger dipole-dipole forces between molecules because the partially positive end of one molecule bonds with the partially negative end of another.

*Nonpolar compounds* have induced dipole-induced dipole intermolecular force, which is weak compared to dipole-dipole force. The strength of induced dipole-induced dipole interaction correlates with an atom's number of electrons and protons.

 

**82. B is correct.**

Valence shell is an atom's outermost shell (i.e., highest principal quantum number $n$). Valence electrons are in the outermost electron shell that can participate in a chemical bond.

The number of valence electrons on the periodic table can be determined by its group (i.e., vertical column). Except for the transition metals (i.e., groups 3-12), the group number identifies how many valence electrons are associated with a particular element: elements of the same group have the same number of valence electrons.

Oxygen has 8 electrons.

The first shell ($1s$) holds 2 electrons, and the remaining 6 electrons are located within the orbitals of the next shell ($2s$, $2p$).

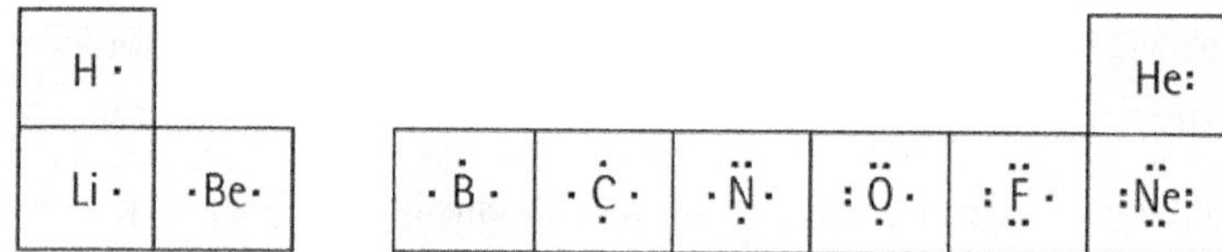


*Sample Lewis dot structures for some elements*

**83. C is correct.**

Nitrate ion is $NO_3^-$ and has the following resonance forms:

*Nitrate exists as a superposition of these three resonance forms*

**84. A is correct.**

Ionic bonds form when the difference in electronegativity of atoms in a compound is greater than 1.7 Pauling units.

Ionic bonds transfer an electron from the electropositive element (along the left-hand column/group) to the electronegative element (along the right-hand column/group) on the periodic table.

 

**85. C is correct.**

Electronegativity is a chemical property of an atom's tendency to attract electrons to itself.

The most common use of electronegativity pertains to polarity along the *sigma* (single) bond.

The most electronegative atom is F, while the least is Fr.

The trend for increasing electronegativity within the periodic table is up and toward the right (i.e., fluorine).

**86. B is correct.**

Octet rule states that atoms of main-group elements tend to combine, so each atom has eight electrons in its valence shell. This occurs because electron arrangements involving eight valence electrons are highly stable, as with noble gasses.

Valence shell is an atom's outermost shell (i.e., highest principal quantum number *n*).

Valence electrons are in the outermost electron shell that can participate in a chemical bond.

The number of valence electrons for an element can be determined by its group (i.e., vertical column) on the periodic table. Except for the transition metals (i.e., groups 3-12), the group number identifies how many valence electrons are associated with a particular element: elements of the same group have the same number of valence electrons.

Atoms with a complete valence shell (i.e., containing the maximum number of electrons), such as noble gases, are the most non-reactive elements.

The most reactive elements are atoms with only one electron in their valence shells (alkali metals) or those missing just one electron from a complete valence shell (halogens).

**87. B is correct.**

To help determine the polarity of molecules, draw the Lewis structures of the molecules.

With $SO_3$, $CO_2$, $CH_4$, and $CCl_4$, the electrons in the central atom are bonded and symmetrical. Therefore, they are nonpolar.

$SO_2$ has a pair of nonbonding electrons on the central atom (S), which means the molecule is polar.

**88. C is correct.**

*Intermolecular forces* are between molecules.

*Hydrogen bonds* are the strongest intermolecular forces (i.e., between molecules), followed by *dipole-dipole*, *dipole-induced dipole*, and *van der Waals forces* (i.e., London dispersion).

*Dipole-dipole interactions* primarily occur between different molecules, whereas ionic bonds, covalent and coordinate covalent bonds hold atoms in a molecule (i.e., intramolecular forces).

*Notes for active learning*

*Notes for active learning*

# 6 – Stoichiometry: Detailed Explanations

===========================================================

**Practice Set 1: Questions 1–20**

===========================================================

**1. C is correct.**

One mole of an ideal gas occupies 22.71 L at STP.

3 mole × 22.71 L = 68.13 L

**2. C is correct.**

Use the mnemonic OIL RIG: Oxidation Is Loss, Reduction Is Gain (of electrons).

Oxidation is the loss of electrons, while reduction is the gain of electrons.

An oxidizing agent undergoes reduction, while a reducing agent undergoes oxidation.

Check the oxidation numbers of each atom.

Oxidation number for S is +6 as reactant and +4 as product, which means that it is reduced.

If a substance is reduced in a redox reaction, it is the oxidizing agent because it causes the other reagent (HI) to be oxidized.

**3. C is correct.**

Balanced equation (combustion): $2\ C_6H_{14} + 19\ O_2 \rightarrow 12\ CO_2 + 14\ H_2O$

**4. D is correct.**

Oxidation number is a charge on an atom not present in the elemental state when the element is neutral.

Electronegativity refers to the attraction an atom has for additional electrons.

**5. C is correct.**

$CH_3COOH$ has 2 carbons, 4 hydrogens, and 2 oxygens: $C_2O_2H_4$

Reduce to the smallest coefficients by dividing by two: $COH_2$

**6. B is correct.**

Oxidation numbers of Cl in each compound:

| | |
|---|---|
| $NaClO_2$ | +3 |
| $Al(ClO_4)_3$ | +7 |
| $Ca(ClO_3)_2$ | +5 |
| $LiClO_3$ | +5 |

**7. A is correct.**

Formula mass is a synonym for molecular mass/molecular weight (MW).

Molecular mass = (atomic mass of C) + (2 × atomic mass of O)

Molecular mass = (12.01 g/mole) + (2 × 16.00 g/mole)

Molecular mass = 44.01 g/mole

Note: 1 amu = 1 g/mole

**8. C is correct.**

Cadmium is in group II, so its oxidation number is +2.

Sulfur tends to form sulfide ions (oxidation number –2).

The net oxidation number is zero; therefore, the compound is CdS.

**9. B is correct.**

Balanced reaction (single replacement): $Cl_2$ (*g*) + 2 NaI (*aq*) → $I_2$ (*s*) + 2 NaCl (*aq*)

**10. C is correct.**

Use mnemonic OIL RIG: Oxidation Is Loss, Reduction Is Gain (of electrons).

Oxidation is the loss of electrons, while reduction is the gain of electrons.

Oxidizing agent undergoes reduction, while a reducing agent undergoes oxidation.

When there are 2 reactants in a redox reaction, one is oxidized, and the other is reduced. The oxidized species is the reducing agent and vice versa.

*continued...*

 

Determine the oxidation number of all atoms:

I changes its oxidation state from –1 as a reactant to 0 as a product.

I is oxidized; therefore, the compound that contains I (NaI) is the reducing agent. Species in their elemental state have an oxidation number of 0.

Since the reaction is spontaneous, NaI is the *strongest* reducing agent because it can reduce another species spontaneously without needing external energy for the reaction to proceed.

**11. C is correct.**

To determine the remainder of a reactant, calculate how many moles of that reactant are required to produce the specified amount of product.

Use the coefficients to calculate the moles of the reactant:

$$\text{moles } N_2 = (1 / 2) \times 18 \text{ moles} = 9 \text{ moles } N_2$$

Therefore, the remainder is:

$$(14.5 \text{ moles } N_2) - (9 \text{ moles } N_2) = 5.5 \text{ moles } N_2 \text{ remaining}$$

**12. D is correct.**

$CaCO_3 \rightarrow CaO + CO_2$ is a decomposition reaction because one complex compound is broken down into two or more parts. However, the redox part is incorrect because the reactants and products are compounds (no single elements), and there is no change in the oxidation state (i.e., no gain or loss of electrons).

A: $AgNO_3 + NaCl \rightarrow AgCl + NaNO_3$ is correctly classified as a double-replacement reaction (sometimes called double-displacement) because parts of two ionic compounds are exchanged to make two new compounds. It is a non-redox reaction because the reactants and products are compounds (no single elements), and there is no change in oxidation state (i.e., no gain or loss of electrons).

B: $Cl_2 + F_2 \rightarrow 2\ ClF$ is correctly classified as a synthesis reaction because two species are combining to form a more complex chemical compound as the product. It is a redox reaction because the F atoms in $F_2$ are reduced (i.e., gain electrons), and the Cl atoms in $Cl_2$ are oxidized (i.e., lose electrons), forming a covalent compound.

C: $H_2O + SO_2 \rightarrow H_2SO_3$ is correctly classified as a synthesis reaction because two species are combining to form a more complex chemical compound as the product. It is a non-redox reaction because the reactants and products are compounds (no single elements), and there is no change in oxidation state (i.e., no gain or loss of electrons).

**13. C is correct.**

Hydrogen is being oxidized (or ignited) to produce water.

The balanced equation that describes the ignition of hydrogen gas in the air to produce water is:

$$\frac{1}{2} O_2 + H_2 \rightarrow H_2O$$

This equation suggests that 1 mole of hydrogen and ½ mole of oxygen gas are needed for every mole of water produced in the reaction.

Maximum amount of water produced in the reaction is determined by the amount of the limiting reactant available. This requires identifying which reactant is the limiting reactant, which can be done by comparing the number of moles of oxygen and hydrogen.

From the question, there are 10 grams of oxygen and 1 gram of hydrogen gas. This is equivalent to 0.3125 moles of oxygen and 0.5 moles of hydrogen.

Because the reaction requires twice as much hydrogen as oxygen gas, the limiting reactant is hydrogen gas (0.3215 moles of $O_2$ requires 0.625 moles of hydrogen, but only 0.5 moles of $H_2$ are available).

Since one equivalent of water is produced for every equivalent of hydrogen that is burned, the amount of water produced is:

$$(18 \text{ grams/mol } H_2O) \times (0.5 \text{ mol } H_2O) = 9 \text{ grams of } H_2O$$

**14. D is correct.**

Calculate the moles of $Cl^-$ ion:

$$\text{Moles of } Cl^- \text{ ion} = \text{number of } Cl^- \text{ atoms / Avogadro's number}$$

$$\text{Moles of } Cl^- \text{ ion} = 6.8 \times 10^{22} \text{ atoms} / 6.02 \times 10^{23} \text{ atoms/mole}$$

$$\text{Moles of } Cl^- \text{ ion} = 0.113 \text{ moles}$$

In a solution, $BaCl_2$ dissociates:

$$BaCl_2 \rightarrow Ba^{2+} + 2\ Cl^-$$

The moles from $Cl^-$ previous calculation can be used to calculate the moles of $Ba^{2+}$:

$$\text{Moles of } Ba^{2+} = (\text{coefficient of } Ba^{2+} / \text{coefficient } Cl^-) \times \text{moles of } Cl^-$$

$$\text{Moles of } Ba^{2+} = (\tfrac{1}{2}) \times 0.113 \text{ moles}$$

$$\text{Moles of } Ba^{2+} = 0.0565 \text{ moles}$$

*continued...*

 

Calculate the mass of $Ba^{2+}$:

Mass of $Ba^{2+}$ = moles of $Ba^{2+}$ × atomic mass of Ba

Mass of $Ba^{2+}$ = 0.0565 moles × 137.33 g/mole

Mass of $Ba^{2+}$ = 7.76 g

**15. D is correct.**

| Na | Br | $O_3$ |
|---|---|---|
| +1 | $x$ | $(3 \times -2)$ |
| +1 | $x$ | –6 |

Sum of charges in a neutral molecule is zero:

1 + oxidation number of Br:

$1 + x + (-6) = 0$

$-5 + x = 0$

$x = +5$

oxidation number of Br = +5

**16. A is correct.**

Balanced equation: $2\ Al + Fe_2O_3 \rightarrow 2\ Fe + Al_2O_3$

The sum of the coefficients of the products = 3.

**17. B is correct.**

$2\ H_2O_2\ (s) \rightarrow 2\ H_2O\ (l) + O_2\ (g)$ is incorrectly classified. The decomposition part of the classification is correct because one complex compound ($H_2O_2$) is broken down into two or more parts ($H_2O$ and $O_2$).

However, it is a redox reaction because oxygen is being lost from $H_2O_2$, one of the three indicators of a redox reaction (i.e., electron loss/gain, hydrogen loss/gain, oxygen loss/gain). Additionally, one of the products is a single element ($O_2$), which is a clue that it is a redox reaction.

$AgNO_3\ (aq) + KOH\ (aq) \rightarrow KNO_3\ (aq) + AgOH\ (s)$ is correctly classified as a non-redox reaction because the reactants and products are compounds (no single elements), and there is no change in oxidation state (i.e., no gain or loss of electrons). It is a precipitation reaction because the chemical reaction occurs in an aqueous solution, and one of the products formed (AgOH) is insoluble, making it a precipitate.

*continued...*

 

$Pb(NO_3)_2$ (*aq*) + 2 Na (*s*) → Pb (*s*) + 2 $NaNO_3$ (*aq*) is correctly classified as a redox reaction because the nitrate ($NO_3^-$), which dissociates, is oxidized (loses electrons), the $Na^+$ is reduced (gains electrons). The two combine to form the ionic compound $NaNO_3$. It is a single-replacement reaction because one element is substituted for another element in a compound, making a new compound ($2NaNO_3$) and an element (Pb).

$HNO_3$ (*aq*) + LiOH (*aq*) → $LiNO_3$ (*aq*) + $H_2O$ (*l*) is correctly classified as a non-redox reaction because the reactants and products are compounds (no single elements), and there is no change in oxidation state (i.e., no gain or loss of electrons). It is a double-replacement reaction because parts of two ionic compounds are exchanged to make two new compounds.

**18. C is correct.**

At STP, 1 mole of gas has a volume of 22.4 L.

Calculate moles of $O_2$:

moles of $O_2$ = (15.0 L) / (22.4 L/mole)

moles of $O_2$ = 0.67 mole

Calculate the moles of $O_2$, using the fact that 1 mole of $O_2$ has $6.02 \times 10^{23}$ $O_2$ molecules:

molecules of $O_2$ = 0.67 mole × ($6.02 \times 10^{23}$ molecules/mole)

molecules of $O_2$ = $4.03 \times 10^{23}$ molecules

**19. D is correct.**

Guideline for balancing redox equations by the oxidation number method:

1) verify that the number of atoms and the ionic charge is the same for reactants and products,

2) in front of the substance reduced, place a coefficient that corresponds to the number of electrons lost by the substance oxidized, and

3) in front of the substance oxidized, place a coefficient that corresponds to the number of electrons gained by the substance reduced.

**20. D is correct.**

Use the mnemonic OIL RIG: Oxidation Is Loss, Reduction Is Gain (of electrons).

Oxidation is the loss of electrons, while reduction is the gain of electrons. An oxidizing agent undergoes reduction, while a reducing agent undergoes oxidation.

$Co^{2+}$ is the starting reactant because it loses electrons and produces an ion with a higher oxidation number.

 

==========================================================

**Practice Set 2: Questions 21–40**

==========================================================

**21. D is correct.**

Double replacement reaction:

HCl has a molar mass of 36.5 g/mol:

365 grams HCl = 10 moles HCl

10 moles are 75% of the total yield.

10 moles $PCl_3$ / 0.75 = 13.5 moles of HCl.

Apply the mole ratios:

Coefficients: 3 HCl = $PCl_3$

13.5 / 3 = moles of $PCl_3$

13.5 moles of HCl is produced by 4.5 moles of $PCl_3$

**22. A is correct.**

All elements in their free state have an oxidation number of zero.

**23. B is correct.**

Balanced chemical equation:

$C_3H_8 + 5\ O_2 \rightarrow 3\ CO_2 + 4\ H_2O$

By using the coefficients from the balanced equation, apply dimensional analysis to solve for the volume of $H_2O$ produced from the 2.6 L $C_3H_8$ reacted:

$V_{H2O} = V_{C3H8} \times (\text{mol } H_2O / \text{mol } C_3H_8)$

$V_{H2O} = (2.6\text{ L}) \times (4\text{ mol} / 1\text{ mol})$

$V_{H2O} = 10.4\text{ L}$

**24. C is correct.**

Use mnemonic OIL RIG: Oxidation Is Loss, Reduction Is Gain (of electrons).

Oxidation is the loss of electrons, while reduction is the gain of electrons.

An oxidizing agent undergoes reduction, while a reducing agent undergoes oxidation.

Oxidation number of Hg decreases (i.e., reduction) from +2 as a reactant ($HgCl_2$) to +1 as a product ($Hg_2Cl_2$). This means that Hg gained one electron.

**25. C is correct.**

Balanced equation (synthesis):

$$4\ P\ (s) + 5\ O_2\ (g) \rightarrow 2\ P_2O_5\ (s)$$

**26. D is correct.**

In basic solutions, the following are guidelines for balancing a redox equation using the half-reaction method:

1) add the two half-reactions and cancel identical species on each side of the equation,

2) multiply each half-reaction by a whole number so that the number of electrons lost by the substance oxidized is equal to the electrons gained by the substance reduced, and

3) write a balanced half-reaction for the substance oxidized and the substance reduced.

**27. A is correct.**

Formula mass is a synonym for molecular mass/molecular weight (MW).

Start by calculating the mass of the formula unit ($CH_2O$)

$$CH_2O = (12.01\ \text{g/mol} + 2.02\ \text{g/mol} + 16.00\ \text{g/mol}) = 30.03\ \text{g/mol}$$

Divide the molar mass with the formula unit mass:

$$180\ \text{g/mol} / 30.03\ \text{g/mol} = 5.994$$

Round to the closest whole number: 6

Multiply the formula unit by 6:

$$(CH_2O)_6 = C_6H_{12}O_6$$

 

**28. B is correct.**

Calculate the moles of LiI present. The molecular weight of LiI is 133.85 g/mol.

Moles of LiI:

6.45 g / 133.85 g/mole = 0.0482 moles

Each mole of LiI contains $6.02 \times 10^{23}$ molecules.

Number of molecules:

$0.0482 \times 6.02 \times 10^{23} = 2.90 \times 10^{22}$ molecules

1 molecule is equal to 1 formula unit.

**29. D is correct.**

A combustion engine creates various nitrogen oxide compounds, often referred to as $NO_x$ gases, because each gas would have a different value of $x$ in its formula.

**30. C is correct.**

Balanced equation (synthesis):

$4\ P\ (s) + 3\ O_2\ (g) \rightarrow 2\ P_2O_3\ (s)$

**31. A is correct.**

Use the mnemonic OIL RIG: Oxidation Is Loss, Reduction Is Gain (of electrons).

Oxidation is the loss of electrons, while reduction is the gain of electrons.

An oxidizing agent undergoes reduction, while a reducing agent undergoes oxidation.

Because the forward reaction is spontaneous, the reverse reaction is not spontaneous.

Sn cannot reduce $Mg^{2+}$; therefore, Sn is the weakest reducing agent.

**32. C is correct.**

*Balancing Redox Equations*

From the balanced equation, the coefficient for the proton can be determined. Balancing a redox equation requires balancing the atoms, but the charges must be balanced.

The equations must be separated into two different half-reactions. One of the half-reactions addresses the oxidizing component, and the other addresses the reducing component.

*continued...*

Magnesium is being oxidized; therefore, the unbalanced oxidation half-reaction is:

$Mg\ (s) \rightarrow Mg^{2+}\ (aq)$

Furthermore, the nitrogen is reduced; therefore, the unbalanced reduction half-reaction is:

$NO_3^-\ (aq) \rightarrow NO_2\ (aq)$

Each half-reaction must be balanced for each atom at this stage, and the net electric charge on each side of the equations must be balanced.

*Order of operations for balancing half-reactions:*

1) Balance atoms except for oxygen and hydrogen.
2) Balance the oxygen atom count by adding water.
3) Balance the hydrogen atom count by adding protons.
   a) If in basic solution, add equal amounts of hydroxide to each side to cancel the protons.
4) Balance the electric charge by adding electrons.
5) If necessary, multiply the coefficients of one half-reaction equation by a factor that cancels the electron count when both equations are combined.
6) Cancel ions or molecules that appear on each side of the overall equation.

After determining the balanced overall redox reaction, stoichiometry indicates the moles of involved protons.

*For magnesium:*

$Mg\ (s) \rightarrow Mg^{2+}\ (aq)$

Magnesium is balanced with a coefficient of 1. There are no hydrogen or oxygen atoms present in the equation. The magnesium cation has a +2 charge, so 2 moles of electrons should be added to the right side to balance the charge.

Balanced half-reaction for oxidation:

$Mg\ (s) \rightarrow Mg^{2+}\ (aq) + 2\ e^-$

*For nitrogen:*

$NO_3^-\ (aq) \rightarrow NO_2\ (aq)$

*continued...*

 

Equation is balanced for nitrogen because one nitrogen atom appears on each reaction side. The nitrate reactant has three oxygen atoms, while the nitrite product has two oxygen atoms.

To balance the oxygen, one mole of water should be added to the right side:

$$NO_3^- \,(aq) \rightarrow NO_2\,(aq) + H_2O$$

Adding water to the right side of the equation introduces hydrogen atoms to that side. Therefore, the hydrogen atom count needs to be balanced.

Water possesses two hydrogen atoms; therefore, two protons need to be added to the left side of the reaction:

$$NO_3^- \,(aq) + 2\,H^+ \rightarrow NO_2\,(aq) + H_2O$$

The reaction occurs in acidic conditions. If the reaction were basic, $OH^-$ would need to be added to each side to cancel the protons.

Next, the net charge needs to be balanced. The left side has a net charge of +1 (+2 from the protons and –1 from the electron), while the right side is neutral. Therefore, one electron should be added to the left side:

$$NO_3^- \,(aq) + 2\,H^+ + e^- \rightarrow NO_2\,(aq) + H_2O$$

When half-reactions are recombined, the electrons in the overall reaction must cancel.

The reduction half-reaction contributes one electron to the left side of the overall equation, while the oxidation half-reaction contributes two electrons to the product side.

Therefore, the coefficients of the reduction half-reaction should be doubled:

$$2 \times [NO_3^- \,(aq) + 2\,H^+ + e^- \rightarrow NO_2\,(aq) + H_2O]$$

$$= 2\,NO_3^- \,(aq) + \mathbf{4\,H^+} + 2\,e^- \rightarrow 2\,NO_2\,(aq) + 2\,H_2O$$

The answer is 4 at this step.

Combining both half-reactions gives:

$$Mg\,(s) + 2\,NO_3^- \,(aq) + \mathbf{4\,H^+} + 2\,e^- \rightarrow Mg^{2+}\,(aq) + 2\,e^- + 2\,NO_2\,(aq) + 2\,H_2O$$

Cancel the electrons that appear on each side of the reaction in the following balanced net equation:

$$Mg\,(s) + 2\,NO_3^- \,(aq) + \mathbf{4\,H^+} \rightarrow Mg^{2+}\,(aq) + 2\,NO_2\,(aq) + 2\,H_2O$$

This equation is now fully balanced for mass, oxygen, hydrogen and electric charge.

 

**33. B is correct.**

To obtain moles, divide the sample mass by the molecular mass.

Because all options have the same mass, comparing molecular mass provides the answer without performing the actual calculation.

The molecule with the smallest molecular mass has the greatest number of moles.

**34. A is correct.**

*Law of constant composition* states that *samples of a given chemical compound have the same chemical composition by mass*. This is true for all compounds, including ethyl alcohol.

This law is referred to as the *law of definite proportions* or *Proust's Law* because this observation was first made by the French chemist Joseph Proust.

**35. A is correct.**

Coefficients in balanced reactions refer to moles (or molecules) but not grams.

1 mole of $N_2$ gas reacts with 3 moles of $H_2$ gas to produce 2 moles of $NH_3$ gas.

**36. C is correct.**

Calculate the number of moles of $Al_2O_3$:

$$2\ Al / 1\ Al_2O_3 = 0.2 \text{ moles } Al / x \text{ moles } Al_2O_3$$

$$x \text{ moles } Al_2O_3 = 0.2 \text{ moles } Al \times 1\ Al_2O_3 / 2\ Al$$

$$x = 0.1 \text{ moles } Al_2O_3$$

Therefore:

$$(0.1 \text{ mole } Al_2O_3)\cdot(102 \text{ g/mole } Al_2O_3) = 10.2 \text{ g } Al_2O_3$$

**37. D is correct.**

Because Na is a metal, Na is more electropositive than H. Therefore, the positive charge is on Na (i.e., +1), which means that the charge on H is –1.

**38. A is correct.**

$PbO + C \rightarrow Pb + CO$ is a single replacement reaction because only one element is transferred from one reactant to another.

*continued...*

 

However, it is a redox reaction because the lead ($Pb^{2+}$), which dissociates, is reduced (gains electrons), and the oxygen ($O^{2-}$) is oxidized (loses electrons). Additionally, one of the products is a single element (Pb), which is a clue that it is a redox reaction.

C: the reaction is double-replacement and not combustion because the combustion reactions must have hydrocarbon and $O_2$ as reactants. Double-replacement reactions often result in the formation of a solid, water, and gas.

**39. B is correct.**

Sum of the oxidation numbers in a neutral molecule must be zero.

Oxidation number of each H is +1, and the oxidation number of each O is –2.

If $x$ denotes the oxidation number of S in $H_2SO_4$, then:

$$2(+1) + x + 4(-2) = 0$$

$$+2 + x + -8 = 0$$

$$x = +6$$

In $H_2SO_4$, the oxidation state of sulfur is +6.

**40. A is correct.**

Calculate the number of molecules using the moles of gas using the ideal gas equation:

$$PV = nRT$$

Rearrange to isolate the number of moles:

$$n = PV / RT$$

Because the gas constant R is in L·atm $K^{-1}$ $mol^{-1}$, the pressure must be converted into atm:

$$780 \text{ torr} \times (1 \text{ atm} / 760 \text{ torr}) = 1.026 \text{ atm}$$

Convert the volume to liters:

$$500 \text{ mL} \times 0.001 \text{ L/mL} = 0.5 \text{ L}$$

Substitute into the ideal gas equation:

$$n = PV / RT$$

$$n = 1.026 \text{ atm} \times 0.5 \text{ L} / (0.08206 \text{ L·atm K}^{-1} \text{ mol}^{-1} \times 320 \text{ K})$$

$$n = 0.513 \text{ L·atm} / (26.259 \text{ L·atm mol}^{-1})$$

$$n = 0.0195 \text{ mol}$$

==================================================================

**Practice Set 3: Questions 41–60**

==================================================================

**41. B is correct.**

Use the mnemonic OIL RIG: Oxidation Is Loss, Reduction Is Gain (of electrons).

In a redox reaction, oxidizing and reducing agents are reactants, not products.

The oxidizing agent is the species that gets reduced (i.e., gains electrons) and causes oxidation of the other species.

Reducing agents are oxidized species (i.e., lose electrons) and reduce the other species.

**42. D is correct.**

Balancing a redox equation in an acidic solution by the half-reaction method involves the steps:

1) Multiply each half-reaction by a whole number so that the number of electrons lost by the substance oxidized is equal to the electrons gained by the substance reduced,

2) add the two half-reactions and cancel identical species from each side of the equation, and

3) write a half-reaction for the substance oxidized and the substance reduced.

**43. D is correct.**

Formula mass is a synonym for molecular mass/molecular weight (MW).

MW of $C_6H_{12}O_6$ = (6 × atomic mass C) + (12 × atomic mass H) + (6 × atomic mass O)

MW of $C_6H_{12}O_6$ = (6 × 12.01 g/mole) + (12 × 1.01 g/mole) + (6 × 16.00 g/mole)

MW of $C_6H_{12}O_6$ = (72.06 g/mole) + (12.12 g/mole) + (96.00 g/mole)

MW of $C_6H_{12}O_6$ = 180.18 g/mole

**44. A is correct.**

Molecular mass of oxygen is 16 g/mole.

Atomic mass unit (amu) or dalton (Da) is the standard unit for indicating mass on an atomic or molecular scale (atomic mass). One amu is approximately the mass of one nucleon (a single proton or neutron) and is numerically equivalent to 1 g/mol.

 

**45. D is correct.**

Balanced equation (combustion): $2\ C_2H_6 + 7\ O_2 \rightarrow 4\ CO_2 + 6\ H_2O$

**46. B is correct.**

Use the mnemonic OIL RIG: Oxidation Is Loss, Reduction Is Gain (of electrons).

Oxidation is the loss of electrons, while reduction is the gain of electrons.

An oxidizing agent undergoes reduction, while a reducing agent undergoes oxidation.

Evaluate the oxidation number in each atom.

S changes from +6 (reactant) to –2 (product), reducing it because oxidation number decreases.

**47. C is correct.**

Balanced equation (double replacement):

$$C_2H_5OH\ (g) + 3\ O_2\ (g) \rightarrow 2\ CO_2\ (g) + 3\ H_2O\ (g)$$

**48. A is correct.**

Note: 1 amu (atomic mass unit) = 1 g/mole.

Moles of C = mass of C / atomic mass of C

Moles of C = (4.50 g) / (12.01 g/mole)

Moles of C = 0.375 mole

**49. C is correct.**

According to the law of mass conservation, there should be equal amounts of carbon in the product ($CO_2$) and reactant (the hydrocarbon sample).

Start by calculating the amount of carbon in the product ($CO_2$).

First, calculate the mass % of carbon in $CO_2$:

Molecular mass of $CO_2$ = atomic mass of carbon + (2 × atomic mass of oxygen)

Molecular mass of $CO_2$ = 12.01 g/mole + (2 × 16.00 g/mole)

Molecular mass of $CO_2$ = 44.01 g/mole

Mass % of carbon in $CO_2$ = (mass of carbon / molecular mass of $CO_2$) × 100%

Mass % of carbon in $CO_2$ = (12.01 g/mole / 44.01 g/mole) × 100%

*continued…*

Mass % of carbon in $CO_2$ = 27.3%

Calculate the mass of carbon in the $CO_2$:

Mass of carbon = mass of $CO_2$ × mass % of carbon in $CO_2$

Mass of carbon = 8.98 g × 27.3%

Mass of carbon = 2.45 g

Mass of carbon in the starting reactant (the hydrocarbon sample) should be 2.45 g.

Mass % carbon in hydrocarbon = (mass carbon in sample / total mass sample) × 100%

Mass % of carbon in hydrocarbon = (2.45 g / 6.84 g) × 100%

Mass % of carbon in hydrocarbon = 35.8%

**50. B is correct.**

The number of each atom on the reactants' side must be equal to the number of atoms on the product side of the reaction.

**51. D is correct.**

Using ½ reactions, the balanced reaction is:

$$6\,(Fe^{2+} \rightarrow Fe^{3} + 1\,e^-)$$

$$Cr_2O_7^{2-} + 14\,H^+ + 6\,e^- \rightarrow 2\,Cr^{3+} + 7\,H_2O$$

$$Cr_2O_7^{2-} + 14\,H^+ + 6\,Fe^{2+} \rightarrow 2\,Cr^{3+} + 6\,Fe^{3+} + 7\,H_2O$$

The sum of coefficients in the balanced reaction is 36.

**52. A is correct.**

Balanced reaction: $4\,RuS\,(s) + 9\,O_2 + 4\,H_2O \rightarrow 2\,Ru_2O_3\,(s) + 4\,H_2SO_4$

Calculate the moles of $H_2SO_4$:

49 g × 1 mol / 98 g = 0.5 mole $H_2SO_4$

If 0.5 mole $H_2SO_4$ is produced, set a ratio for $O_2$ consumed.

$9\,O_2 / 4\,H_2SO_4 = x$ moles $O_2$ / 0.5 mole $H_2SO_4$

$x$ moles $O_2 = 9\,O_2 / 4\,H_2SO_4 \times 0.5$ mole $H_2SO_4$

$x$ = 1.125 moles $O_2$ × 22.4 liters / 1 mole

$x$ = 25.2 liters of $O_2$

**53. C is correct.**

Identify the limiting reactant:

Al is the limiting reactant because 2 atoms combine with 1 molecule of ferric oxide to produce the products.

Al is depleted if the reaction starts with equal numbers of moles of each reactant.

Use moles of Al and coefficients to calculate moles of other products/reactants.

Coefficient of Al = coefficient of Fe

Moles of iron produced = 0.20 moles.

**54. B is correct.**

$S_2$ $O_8$

$2x$ $(8 \times -2)$

Sum of charges on the molecule is –2:

$2x + (8 \times -2) = -2$

$2x + (-16) = -2$

$2x = +14$

oxidation number of S = +7

**55. D is correct.**

Na is in group IA, so its oxidation number is +1.

Oxygen is usually −2, with some exceptions, such as peroxide ($H_2O_2$), where it is −1.

Cr is a transition metal with more than one possible oxidation number.

To determine Cr's oxidation number, use the known oxidation numbers:

$2(+1) + Cr + 4(-2) = 0$

$2 + Cr - 8 = 0$

$Cr = 6$

**56. B is correct.**

Net charge of $H_2SO_4 = 0$. Hydrogen has a common oxidation number of +1, whereas each oxygen atom has an oxidation number of –2.

| $H_2$ | S | $O_4$ |
|---|---|---|
| $(2 \times 1)$ | $x$ | $(4 \times -2)$ |

$2 + x + -8 = 0$

$x = +6$

In $H_2SO_4$, the oxidation of sulfur is +6.

**57. C is correct.**

Use the mnemonic OIL RIG: Oxidation Is Loss, Reduction Is Gain (of electrons).

Oxidation is the loss of electrons, while reduction is the gain of electrons.

An oxidizing agent undergoes reduction, while a reducing agent undergoes oxidation.

Oxidation number for Sn is +2 as a reactant and +4 as a product; an increase in oxidation means that $Sn^{2+}$ is oxidized in this reaction.

**58. D is correct.**

$Na_2Cl_2$ and 2 NaCl are not the same.

$Na_2Cl_2$ is a molecule of 2 Na and 2 Cl.

NaCl is a compound that consists of 1 Na and 1 Cl.

**59. D is correct.**

I: represents the volume of 1 mol of an ideal gas at STP

II: mass of 1 mol of $PH_3$

III: number of molecules in 1 mol of $PH_3$

**60. D is correct.**

Use the mnemonic OIL RIG: Oxidation Is Loss, Reduction Is Gain (of electrons).

Oxidation is the loss of electrons, while reduction is the gain of electrons.

An oxidizing agent undergoes reduction, while a reducing agent undergoes oxidation.

$Co^{2+}$ is the starting reactant because it loses electrons and produces an ion with a higher oxidation number.

 

==========================================================

**Practice Set 4: Questions 61–80**

==========================================================

**61. B is correct.**

Use mnemonic OIL RIG: Oxidation Is Loss, Reduction Is Gain (of electrons).

Oxidizing and reducing agents are reactants, not products, in a redox reaction.

Oxidizing agent is the reduced species (i.e., gains electrons).

Reducing agent is the oxidized species (i.e., loses electrons).

**62. D is correct.**

Calculations:

% mass Cl = (molecular mass Cl) / (molecular mass of compound) × 100%

% mass Cl = 4(35.5 g/mol) / [12 g/mol + 4(35.5 g/mol)] × 100%

% mass Cl = (142 g/mol) / (154 g/mol) × 100%

% mass Cl = 0.922 × 100% = 92%

**63. D is correct.**

I: ionic charge of reactants must equal the ionic charge of products, which means that no electrons can "disappear" from the reaction; they can only be transferred from one atom to another. Therefore, the total charge will always be the same.

II: atoms of each reactant must equal atoms of product, which is true for chemical reactions because atoms in chemical reactions cannot be created or destroyed.

III: electrons gained by one atom must be lost by another.

**64. B is correct.**

Determine the molecular weight of each compound.

Note, unlike % by mass, there is no division.

$Cl_2C_2H_4$ = 2(35.5 g/mol) + 2(12 g/mol) + 4(1 g/mol)

$Cl_2C_2H_4$ = 71 g/mol + 24 g/mol + 4 g/mol

$Cl_2C_2H_4$ = 99 g/mol

 

**65. C is correct.**

Formula mass is a synonym for molecular mass/molecular weight (MW).

Start by calculating the mass of the formula unit (CH):

$$CH = (12.01 \text{ g/mol} + 1.01 \text{ g/mol}) = 13.02 \text{ g/mol}$$

Divide the molar mass by the formula unit mass:

$$78 \text{ g/mol} / 13.02 \text{ g/mol} = 5.99$$

Round to the closest whole number: 6

Multiply the formula unit by 6:

$$(CH)_6 = C_6H_6$$

**66. D is correct.**

There is 1 oxygen atom from GaO plus 6 oxygen atoms from $(NO_3)_2$, so the total is 7.

**67. B is correct.**

Ethanol ($C_2H_5OH$) combines with oxygen to produce carbon dioxide and water.

**68. C is correct.**

Balanced reaction (combustion):

$$2\ C_3H_7OH + 9\ O_2 \rightarrow 6\ CO_2 + 8\ H_2O$$

**69. C is correct.**

Balanced equation (double replacement):

$$Co_2O_3\ (s) + 3\ CO\ (g) \rightarrow 2\ Co\ (s) + 3\ CO_2\ (g)$$

**70. B is correct.**

*Molar volume* ($V_m$) is the volume occupied by one mole of a substance (i.e., element or compound) at a given temperature and pressure.

 

**71. A is correct.**

Use the mnemonic OIL RIG: Oxidation Is Loss, Reduction Is Gain (of electrons).

Oxidation is the loss of electrons, while reduction is the gain of electrons. An oxidizing agent undergoes reduction, while a reducing agent undergoes oxidation.

Because the forward reaction is spontaneous, the reverse reaction is not spontaneous. $Mg^{2+}$ cannot oxidize Sn; therefore, $Mg^{2+}$ is the weakest oxidizing agent.

**72. D is correct.**

$H_2$ is the limiting reactant.

24 moles of $H_2$ should produce: $24 \times (2 / 3) = 16$ moles of $NH_3$

If 13.5 moles are produced, the yield:

13.5 moles / 16 moles × 100% = 84%

**73. B is correct.**

Balanced reaction:

$H_2 + ½\ O_2 \rightarrow H_2O$

Multiply the equations by 2 to remove the fraction:

$2\ H_2 + O_2 \rightarrow 2\ H_2O$

Find the limiting reactant by calculating the moles of each reactant:

Moles of hydrogen = mass of hydrogen / (2 × atomic mass of hydrogen)

Moles of hydrogen = 25 g / (2 × 1.01 g/mole)

Moles of hydrogen = 12.38 moles

Moles of oxygen = mass of oxygen / (2 × atomic mass of oxygen)

Moles of oxygen = 225 g / (2 × 16.00 g/mole)

Moles of oxygen = 7.03 moles

Divide the number of moles of each reactant by its coefficient.

The reactant with a smaller number of moles is the limiting reactant.

Hydrogen = 12.38 moles / 2

Hydrogen = 6.19 moles

*continued...*

Oxygen = 7.03 moles / 1

Oxygen = 7.03 moles

Because hydrogen has a smaller number of moles after the division, hydrogen is the limiting reactant, and hydrogen will be depleted in the reaction.

$2\ H_2 + O_2 \rightarrow 2\ H_2O$

Hydrogen and water have coefficients of 2; therefore, these have the same number of moles.

There are 12.38 moles of hydrogen, which means this reaction produces 12.38 moles of water.

Molecular mass of $H_2O$ = (2 × molecular mass of hydrogen) + atomic mass of oxygen

Molecular mass of $H_2O$ = (2 × 1.01 g/mole) + 16.00 g/mole

Molecular mass of $H_2O$ = 18.02 g/mole

Mass of $H_2O$ = moles of $H_2O$ × molecular mass of $H_2O$

Mass of $H_2O$ = 12.38 moles × 18.02 g/mole

Mass of $H_2O$ = 223 g

**74. C is correct.**

| Li | Cl | $O_2$ |
|---|---|---|
| +1 | $x$ | (2 × –2) |

The sum of charges in a neutral molecule is zero:

$1 + x + (2 \times -2) = 0$

$1 + x + (-4) = 0$

$x = -1 + 4$

oxidation number of Cl = +3

 

**75. D is correct.**

Using half-reactions, the balanced reaction is:

Balance hydrogen ↓ Balance oxygen ↓

$$4[5\ e^- + 8\ H^+ + MnO_4^- \rightarrow Mn^{2+} + 4\ (H_2O)]$$

$$5[(H_2O) + C_3H_7OH \rightarrow C_2H_5CO_2H + 4\ H^+ + 4\ e^-]$$

$$12\ H^+ + 4\ MnO_4^- + 5\ C_3H_7OH \rightarrow 11\ H_2O + 4\ Mn^{2+} + 5\ C_2H_5CO_2H$$

Multiply by a common multiple of 4 and 5:

The sum of the product's coefficients: 11 + 4 + 5 = 20.

**76. A is correct.**

PbO (*s*) + C (*s*) → Pb (*s*) + CO (*g*) is a single-replacement reaction because one element (i.e., oxygen) is being transferred from one reactant to another.

**77. D is correct.**

Chemical reactants are on the left of the reaction arrow; the products are on the right.

In this reaction, the reactants are $C_6H_{12}O_6$, $H_2O$ and $O_2$.

There is a distinction between the terms *reactant* and *reagent.*

*Reactant* is a substance consumed during a chemical reaction.

*Reagent* is a substance (e.g., solvent) added to a system to cause a chemical reaction.

**78. C is correct.**

Determine the number of moles of He:

$$4\text{ g} \div 4.0\text{ g/mol} = 1\text{ mole}$$

He has 2 electrons since He has atomic number 2 (# electrons = # protons for neutral atoms).

Therefore, 1 mole of He × 2 electrons / mole = 2 mole of electrons

**79. D is correct.**

In this molecule, Br has the oxidation number of –1 because it is a halogen, and the gaining of one electron results in a complete octet for bromine.

Sum of charges in a neutral molecule = 0

0 = (oxidation state of Fe) + (3 × oxidation state of Br)

0 = (oxidation state of Fe) + (3 × –1)

0 = oxidation state of Fe – 3

oxidation state of Fe = +3

**80. B is correct.**

*Mole* is a unit of measurement used to express amounts of a chemical substance.

Number of molecules in a mole is $6.02 \times 10^{23}$, which is Avogadro's number.

However, *Avogadro's number* relates to the number of molecules, not the amount of substance.

*Molar mass* refers to the mass per mole of a substance.

*Formula mass* is a term that is sometimes used to mean *molecular mass* or *molecular weight*, and it refers to the mass of a particular molecule.

 

---

**Practice Set 5: Questions 81–100**

---

**81. B is correct.**

Use mnemonic OIL RIG: Oxidation Is Loss, Reduction Is Gain (of electrons).

To determine which species undergoes oxidation, determine the oxidation number of species involved in the reaction.

Reactants: CuBr

Br ion is –1, which means that Cu is +1.

Products: Cu = 0 (it's an element)

$Br_2$ = 0 (it's an element)

Br is the species that undergoes oxidation. It loses electrons and has its oxidation number increase.

**82. C is correct.**

Use mnemonic OIL RIG: Oxidation Is Loss, Reduction Is Gain (of electrons).

Oxidation is the loss of electrons, while reduction is the gain of electrons. An oxidizing agent undergoes reduction, while a reducing agent undergoes oxidation.

Each formula unit of $CuBr_2$ is converted into Cu.

Br ion does not change (i.e., neither oxidation nor reduction); it just moves to Zn to form another compound.

In $CuBr_2$, the oxidation number of Cu is +2. Therefore, when reduced, it gains 2 electrons to form Cu.

**83. B is correct.**

Calculate molecular mass of $CO_2$:

Molecular mass of $CO_2$ = atomic mass of C + (2 × atomic mass of O)

Molecular mass of $CO_2$ = 12.01 g/mole + (2 × 16.00 g/mole)

Molecular mass of $CO_2$ = 44.01 g/mole

*continued...*

Calculate moles of $CO_2$:

Moles of $CO_2$ = mass of $CO_2$ / molecular mass of $CO_2$

Moles of $CO_2$ = 168 g / (44.01 g/mole)

Moles of $CO_2$ = 3.82 moles

One mole of atoms has $6.02 \times 10^{23}$ atoms (Avogadro's number, or $N_A$).

Number of $CO_2$ atoms = moles of $CO_2$ × Avogadro's number

Number of $CO_2$ atoms = 3.82 moles × ($6.02 \times 10^{23}$ atoms/mole)

Number of $CO_2$ atoms = $2.30 \times 10^{24}$ atoms

**84. A is correct.**

For calculations, assume 100 g of the compound:

64 g of Ag, 8 g of N and 28 g of O.

Find the number of moles of each of these elements:

Ag: (64 g) / (108 g/mol) = 0.6 mol

N: (8 g) / (14 g/ mol) = 0.6 mol

O: (28 g) / (16 g/mol) = 1.8 mol

Reduce to the smallest coefficients by dividing by 0.6: $AgNO_3$

**85. D is correct.**

To solve the density of the gas, a modification of the ideal gas law must be made.

Ideal gas law: PV = nRT

where n is the number of moles equal to mass/molecular weight (MW). Substitute that expression into the ideal gas equation:

PV = (mass / MW) × RT

density ($\rho$) = mass / volume

Rearrange the equation to get mass / volume on one side:

mass / volume = (P × MW) / RT

density ($\rho$) = (P × MW) / RT

*continued...*

 

Rearrange the equation to solve for molecular weight (MW):

$$MW = \rho RT / P$$

The gas is in STP condition, with T of 0 °C = 273.15 K and P of 1 atm.

Calculate MW:

$$MW = 1.34 \text{ g/L} \times 0.0821 \text{ L atm K mol}^{-1} \times 273.15 \text{ K} / 1 \text{ atm}$$

$$MW = 30.1 \text{ g}$$

**86. C is correct.**

Balanced equation (double replacement):

$$C_3H_8\ (g) + 5\ O_2\ (g) \rightarrow 3\ CO_2\ (g) + 4\ H_2O\ (g)$$

**87. D is correct.**

To find what the reaction yields, look at the right side of the equation.

There are 6 $CO_2$ atoms: 6 × 2 = 12

There are 12 $H_2O$ atoms: 12 × 1 = 12

12 + 12 = 24 atoms of oxygen.

**88. C is correct.**

Balanced equation (single replacement):

$$2\ Al_2O_3\ (s) + 6\ Cl_2\ (g) \rightarrow 4\ AlCl_3\ (aq) + 3\ O_2\ (g)$$

**89. C is correct.**

The given half-reaction equation for the reduction of chromium is:

$$Cr_2O_7^{2-}\ (aq) \rightarrow Cr^{3+}\ (aq)$$

where the oxidation state of oxygen in $Cr_2O_7^{2-}$ is –2, and the oxidation state of chromium is +6:

2(charge of metal) + 7(charge of oxygen) = net charge of the molecule or ion

$$2(Cr^{n+}) + 7(-2) = -2$$

$$2(Cr^{n+}) - 14 = -2$$

$$2(Cr^{n+}) = 12$$

$$Cr^{n+} = +6$$

*continued...*

To balance this reduction half-reaction:

1) Balance the metal count.
2) Balance oxygen count.
3) Balance hydrogen count.
4) Add hydroxide if the reaction is done in a base.
5) Balance the charge.
6) Optional step: multiply the coefficients by a factor to cancel electrons if the reaction is to be combined with the oxidation half-reaction

Applying these operations to the half-reaction:

$$Cr_2O_7^{2-} \rightarrow Cr^{3+}$$

1) Balance the metal count on each side of the equation by adding the appropriate coefficients:

$$Cr_2O_7^{2-} \rightarrow 2\ Cr^{3+}$$

2) Balance the oxygen count by adding water:

$$Cr_2O_7^{2-} \rightarrow 2\ Cr^{3+} + 7\ H_2O$$

3) Balance the hydrogen count by adding protons:

$$14\ H^+ + Cr_2O_7^{2-} \rightarrow 2\ Cr^{3+} + 7\ H_2O$$

Skip step 4.

5) Balance the charge by adding electrons (there should be a net charge of +6 on each side):

$$14\ H^+ + Cr_2O_7^{2-} + 6\ e^- \rightarrow 2\ Cr^{3+} + 7\ H_2O$$

**90. B is correct.**

| Na | H | C | $O_3$ |
|---|---|---|---|
| +1 | +1 | $x$ | $(3 \times -2)$ |

The sum of charges in a neutral molecule = 0

$$1 + 1 + x + (3 \times -2) = 0$$

$$1 + 1 + x + (-6) = 0$$

$$x = -1 - 1 + 6$$

$$x = +4$$

oxidation number of C = +4

 

**91. D is correct.**

*Avogadro's law* states that the relationship between the masses of the same volume of the same gases (at the same temperature and pressure) corresponds to their respective molecular weights. Hence, the relative molecular mass of a gas can be calculated from the mass of a sample of known volume.

The flaw in Dalton's theory was corrected in 1811 by Avogadro, who proposed that equal volumes of two gases contain equal numbers of molecules at equal temperature and pressure (i.e., the mass of a gas's particles does not affect the volume that it occupies).

Avogadro's law allowed him to deduce the diatomic nature of numerous gases by studying the volumes at which they reacted. For example, two liters of hydrogen react with just one liter of oxygen to produce two liters of water vapor (at constant pressure and temperature). This means that a single oxygen molecule ($O_2$) splits to form two water particles.

Thus, Avogadro offered a more accurate estimate of the atomic mass of oxygen and various other elements and distinguished between molecules and atoms.

**92. B is correct.**

Balanced reaction: $BaCl_2$ (*aq*) + $K_2SO_4$ (*aq*) → $BaSO_4$ and 2 KCl

*Double replacement reaction* indicates an exchange of cations and anions between the reactants. Separate the reactants into ions and then exchange the cation and anion pairings.

**93. B is correct.**

% mass Cl = (mass of chlorine in molecule / molecular mass) × 100%

% mass Cl = (2 × 35.45 g/mol) / 159.09 g/mol × 100%

% mass Cl = 0.446 × 100% = 44.6%

**94. A is correct.**

Use mnemonic OIL RIG: Oxidation Is Loss, Reduction Is Gain (of electrons).

Oxidation is the loss of electrons, while reduction is the gain of electrons.

An oxidizing agent undergoes reduction, while a reducing agent undergoes oxidation.

Because the forward reaction is spontaneous, the reverse reaction is not spontaneous.

$I_2$ cannot oxidize $FeCl_2$; therefore, $I_2$ is the weakest oxidizing agent.

NaCl is not the answer because Na and Cl ions do not participate in the redox reaction; each has an identical oxidation number on each side of the reaction.

**95. C is correct.**

Balanced reaction:

$4\ RuS\ (s) + 9\ O_2 + 4\ H_2O \rightarrow 2\ Ru_2O_3\ (s) + 4\ H_2SO_4$

Find moles of RuS:

22.4 liters $O_2$ at STP = 1 mole $O_2$

67 g RuS × 1 mole/133 g = 0.5 mole RuS

Since 4 RuS molecules combine with 9 $O_2$ molecules, given 1 mole of $O_2$ and 0.5 mole RuS, $O_2$ is the limiting reactant.

Ratio for the limiting reactant:

molar ratio: 9 $O_2$ / 4 RuS = 1 mole $O_2$ / $x$ mole RuS

4 RuS / 9 $O_2$ = $x$ mole RuS / 1 mole $O_2$

$x$ mole RuS / 1 mole $O_2$ = 4 RuS / 9 $O_2$

$x$ mole RuS = 4 RuS / 9 $O_2$ × 1 mole $O_2$

$x = (4 / 9)$ mole RuS

Since RuS and $H_2SO_4$ have the same ratio, 4 / 9 $H_2SO_4$ is produced.

(4 / 9) mole $H_2SO_4$ × 98 g/mole = 44 g

**96. A is correct.**

| Ca | S | $O_4$ |
|---|---|---|
| +2 | $x$ | $(4 \times -2)$ |

The sum of charges in a neutral molecule is zero:

$2 + x + (4 \times -2) = 0$

$2 + x + (-8) = 0$

$x = -2 + 8$

oxidation number of S = +6

 

**97. C is correct.**

Balanced double replacement reaction:

$$C_6H_{12}O_6\,(s) + 6\ O_2\,(g) \rightarrow 6\ CO_2\,(g) + 6\ H_2O\,(g)$$

**98. B is correct.**

$SO_3 + H_2O \rightarrow H_2SO_4$ is a synthesis reaction because two compounds join to produce one product.

$3\ CuSO_4 + 2\ Al \rightarrow Al_2(SO_4)_3 + 3\ Cu$ is a single-replacement reaction because only one element (i.e., $SO_4$) is transferred from one reactant to another.

$2\ NaHCO_3 \rightarrow Na_2CO_3 + CO_2 + H_2O$ is a decomposition reaction where a compound is broken down into constituent elements.

$C_3H_8 + 5\ O_2 \rightarrow 3\ CO_2 + 4\ H_2O$ is a combustion reaction because a hydrocarbon (i.e., propane) reacts with oxygen to produce carbon dioxide and water.

**99. A is correct.**

The coefficients give the stoichiometric number (i.e., relative quantities) of molecules (or atoms) involved in the reaction.

**100. D is correct.**

Calculate molecular mass of NO:

Molecular mass of NO = atomic mass of N + atomic mass of O

Molecular mass of NO = N: 14.01 g/mole + O: 16.00 g/mole

Molecular mass of NO = 30.01 g/mole

Calculate moles of NO:

Moles of NO = mass of NO / molecular mass of NO

Moles of NO = 17.0 g / 30.01 g/mole

Moles of NO = 0.57 mole

For STP conditions, 1 mole of gas has a volume of 22.4 L:

Volume of NO = moles of NO × 22.4 L/mole

Volume of NO = 0.57 mole × 22.4 L

Volume of NO = 12.7 L

==========================================================================

**Practice Set 6: Questions 101–120**

==========================================================================

**101. B is correct.**

Use mnemonic OIL RIG: Oxidation Is Loss, Reduction Is Gain (of electrons).

Oxidizing agents undergo reduction (gain electrons) in a redox reaction.

$Cl^-$ has a full octet electron configuration and has a slight tendency to pick up more electrons.

$Na^+$ has a full octet electron configuration. Sodium metal is very reactive, and it is a strong reducing agent. As a result, the conjugate ion ($Na^+$) will be a weak oxidizing agent.

$Cl_2$ is in a stable electron configuration but can be reduced to $Cl^-$.

$Cl_2$ is very reactive and a strong oxidizing agent.

**102. B is correct.**

Use mnemonic OIL RIG: Oxidation Is Loss, Reduction Is Gain (of electrons).

Oxidation is the loss of electrons, while reduction is the gain of electrons.

An oxidizing agent undergoes reduction, while a reducing agent undergoes oxidation.

If the substance loses electrons (being oxidized), another substance must be gaining electrons (being reduced).

The original substance is referred to as a reducing agent, even though it is being oxidized.

**103. A is correct.**

*Molecular formula* expresses the number of atoms of each element in a molecule.

*Empirical formula* is the simplest formula for a compound. For example, if the molecular formula of a compound is $C_6H_{16}$, the empirical formula is $C_3H_8$.

*Elemental formula* and *atomic formula* are *not* valid chemistry terms.

**104. D is correct.**

Use mnemonic OIL RIG: Oxidation Is Loss, Reduction Is Gain (of electrons).

Oxidation is the loss of electrons, while reduction is the gain of electrons.

*continued...*

 

Oxidizing agent undergoes reduction, while a reducing agent undergoes oxidation.

Because the forward reaction is spontaneous, the reverse reaction is not spontaneous.

$FeCl_2$ cannot reduce $I_2$; therefore, $FeCl_2$ is the weakest reducing agent.

NaCl is not the answer because Na and Cl ions do not participate in the redox reaction; each has an identical oxidation number on each side of the reaction.

**105. C is correct.**

Use atomic mass from the periodic table:

$$(C = 6 \times 12 \text{ grams}) + (H = 12 \times 1 \text{ gram}) + (O = 6 \times 16 \text{ grams}) = 180 \text{ g/mole}$$

$$180 \text{ g/mole} \times 3.5 \text{ moles} = 630 \text{ grams}$$

**106. B is correct.**

The number of oxygens on each side of the reaction equation must be equal.

On the right side, there are 8 $CO_2$ molecules; since each $CO_2$ molecule contains 2 oxygens, multiply $8 \times 2 = 16$.

On the right side are 10 $H_2O$ molecules:

$$10 \times 1 = 10$$

Add 16 and 10 to get the number of oxygens on the right side:

$$16 + 10 = 26$$

Since there are 26 oxygens on the right side, there should be 26 oxygens on the left side.

Each $O_2$ molecule contains 2 oxygens:

$$26 / 2 = 13$$

Therefore, coefficient 13 is needed to balance the equation.

$$2\ C_4H_{10}\ (g) + 13\ O_2\ (g) \rightarrow 8\ CO_2\ (g) + 10\ H_2O\ (g)$$

**107. D is correct.**

A mole is a unit of measurement used to express amounts of a chemical substance.

Avogadro's number is the number of constituent particles (i.e., often atoms or molecules) that equals approximately $6.022 \times 10^{23}$.

Avogadro's number equals the number of atoms in 1 mole (g atomic weight) of an element.

The number of molecules in a mole is $6.02 \times 10^{23}$, which is Avogadro's number.

 

**108. A is correct.**

Balanced equation is:

$$2\ RuS + 9/2\ O_2 + 2\ H_2O \rightarrow Ru_2O_3 + 2\ H_2SO_4$$

To avoid fractional coefficients, multiply coefficients by 2:

$$4\ RuS + 9\ O_2 + 4\ H_2O \rightarrow 2\ Ru_2O_3 + 4\ H_2SO_4$$

The sum of the coefficients:

$$4 + 9 + 4 + 2 + 4 = 2$$

**109. C is correct.**

The % by mass alone is insufficient because it provides only the empirical formula.

The molecular weight alone is insufficient because two different compounds might have similar molecular weights.

If molecular mass and % by mass are known, the empirical formula is multiplied by a factor to determine the molecular weight.

**110. C is correct.**

Balanced equation (synthesis) for reaction II:

$$4\ Al\ (s) + 3\ Br_2\ (l) \rightarrow 2\ Al_2Br_3\ (s)$$

**111. B is correct.**

First, calculate the molecular mass (MW) of $MgSO_4$:

MW of $MgSO_4$ = atomic mass of Mg + atomic mass of S + (4 × atomic mass of O)

MW of $MgSO_4$ = 24.3 g/mole + 32.1 g/mole + (4 × 16.0 g/mole)

MW of $MgSO_4$ = 120.4 g/mole

Then, use the molecular mass to determine the number of moles:

Moles of $MgSO_4$ = mass of $MgSO_4$ / MW of $MgSO_4$

Moles of $MgSO_4$ = 60.2 g / 120.4 g/mole

Moles of $MgSO_4$ = 0.5 mole

 

**112. A is correct.**

Grams of a compound often contain a very large number of atoms.

72.9 g of Mg × (1 mol/24.3 g) = 3 moles Mg

3 moles Mg × (6.02 × $10^{23}$ atoms/mol) = 1.81 × $10^{24}$ atoms

**113. A is correct.**

To obtain the percent mass composition of oxygen, calculate the mass of oxygen in the molecule, divide it by the molecular mass and multiply by 100%.

% mass oxygen = (mass of oxygen in molecule / molecular mass) × 100%

% mass oxygen = [(4 moles × 16.00 g/mole) / (303.39 g/mole)] × 100%

% mass oxygen = 0.211 × 100% = 21.1%

**114. A is correct.**

Balanced reaction:

4 RuS (*s*) + 9 $O_2$ + 4 $H_2O$ → 2 $Ru_2O_3$ (*s*) + 4 $H_2SO_4$

Set up a ratio and solve:

4 RuS / 2 $Ru_2O_3$ = 7 moles RuS / *x* moles $Ru_2O_3$

*x* moles $Ru_2O_3$ = 4 RuS × 7 moles RuS / 2 $Ru_2O_3$

*x* moles $Ru_2O_3$ = 7 moles RuS × 2 $Ru_2O_3$ / 4 RuS

*x* = 3.5 moles

**115. C is correct.**

Oxidation numbers of S in compounds:

| | |
|---|---|
| $SO_4^{2-}$ | +6 |
| $S_2O_3^{2-}$ | +2 |
| $S^{2-}$ | –2 |

**116. B is correct.**

Li is in group IA, and its oxidation number is +1. Therefore, in $Li_2O_2$ the oxidation number of oxygen is −1.

Oxygen has an oxidation number of −1 (like in peroxides, $H_2O_2$).

**117. C is correct.**

*Double-replacement reaction* occurs when parts of two ionic compounds (e.g., HBr and KOH) are exchanged to make two new compounds ($H_2O$ and KBr).

2 HI → $H_2$ + $I_2$ is a *decomposition reaction* where a compound breaks down into constituent elements.

$SO_2$ + $H_2O$ → $H_2SO_4$ is a *synthesis* (or composition) reaction when two species combine to form a more complex chemical product.

CuO + $H_2$ → Cu + $H_2O$ is a *single-replacement reaction* when one element is transferred from one reactant to another (i.e., the oxygen atom leaves the Cu and joins $H_2$).

**118. B is correct.**

Density = mass / volume

Density = (5 μg × 1g /1,000,000 μg) / (25 μL × 1 L / 1,000,000 μL)

Density = $(5 \times 10^{-6}$ g) / $(2.5 \times 10^{-5}$ L)

Density = 0.2 g/L

Perform the above calculation (i.e., density = mass/volume) for each answer to determine the lowest density.

**119. D is correct.**

Balanced double replacement reaction:

$$2\ C_6H_{14}\ (g) + 19\ O_2\ (g) \rightarrow 12\ CO_2\ (g) + 14\ H_2O\ (g)$$

**120. A is correct.**

Mole is a unit of measurement used to express amounts of a chemical substance.

Number of molecules in a mole is $6.02 \times 10^{23}$, which is Avogadro's number.

 

===============================================================

**Practice Set 7: Questions 121–140**

===============================================================

**121. A is correct.**

Use mnemonic OIL RIG: Oxidation Is Loss, Reduction Is Gain (of electrons).

Oxidizing agents undergo reduction (gain electrons) in a redox reaction.

$Cl^-$ has a full octet electron configuration and has little tendency to pick up more electrons.

$Na^+$ has a full octet electron configuration. Sodium metal is very reactive, and it is a strong reducing agent. As a result, the conjugate ion ($Na^+$) will be a weak oxidizing agent.

$Cl_2$ is already in a stable electron configuration but can be easily reduced to $Cl^-$. $Cl_2$ is very reactive and a strong oxidizing agent.

**122. A is correct.**

Use mnemonic OIL RIG: Oxidation Is Loss, Reduction Is Gain (of electrons).

Oxidation is the loss of electrons, while reduction is the gain of electrons.

Oxidizing agent undergoes reduction, while a reducing agent undergoes oxidation.

Therefore, the substance that is reduced always gains electrons. It does not have to contain an element that increases in oxidation number.

It is not the reducing agent; the reduced substance is the oxidizing agent.

**123. A is correct.**

*Molecular formula* expresses the number of atoms of each element in a molecule.

*Empirical formula* is the simplest formula for a compound.

For example, if the molecular formula of a compound is $C_6H_{16}$, the empirical formula is $C_3H_8$.

Elemental formula and atomic formula are *not* chemistry terms.

**124. C is correct.**

Multiply the values in the bracket by the multiplier (i.e., subscript) outside the bracket.

For $C_6H_3(C_3H_7)_2(C_2H_5)$: $3 + (2 \times 7) + 5 = 22$ hydrogens

**125. D is correct.**

*Law of conservation of mass* states that for a closed system, the mass of the system must remain constant over time.

The law implies that during any chemical reaction, the mass of the reactants or starting materials must equal the mass of the products. From an understanding of this principle, a scientist could undertake quantitative studies of the transformations of substances.

**126. D is correct.**

The number of oxygens on both sides of the reaction equation must be equal.

On the right side, there are 8 $CO_2$ molecules; since each $CO_2$ molecule contains 2 oxygens, multiply $8 \times 2 = 16$.

On the right side are 10 $H_2O$ molecules: $10 \times 1 = 10$.

Add 16 and 10 to get the number of oxygens on the right side: $16 + 10 = 26$.

Since there are 26 oxygens on the right side, there should be 26 oxygens on the left side.

Each $O_2$ molecule contains 2 oxygens: $26 / 2 = 13$.

Therefore, coefficient 13 is needed to balance the equation.

$$2\ C_4H_{10}\ (g) + 13\ O_2\ (g) \rightarrow 8\ CO_2\ (g) + 10\ H_2O\ (g)$$

**127. C is correct.**

Use mnemonic OIL RIG: Oxidation Is Loss, Reduction Is Gain (of electrons).

Oxidation is the loss of electrons, while reduction is the gain of electrons.

Oxidizing agent undergoes reduction, while a reducing agent undergoes oxidation. The oxidation number for F is 0 as a reactant and –1 as a product; a decrease in oxidation number means that F is reduced in this reaction.

**128. B is correct.**

Balanced equation (double replacement): $N_2H_4 + 2\ H_2O_2 \rightarrow N_2 + 4\ H_2O$

 

**129. B is correct.**

First, calculate the molecular mass (MW) of $CH_4$:

Molecular mass of $CH_4$ = atomic mass of C + (4 × atomic mass of H)

Molecular mass of $CH_4$ = 12.01 g/mole + (4 × 1.01 g/mole)

Molecular mass of $CH_4$ = 16.05 g/mole

Use this information to determine the number of moles:

Moles of $CH_4$ = mass of $CH_4$ / molecular mass of $CH_4$

Moles of $CH_4$ = 6.40 g / 16.05 g/mole

Moles of $CH_4$ = 0.4 mole

Then, multiply by Avogadro's number to obtain the number of molecules:

Number of $CH_4$ molecules = moles of $CH_4$ × Avogadro's number

Number of $CH_4$ molecules = 0.4 moles × $6.02 \times 10^{23}$ molecules/mole

Number of $CH_4$ molecules = $2.40 \times 10^{23}$ molecules

**130. C is correct.**

To determine the limiting reactant, divide the quantity by the coefficient of each reactant.

Al: 0.2 mole / 2 = 0.1

$Fe_2O_3$: = 0.4 mole / 1 = 0.4

Whichever reactant has a smaller number is the limiting reactant.

**131. B is correct.**

Start by balancing the number of nitrogens by adding 2 to $NH_3$:

__$H_2$ + __$N_2$ + → 2 $NH_3$

Now balance the hydrogen by adding 3 to $H_2$:

3 $H_2$ + __$N_2$ → 2 $NH_3$

Balanced equation:

3 $H_2$ + $N_2$ → 2 $NH_3$

**132. B is correct.**

In order to calculate the number of atoms, start by calculating the molecular mass (MW) of fructose:

MW of fructose = (6 × atomic mass of C) + (12 × atomic mass of H) + (6 × atomic mass of O)

MW of fructose = (6 × 12.01 g/mole) + (12 × 1.01 g/mole) + (6 × 16.00 g/mole)

MW of fructose = 180.18 g/mole

Using the molecular mass, calculate the number of moles in the fructose sample:

Moles of fructose = mass of fructose / MW of fructose

Moles of fructose = 5.40 g / 180.18 g/mole

Moles of fructose = 0.03 mole

Because there are 6 oxygen atoms for each molecule of fructose:

Moles of oxygen = 6 × moles of fructose

Moles of oxygen = 6 × 0.03 mole

Moles of oxygen = 0.18 mole

One mole of atoms has $6.02 \times 10^{23}$ atoms; this is Avogadro's Number ($N_A$).

Number of oxygen atoms = moles of oxygen × Avogadro's number

Number of oxygen atoms = 0.18 mole × ($6.02 \times 10^{23}$ atoms/mole)

Number of oxygen atoms = $1.08 \times 10^{23}$ atoms

**133. A is correct.**

Molecular formula of iron (III) oxide is $Fe_3O_4$; therefore, only choices A and B are correct equations for this reaction.

$2\ Fe_2O_3 + 3\ C\ (s) \rightarrow 4\ Fe\ (l) + 3\ CO_2\ (g)$ is properly balanced because the coefficients are reduced to the lowest values.

**134. C is correct.**

In the balanced equation, the molar ratio of $Sn^{4+}$ and $Sn^{2+}$ are equal because the number of moles of $Sn^{2+}$ (reactant) = the number of moles of $Sn^{4+}$ (product).

 

**135. A is correct.**

Calculate the oxidation number for each species.

A: Mn = +7

B: Br = +5

C: C = +3

D: S = +6

E: K = +1

**136. B is correct.**

$2\ H_2O_2\ (s) \rightarrow 2\ H_2O\ (l) + O_2\ (g)$ is incorrectly classified. The decomposition part of the classification is correct because one complex compound ($H_2O_2$) is broken down into two or more parts ($H_2O$ and $O_2$).

However, it is a redox reaction because oxygen is being lost from $H_2O_2$, one of the three indicators of a redox reaction (i.e., electron loss/gain, hydrogen loss/gain, oxygen loss/gain). Additionally, one of the products is a single element ($O_2$), which is a clue that it is a redox reaction.

$AgNO_3\ (aq) + KOH\ (aq) \rightarrow KNO_3\ (aq) + AgOH\ (s)$ is correctly classified as a non-redox reaction because the reactants and products are compounds (no single elements), and there is no change in oxidation state (i.e., no gain or loss of electrons). It is a precipitation reaction because the chemical reaction occurs in an aqueous solution, and one of the products formed (AgOH) is insoluble, which makes it a precipitate.

$Pb(NO_3)_2\ (aq) + 2\ Na\ (s) \rightarrow Pb\ (s) + 2\ NaNO_3\ (aq)$ is correctly classified as a redox reaction because the nitrate ($NO_3^-$), which dissociates, is oxidized (loses electrons), the $Na^+$ is reduced (gains electrons). The two combine to form the ionic compound $NaNO_3$. It is a single-replacement reaction because one element is substituted for another element in a compound, making a new compound ($2NaNO_3$) and an element (Pb).

$HNO_3\ (aq) + LiOH\ (aq) \rightarrow LiNO_3\ (aq) + H_2O\ (l)$ is correctly classified as a non-redox reaction because the reactants and products are compounds (no single elements), and there is no change in oxidation state (i.e., no gain or loss of electrons). It is a double-replacement reaction because parts of two ionic compounds are exchanged to make two new compounds.

**137. A is correct.**

Formula mass is a synonym for molecular mass/molecular weight (MW).

Start by calculating the mass of the formula unit (CHCl):

$$CHCl = (12.01\ g/mol + 1.01\ g/mol + 35.45\ g/mol) = 48.47\ g/mol$$

Divide the molar mass by formula unit mass:

$$194\ g/mol / 48.47\ g/mol = 4.0025$$

Round it to the closest whole number: 4

Multiply the formula unit by 4:

$$(CHCl)_4 = C_4H_4Cl_4$$

**138. A is correct.**

Moles of Al = mass of Al / atomic mass of Al

Moles of Al = (8.52 g) / (26.98 g/mol)

Moles of Al = 0.316 moles

**139. C is correct.**

Use oxidation values of:

$H = +1$; $O = –2$; $Cl = –1$

| H | $Cr_2$ | $O_4$ | Cl |
|---|---|---|---|
| 1 | $2x$ | $4 \times -2$ | $-1$ |

$$1 + 2x - 8 - 1 = 0$$

$$2x = 8$$

$$x = 4$$

In $HCr_2O_4Cl$, Cr has an oxidation number of +4.

**140. B is correct.**

Molar mass (or *molecular weight*) is the mass in grams of the atoms that make up a mole of a particular molecule.

The unit used to measure molar mass is grams per mole.

Molar mass = grams / mole

 

*Notes for active learning*

*Notes for active learning*

# 7 – Chemical Kinetics and Equilibria: Detailed Explanations

---

**Practice Set 1: Questions 1–20**

---

**1. A is correct.**

General formula for the equilibrium constant of a reaction:

$aA + bB \leftrightarrow cC + dD$

$K_{eq} = ([C]^c \times [D]^d) / ([A]^a \times [B]^b)$

For the reaction above:

$K_{eq} = [C] / [A]^2 \times [B]^3$

**2. B is correct.**

Two gases having the same temperature have the same average molecular kinetic energy.

Therefore,

$(½)m_A v_A^2 = (½)m_B v_B^2$

$m_A v_A^2 = m_B v_B^2$

$(m_B / m_A) = (v_A / v_B)^2$

$(v_A / v_B) = 2$

$(m_B / m_A) = 2^2 = 4$

The only gases listed with a mass ratio of about 4 to 1 are iron (55.8 g/mol) and nitrogen (14 g/mol).

**3. A is correct.**

The slowest reaction would be the reaction with the highest activation energy and the lowest temperature.

**4. A is correct.**

Rate law is calculated experimentally by comparing trials and determining how changes in the reactants' initial concentrations affect the reaction rate.

rate = $k[A]^x \cdot [B]^y$

where $k$ is the rate constant, and the exponents $x$ and $y$ are the partial reaction orders (i.e., determined experimentally). They are not equal to the stoichiometric coefficients.

To determine the order of reactant A, find two experiments where the concentrations of B are identical and concentrations of A are different.

Use data from experiments 2 and 3 to calculate order of A:

$Rate_3 / Rate_2 = ([A]_3 / [A]_2)^{\text{order of A}}$

$0.500 / 0.500 = (0.060 / 0.030)^{\text{order of A}}$

$1 = (2)^{\text{order of A}}$

Order of A = 0

**5. C is correct.**

Balanced equation:

$C_2H_6O + 3\ O_2 \rightarrow 2\ CO_2 + 3\ H_2O$

Rate of reaction/consumption is proportional to the coefficients.

Rate of carbon dioxide production is twice the rate of ethanol consumption:

$2 \times 4.0\ M\ s^{-1} = 8.0\ M\ s^{-1}$

**6. D is correct.**

General formula for the equilibrium constant of a reaction:

$aA + bB \leftrightarrow cC + dD$

$K_{eq} = ([C]^c \times [D]^d) / ([A]^a \times [B]^b)$

For equilibrium constant calculations, only include species in aqueous or gas phases.

$K_{eq} = 1 / [Cl_2]$

 

**7. B is correct.**

$K_{eq}$ = [products] / [reactants]

If the $K_{eq}$ is less than 1 (e.g., $6.3 \times 10^{-14}$), then the numerator (i.e., products) is smaller than the denominator (i.e., reactants), and fewer products have formed relative to the reactants.

Equilibrium lies to the left if the reaction favors reactants compared to products.

**8. A is correct.**

The higher the activation energy, the slower the reaction is. The height of this barrier is independent of the determination of spontaneous (ΔG is negative with products more stable than reactants) or nonspontaneous reactions (ΔG is positive with products less stable than reactants).

**9. D is correct.**

The primary function of catalysts is lowering the reaction's activation energy (energy barrier), thus increasing its rate ($k$). Temperature measures the molecules' average kinetic energy (i.e., KE = ½mv2).

Except for zero-order chemical reactions, increased concentration of reactants increases the probability that the reactants collide with sufficient energy and orientation to overcome the energy of the activation barrier and proceed toward products.

**10. D is correct.**

Every reaction has activation energy: the amount of energy the reactant requires to start reacting.

On the graph, activation energy can be estimated by calculating the distance between the initial energy level (i.e., reactant) and the peak (i.e., transition state).

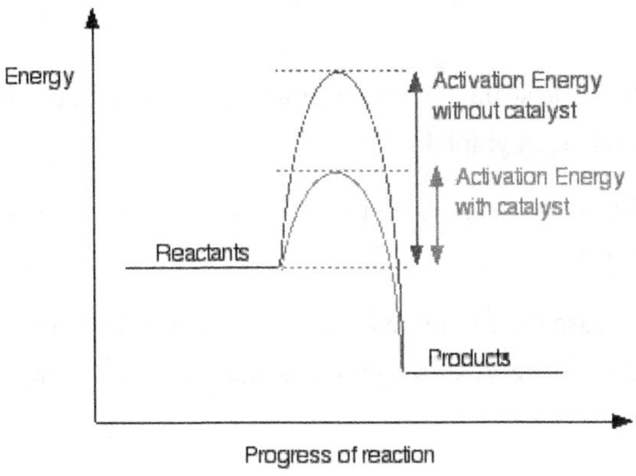


continued...

Activation energy is *not* the difference between energy levels of the initial and final state (reactant and product).

Catalysts provide an alternative pathway for the reaction to proceed to product formation. It lowers the activation energy (i.e., relative energy between reactants and transition state) and speeds the reaction rate.

Catalysts do not affect the Gibbs free energy ($\Delta G$: stability of products vs. reactants) or the enthalpy ($\Delta H$: bond breaking in reactants or bond making in products).

**11. D is correct.**

Activation energy is the barrier that must be overcome to transform reactant(s) into the product(s).

A reaction with higher activation energy (energy barrier) decreases the rate ($k$).

**12. B is correct.**

Activation energy is the barrier that must be overcome to transform reactant(s) into the product(s).

A reaction with lower activation energy (energy barrier) has an increased rate ($k$).

**13. D is correct.**

Activation energy is the barrier that must be overcome to transform reactant(s) into the product(s).

A reaction with higher activation energy (energy barrier) decreases the rate ($k$).

If the graphs are on the same scale, the height of the activation energy (R to the highest peak on the graph) is greatest in graph d.

**14. B is correct.**

If the graphs are on the same scale, the height of the activation energy (R to the highest peak on the graph) is the smallest in graph b.

The main function of catalysts is lowering the reaction's activation energy (energy barrier), thus increasing its rate ($k$).

Although catalysts decrease the energy required to reach the rate-limiting transition state, they do *not* decrease the relative energy of the products and reactants. Therefore, a catalyst does not affect $\Delta G$.

*continued...*

 

A catalyst provides an *alternative pathway* for the reaction to proceed to product formation. It lowers the activation energy (i.e., relative energy between reactants and transition state) and speeds the reaction rate.

Catalysts do *not* affect the Gibbs free energy (ΔG: stability of products vs. reactants) or the enthalpy (ΔH: bond breaking in reactants or bond making in products).

Graph D shows an endergonic reaction with less stable products than reactants.

All other graphs show an exergonic reaction with the same relative difference between the more stable products and the less stable reactants; only the activation energy is different.

**15. B is correct.**

A reaction is at equilibrium when the forward reaction rate equals the rate of the reverse reaction. Achieving equilibrium is common for reactions.

Catalysts, by definition, are regenerated during a reaction and are *not consumed* by the reaction.

**16. D is correct.**

The molecules must collide with sufficient energy, the frequency of collision, and the proper orientation to overcome the activation energy barrier.

**17. B is correct.**

Catalysts lower the energy of activation, which increases the rate of reaction.

The primary function of catalysts is lowering the reaction's activation energy (energy barrier), thus increasing its rate ($k$).

Although catalysts decrease the energy required to reach the rate-limiting transition state, they do *not* decrease the relative energy of the products and reactants. Therefore, a catalyst does not affect ΔG.

**18. D is correct.**

Le Châtelier's principle states that equilibrium shifts to counteract the change when changing the reaction conditions. If the concentration of reactants or products changes, the position of the equilibrium changes.

Adding reactants or removing products shifts the equilibrium to the right.

According to Le Châtelier's principle, adding or removing a solid does not affect the equilibrium position.

*continued...*

However, solid NaOH dissociates into $Na^+$ and $OH^-$ when placed in an aqueous solution. The $OH^-$ combines with $H^+$ to form water. Reducing $H^+$ as a product drives the reaction to the right.

**19. A is correct.**

Le Châtelier's principle states that equilibrium shifts to counteract the change when changing the reaction conditions. If the reaction temperature, pressure, or volume changes, the position of equilibrium changes.

Adding reactants or removing products shifts the equilibrium to the right.

Heat (i.e., energy-related to temperature) is a reactant, and increasing its value drives product formation.

Decreasing the pressure on the reaction vessel drives the reaction toward reactants (i.e., to the left). The relative molar concentration of the reactants (i.e., 2 + 6) is greater than the products (i.e., 4 + 3).

The other choices listed drive the reaction toward reactants (i.e., to the left).

**20. C is correct.**

First, calculate the concentration/molarity of each reactant:

$H_2$ : 0.20 moles / 4.00 L = 0.05 M

$X_2$ : 0.20 moles / 4.00 L = 0.05 M

$HX$: 0.800 moles / 4.00L = 0.20 M

Use the concentrations to calculate $Q$:

$Q = [HX]^2 / [H_2]\cdot[X_2]$

$Q = (0.20)^2 / (0.05 \times 0.05)$

$Q = 16$

Because $Q < K_c$ (i.e., 16 < 24.4), the reaction shifts to the right.

For $Q$ to increase and match $K_c$, the numerator (i.e., $[HX]^2$), which is the product (right side) of the equation, needs to be increased.

 

---

**Practice Set 2: Questions 21–40**

---

**21. B is correct.**

General formula for the equilibrium constant of a reaction:

$aA + bB \leftrightarrow cC + dD$

$K_{eq} = ([C]^c \times [D]^d) / ([A]^a \times [B]^b)$

For equilibrium constant ($K_c$) calculation, only include species in aqueous or gas phases:

$K_c = [CO_2]\,[H_2O]^2 / [CH_4]$

**22. A is correct.**

The rate law is calculated by determining how changes in reactant concentrations affect the reaction's initial rate.

rate = $k[A]^x \cdot [B]^y$

where $k$ is the rate constant, and the exponents $x$ and $y$ are the partial reaction orders; they are not equal to the stoichiometric coefficients.

Whenever the fast (i.e., second) step follows the slow (i.e., first) step, the fast step is assumed to reach equilibrium, and the equilibrium concentrations are used for the rate law of the slow step.

**23. D is correct.**

The primary function of catalysts is lowering the reaction's activation energy (energy barrier), thus increasing its rate ($k$).

Although catalysts decrease the energy required to reach the rate-limiting transition state, they do *not* decrease the relative energy of the products and reactants. Therefore, a catalyst does *not* affect $\Delta G$.

A catalyst provides an alternative pathway for the reaction to proceed to product formation. It lowers the activation energy (i.e., relative energy between reactants and transition state) and speeds the reaction rate.

Catalysts do not affect the Gibbs free energy ($\Delta G$: stability of products vs. reactants) or the enthalpy ($\Delta H$: bond breaking in reactants or bond making in products).

 

**24. B is correct.**

Equilibrium reactions vary between conditions, so any proportion of product/reactant mass is possible.

The question refers to the state *after* a reaction reaches equilibrium.

For a reaction that favors product formation ($K_{eq} > 1$), the amount of product would be more than the reactants.

For a reaction that favors reactants ($K_{eq} < 1$), the amount of product would be less than the reactants.

When $K_{eq} = 1$, the amounts of products and reactants would be equal.

**25. C is correct.**

Rate law is calculated *experimentally* by comparing trials and determining how changes in the reactants' initial concentrations affect the reaction rate.

$$\text{rate} = k[A]^x \cdot [B]^y$$

where $k$ is the rate constant, and the exponents $x$ and $y$ are the partial reaction orders (i.e., determined experimentally). They are not equal to the stoichiometric coefficients.

Rate laws *cannot* be determined from the balanced equation (i.e., used to determine equilibrium) unless the reaction occurs in a single step.

From the data, when the concentration doubled, the rate quadrupled. Since 4 is $2^2$, the rate law is second order; therefore, the rate = $k[H_2]^2$.

**26. D is correct.**

Lowering temperature shifts the equilibrium to the right because heat is a product (lowering the temperature is like removing the product).

Increasing temperature (i.e., heating the system) has the opposite effect and shifts the equilibrium to the left (i.e., toward products).

Removing $H_2$ (i.e., reactant) shifts the equilibrium to the left.

Adding $NH_3$ (i.e., products) shifts the equilibrium to the left.

A catalyst lowers the energy of activation but does not affect the position of the equilibrium.

 

**27. B is correct.**

The order of the reaction must be determined experimentally.

A reaction's order can be identical to its coefficients. For example, the coefficients correlate to the rate law in a single-step reaction or during the slow step of a multi-step reaction.

**28. A is correct.**

General formula for the equilibrium constant of a reaction: $a\text{A} + b\text{B} \leftrightarrow c\text{C} + d\text{D}$

$$K_{eq} = ([\text{C}]^c \times [\text{D}]^d) / ([\text{A}]^a \times [\text{B}]^b)$$

For equilibrium constant calculation, only include species in aqueous or gas phases: $K_{eq} = 1 / [\text{CO}] \times [\text{H}_2]^2$

**29. D is correct.**

Two peaks in this reaction indicate two energy-requiring steps with one intermediate (i.e., C) and each peak (i.e., B and D) as an activated complex (i.e., transition states). The activated complex (i.e., transition state) is undergoing bond-breaking/bond-making events.

**30. A is correct.**

Activation energy for the slow step of a reaction is the distance from the starting material (or an intermediate) to the activated complex (i.e., transition state) with the absolute highest energy (i.e., highest point on the graph).

**31. C is correct.**

The activation energy for the slow step of a reverse reaction is the distance from an intermediate (or product) to the activated complex (i.e., transition state) with the absolute highest energy (i.e., highest point on the graph). The slow step may not have the greatest magnitude for activation energy (e.g., E→ D on the graph).

**32. B is correct.**

Change in energy (or ΔH) is the difference between energy content of reactants and products.

**33. A is correct.**

As the temperature (i.e., average kinetic energy) *increases*, the particles move faster (i.e., increased kinetic energy) and collide more frequently per unit time. This *increases* the *reaction rate*.

**34. A is correct.**

Le Châtelier's principle states that equilibrium shifts to counteract the change when changing the reaction conditions.

If reaction temperature, pressure, or volume changes, the equilibrium position changes.

According to Le Châtelier's principle, endothermic ($+\Delta H$) reactions increase the formation of products at higher temperatures.

Since each side of the reaction has 2 gas molecules, a change in pressure does not affect equilibrium.

**35. B is correct.**

All chemical reactions eventually reach equilibrium, the state at which the reactants and products are present in concentrations with no further tendency to change with time.

Each product's production rate (i.e., forward reaction) equals the rate of their consumption by the reverse reaction.

**36. D is correct.**

Chemical equilibrium refers to a dynamic process whereby the *rate* at which a reactant molecule is transformed into a product is the same as at which a product molecule is transformed into a reactant.

All chemical reactions eventually reach equilibrium, the state at which the reactants and products are present in concentrations with no further tendency to change with time.

**37. A is correct.**

Expression for equilibrium constant:

$$K = [H_2O]^2 \cdot [Cl_2]^2 / [HCl]^4 \cdot [O_2]$$

Solve for $[Cl_2]$:

$$[Cl_2]^2 = (K \times [HCl]^4 \cdot [O_2]) / [H_2O]^2$$

$$[Cl_2]^2 = [46.0 \times (0.150)^4 \times 0.395] / (0.625)^2$$

$$[Cl_2]^2 = 0.0235$$

$$[Cl_2] = 0.153 \text{ M}$$

**38. A is correct.**

To determine the amount of each compound at equilibrium, consider the chemical reaction written in the form:

$$aA + bB \leftrightarrow cC + dD$$

Equilibrium constant ($K_c$) is defined as:

$$K_c = ([C]^c \times [D]^d) / ([A]^a \times [B]^b)$$

or

$$K_c = [products] / [reactants]$$

If the $K_{eq}$ is greater than 1, the numerator (i.e., products) is larger than the denominator (i.e., reactants), and more products have formed relative to the reactants.

If the reaction favors products compared to reactants, the equilibrium lies to the right.

**39. D is correct.**

Gases (as opposed to liquids and solids) are most sensitive to changes in pressure.

Three moles of hydrogen gas as a reactant and three moles of water vapor as a product.

Changes in pressure result in proportionate changes to the forward and reverse reactions.

**40. C is correct.**

When changing the reaction conditions, Le Châtelier's principle states that equilibrium shifts to counteract the change. If the temperature, pressure, volume, or concentration changes, the position of the equilibrium changes.

CO is a reactant, and increasing its concentration shifts the equilibrium toward product formation.

## Practice Set 3: Questions 41–60

**41. A is correct.**

General formula for the equilibrium constant of a reaction:

$$aA + bB \leftrightarrow cC + dD$$

$$K_{eq} = ([C]^c \times [D]^d) / ([A]^a \times [B]^b)$$

For equilibrium constant calculation, only include species in aqueous or gas phases:

$$K_{eq} = [NO]^4 \times [H_2O]^6 / [NH_3]^4 \times [O_2]^5$$

**42. D is correct.**

High levels of $CO_2$ cause a person to hyperventilate to reduce the number of $CO_2$ (reactant).

Hyperventilating drives the reaction toward products.

Removing products (i.e., $HCO_3^-$ and $H^+$) or intermediates ($H_2CO_3$) drives the reaction toward products.

**43. B is correct.**

Units of the rate constants:

Zero-order reaction: M $sec^{-1}$

First order reaction: $sec^{-1}$

Second order reaction: L $mole^{-1}$ $sec^{-1}$

**44. D is correct.**

A reaction proceeds when the reactant(s) have sufficient energy to overcome activation energy and form products.

Increasing the temperature increases the molecule's kinetic energy, the frequency of collision, and the probability that the reactants will overcome the barrier (energy of activation) to form products.

 

**45. B is correct.**

Low activation energy increases the rate because lower energy is required for the reaction to proceed is lower.

High temperature increases the rate because faster-moving molecules have a greater collision probability, facilitating the reaction.

Combined, lower activation energy and a higher temperature result in the highest relative rate for reaction.

**46. C is correct.**

Rate law is calculated by comparing trials and determining how changes in the reactants' initial concentrations affect the reaction rate.

$$\text{rate} = k[A]^x \cdot [B]^y$$

where $k$ is the rate constant, and the exponents $x$ and $y$ are the partial reaction orders (i.e., determined experimentally). They are not equal to the stoichiometric coefficients.

Start by identifying two reactions where XO concentration is constant and $O_2$ is different: experiments 1 and 2. When the concentration of $O_2$ is doubled, the rate is also doubled. This indicates that the order of the reaction for $O_2$ is 1.

Now, determine the order for *X*O.

Find 2 reactions where the concentration of $O_2$ is constant, and *X*O is different: experiments 2 and 3. When the concentration of *X*O is tripled, the rate is multiplied by a factor of 9. $3^2 = 9$, which means the order of the reaction with respect to *X*O is 2.

Therefore, the expression of rate law is:

$$\text{rate} = k[XO]^2 \cdot [O_2]$$

**47. B is correct.**

General formula for the equilibrium constant of a reaction:

$$aA + bB \leftrightarrow cC + dD$$

$$K_{eq} = ([C]^c \times [D]^d) / ([A]^a \times [B]^b)$$

For equilibrium constant calculation, only include species in aqueous or gas phases.

$$K_{eq} = 1 / [CO_2]$$

**48. D is correct.**

$K_{eq}$ = [products] / [reactants]

If the numerator (i.e., products) is smaller than the denominator (i.e., reactants), fewer products have formed relative to the reactants. $K_{eq}$ is less than 1.

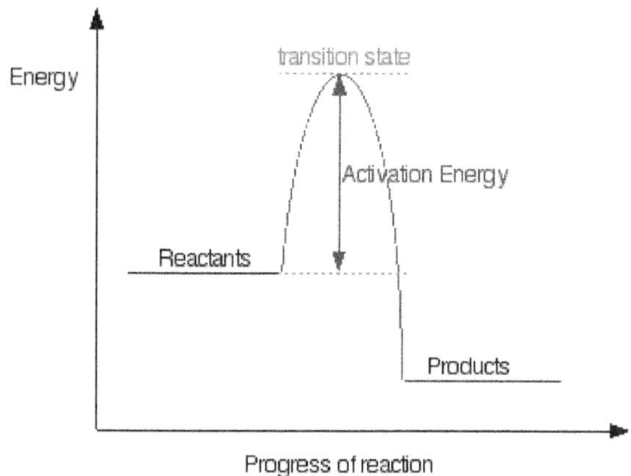


**49. C is correct.**

Activation energy is the energy barrier for a reaction to proceed. The forward reaction is measured from the reactants to the highest energy level in the reaction.

A reaction mechanism with low activation energy proceeds faster than a reaction with high activation energy.

**50. D is correct.**

As the average kinetic energy (i.e., KE = $\frac{1}{2}mv^2$) *increases*, the particles move faster, collide more frequently per unit time, and possess greater energy when they collide. This *increases* the *reaction rate*.

Hence the *reaction rate* of most *reactions increases* with *increasing temperature*.

For reversible reactions, it is common for the rate law to depend on the concentration of the products. The overall rate is negative (except for autocatalytic reactions).

As the concentration of products decreases, their collision frequency decreases, and the reverse reaction rate decreases. Therefore, the overall rate of reaction increases.

Catalysts lower the reaction's activation energy, thus increasing its rate ($k$).

 

**51. C is correct.**

General formula for the equilibrium constant of a reaction:

$aA + bB \leftrightarrow cC + dD$

$K_{eq} = ([C]^c \times [D]^d) / ([A]^a \times [B]^b)$

For the ionization equilibrium ($K_i$) constant calculation, only include species in aqueous or gas phases (which, in this case, is all):

$K_i = [H^+] \cdot [HS^-] / [H_2S]$

**52. B is correct.**

As the temperature (i.e., average kinetic energy) *increases*, the particles move faster, collide more frequently per unit time, and possess greater energy when they collide. This *increases* the *reaction rate*. Hence, the *reaction rate* of most *reactions increases* with *increasing temperature*.

**53. B is correct.**

Reactions are spontaneous when Gibbs free energy (ΔG) is negative, and the reaction is described as exergonic; the products are more stable than the reactants.

Reactions are nonspontaneous when Gibbs free energy (ΔG) is positive, and the reaction is described as endergonic; the products are less stable than the reactants.

**54. C is correct.**

Le Châtelier's principle states that the equilibrium position shifts to counteract the change when changing the reaction conditions. If the reaction temperature, pressure, or volume changes, the equilibrium position changes.

There are 2 moles on each side, so changes in pressure (or volume) do not shift the equilibrium.

**55. A is correct.**

All the choices are possible events of chemical reactions, but they do not need to happen for a reaction, except that reactant particles must collide (i.e., make contact) with each other.

**56. D is correct.**

The enthalpy or heat of reaction (ΔH) is not a function of temperature.

Since the reaction is exothermic, Le Châtelier's principle states that increasing the temperature decreases the forward reaction.

 

**57. C is correct.**

Le Châtelier's principle states that equilibrium shifts to counteract the change when changing the reaction conditions. If the reaction's concentration changes, the position of the equilibrium changes.

According to Le Châtelier's Principle, adding or removing a solid at equilibrium does not affect equilibrium.

However, adding solid $KC_2H_3O_2$ is equivalent to adding $K^+$ and $C_2H_3O_2^-$ (i.e., product) as it dissociates in an aqueous solution. The increased concentration of products shifts the equilibrium toward reactants (i.e., to the left).

Increasing the pH decreases the $[H^+]$ and shifts the reaction to products (i.e., right).

**58. B is correct.**

Le Châtelier's principle states that equilibrium shifts to counteract the change when changing the reaction conditions. If the reaction temperature, pressure, or volume changes, the equilibrium position changes.

$K_{eq} = 2.8 \times 10^{-21}$, which indicates a low concentration of products compared to reactants:

$$K_{eq} = [\text{products}] / [\text{reactants}]$$

**59. D is correct.**

Chemical equilibrium refers to a dynamic process whereby the rate at which a reactant molecule is transformed into a product is the same as at which a product molecule is transformed into a reactant.

The rate of the forward reaction is equal to the rate of the reverse reaction.

**60. C is correct.**

All chemical reactions eventually reach equilibrium, the state at which the reactants and products are present in concentrations with no further tendency to change with time.

Catalysts speed up reactions and increase the rate at which equilibrium is reached. However, they do not alter the thermodynamics of a reaction and, therefore, do not change free energy $\Delta G$.

Catalysts do not alter the equilibrium constant.

 

---

## Practice Set 4: Questions 61–80

---

**61. A is correct.**

To determine the order with respect to W, compare the data for trials 2 and 4.

Concentrations of X and Y do not change when comparing trials 2 and 4, but the concentration of W changes from 0.015 to 0.03, corresponding to an increase by a factor of 2.

However, rate did not increase (0.08 remains 0.08); therefore, the order with respect to W is 0.

Order for X is determined by comparing the data for trials 1 and 3.

Concentrations for W and Z are constant, and the concentration for X increases by a factor of 3 (from 0.05 to 0.15). The reaction rate increased by a factor of 9 (0.04 to 0.36).

Since $3^2 = 9$, the order of the reaction with respect to X is two.

Order with respect to Y is found by comparing data from trials 1 and 5. [W] and [X] do not change, but [Y] goes up by a factor of 4. The rate from trials 1 to 5 goes up by a factor of 2.

Since $4^{1/2} = 2$, the order of the reaction with respect to Y is ½.

The overall order is found by the sum of the orders for X, Y, and Z: 0 + 2 + ½ = 2½.

**62. B is correct.**

Since the order with respect to W is zero, the rate of formation of Z does not depend on the concentration of W.

**63. D is correct.**

The rate law is calculated by comparing trials and determining how changes in the reactants' initial concentrations affect the reaction rate.

$$\text{rate} = k[A]^x \cdot [B]^y$$

where $k$ is the rate constant, and the exponents $x$ and $y$ are the partial reaction orders (i.e., determined experimentally). They are not equal to the stoichiometric coefficients.

Use the partial orders determined in question **61** to express the rate law:

$$\text{rate} = k[W]^0[X]^2[Z]^{1/2}$$

*continued...*

Substitute the values for trial #1:

$$0.04 = k\,[0.01]^0[0.05]^2[0.04]^{1/2}$$
$$0.04 = (k)\cdot(1)\cdot(2.5 \times 10^{-3})\cdot(2 \times 10^{-1})$$

$$k = 80$$

**64. A is correct.**

When changing the conditions of a reaction, Le Châtelier's principle states that the equilibrium position shifts to counteract the change.

If the reaction temperature, pressure, or volume changes, the position of the equilibrium changes.

Decreasing the temperature of the system favors the production of more heat. It shifts the reaction towards the exothermic side.

Because $\Delta H > 0$ (i.e., heat is a reactant), the reaction is endothermic.

Therefore, decreasing the temperature shifts the equilibrium toward the reactants.

**65. D is correct.**

General formula for the equilibrium constant of a reaction:

$$a\text{A} + b\text{B} \leftrightarrow c\text{C} + d\text{D}$$

$$K_{eq} = ([\text{C}]^c \times [\text{D}]^d) / ([\text{A}]^a \times [\text{B}]^b)$$

For the equilibrium constant ($K_i$), only include species in aqueous phases:

$$K_i = [H^+] \times [H_2PO_4^-] / [H_3PO_4]$$

**66. C is correct.**

Catalysts speed up reactions and increase the rate at which equilibrium is reached.

However, they do not alter the thermodynamics of a reaction and, therefore, do not change free energy $\Delta G$.

A catalyst provides an *alternative pathway* for the reaction to proceed to product formation. It lowers the activation energy (i.e., relative energy between reactants and transition state) and increases the reaction rate.

*continued...*

 

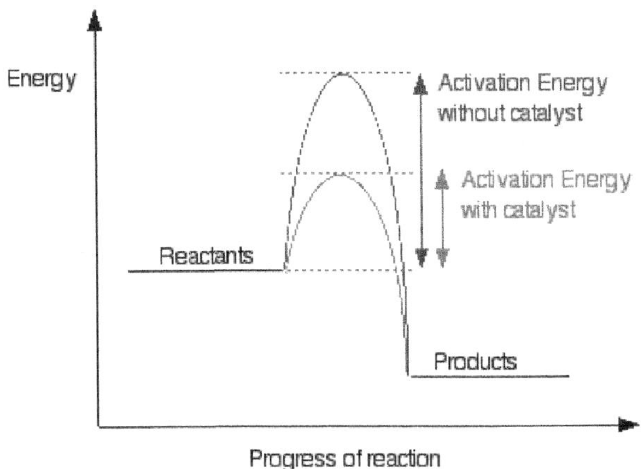


Catalysts do not affect the Gibbs free energy (ΔG: stability of products vs. reactants) or the enthalpy (ΔH: bond breaking in reactants or bond making in products).

**67. D is correct.**

General formula for the equilibrium constant of a reaction:

$aA + bB \leftrightarrow cC + dD$

$K_{eq} = ([C]^c \times [D]^d) / ([A]^a \times [B]^b)$

For the ionization equilibrium constant ($K_i$) calculation, only include species in aqueous or gas phases (which, in this case, is all of them):

$K_i = [H^+]\cdot[HSO_3^-] / [H_2SO_3]$

**68. D is correct.**

$K_{eq}$ = [products] / [reactants]

If the $K_{eq}$ is greater than 1, then the numerator (i.e., products) is greater than the denominator (i.e., reactants), and more products have formed relative to the reactants.

**69. B is correct.**

Activation energy is the energy barrier for a reaction to proceed.

Forward reaction is measured from the energy of the reactants to the highest energy level in the reaction.

**70. A is correct.**

Lowering the temperature decreases the reaction rate because the molecules involved move and collide more slowly, so the reaction occurs slower.

Increasing the concentration of reactants due to the increased probability that the reactants collide with sufficient energy and orientation to overcome the energy of the activation barrier and proceed toward products.

Catalysts lower the reaction's activation energy, thus increasing its rate ($k$).

**71. B is correct.**

Activation energy is the barrier that must be overcome to transform reactant(s) into the product(s).

Molecules must collide with the proper orientation and energy (i.e., kinetic energy is the average temperature) to overcome the energy of the activation barrier.

**72. B is correct.**

Activation energy is the barrier that must be overcome to transform reactant(s) into the product(s). The molecules must collide with the proper orientation and energy (i.e., kinetic energy is the average temperature) to overcome the energy of the activation barrier.

Solids describe molecules with limited relative motion. Liquids are molecules with more relative motion than solids, while gases exhibit the most relative motion.

**73. B is correct.**

Rate law is calculated experimentally by comparing trials and determining how changes in the initial concentrations of the reactants affect the rate.

$$\text{rate} = k[\text{A}]^x \cdot [\text{B}]^y$$

where $k$ is the rate constant, and the exponents $x$ and $y$ are the partial reaction orders (i.e., determined experimentally). They are not equal to the stoichiometric coefficients.

Comparing Trials 1 and 3, [A] increased by a factor of 3, as did the reaction rate; thus, the reaction is first order with respect to A.

Comparing Trials 1 and 2, [B] increased by a factor of 4, and the reaction rate increased by a factor of $16 = 4^2$.

Thus, the reaction is second order with respect to B. Therefore, the rate = $k[\text{A}] \cdot [\text{B}]^2$

 

**74. D is correct.**

The molecules must collide with sufficient energy, the frequency of collision, and the proper orientation to overcome the activation energy barrier.

**75. D is correct.**

Le Châtelier's principle states that equilibrium shifts to counteract the change when changing the reaction conditions. If the reaction temperature, pressure, or volume changes, the equilibrium position changes.

Removing $NH_3$ and adding $N_2$ shifts the equilibrium to the right.

Removing $N_2$ or adding $NH_3$ shifts the equilibrium to the left.

**76. B is correct.**

Increasing $[HC_2H_3O_2]$ and $[H^+]$ affects equilibrium because both species are part of the equilibrium equation.

$K_{eq}$ = [products] / [reactants]

$K_{eq} = [C_2H_3O_2^-] / [HC_2H_3O_2]·[H^+]$

Solid $NaC_2H_3O_2$ is added to aqueous solutions, and $NaC_2H_3O_2$ dissociates into $Na^+$ (*aq*) and $C_2H_3O_2^-$ (*aq*). Therefore, the overall concentration of $C_2H_3O_2^-$ (*aq*) increases, affecting the equilibrium.

Adding solid $NaNO_3$: this is a salt and does not react with reactant or product molecules; therefore, it will not affect the equilibrium.

Adding solid NaOH: the reactant is an acid because it dissociates into hydrogen ions and anions in aqueous solutions. It reacts with bases (e.g., NaOH) to create water and salt. Thus, NaOH reduces the concentration of the reactant, which affects the equilibrium.

Increasing $[HC_2H_3O_2]$ and $[H^+]$ increases reactants and drives the reaction toward product formation.

**77. D is correct.**

Calculate the value of the equilibrium expression: hydrogen iodide concentration decreases for equilibrium to be reached.

$K$ = [products] / [reactants]

$K = [HI]^2 / [H_2]·[I_2]$

*continued…*

To determine the state of a reaction, calculate its reaction quotient ($Q$). $Q$ uses the same calculation method as the equilibrium constant ($K$); the difference is that $K$ must be calculated at the equilibrium point, whereas $Q$ can be calculated at any time.

If $Q > K$, reaction shifts towards reactants (more reactants, fewer products)

If $Q = K$, reaction is at equilibrium

If $Q < K$, reaction shifts towards products (more products, fewer reactants)

$$Q = [HI]^2 / [H_2]\cdot[I_2]$$

$$Q = (3)^2 / 0.4 \times 0.6$$

$$Q = 37.5$$

Because $Q > K$ (i.e., (37.5 > 35), the reaction shifts toward reactants (larger denominator), and the amount of product (HI) decreases.

**78. C is correct.**

Decreasing the concentration of reactants (or decreasing pressure) decreases the reaction rate because there is a decreased probability that two reactant molecules collide with sufficient energy to overcome the energy of activation and form products.

**79. B is correct.**

Chemical equilibrium expression depends on the stoichiometry of the reaction, more specifically, the coefficients of each species involved in the reaction.

The mechanism refers to the pathway of product formation (e.g., $S_N1$ or $S_N2$) but does not affect the relative energies of the reactants and products (i.e., a position for the equilibrium).

The rate refers to the time needed to achieve equilibrium but does not affect the equilibrium position.

**80. D is correct.**

$$K = [\text{products}] / [\text{reactants}]$$

$$K = [CH_3OH] / [H_2O]^2\cdot[CO]$$

None of the stated changes tend to decrease the *magnitude* of the equilibrium constant $K$.

For exothermic reactions ($\Delta H < 0$), decreasing the temperature increases the magnitude of $K$.

Volume, pressure, or concentration changes do not affect the magnitude of $K$.

 

---

**Practice Set 5: Questions 81–100**

---

**81. B is correct.**

When changing the reaction conditions, Le Châtelier's principle states that the position of equilibrium shifts to counteract the change. If the reaction's concentration changes, the position of the equilibrium changes.

Adding reactants or removing products shifts the equilibrium to the right (i.e., toward product formation).

**82. A is correct.**

When two or more reactions are combined to create a new reaction, the $K_c$ of the resulting reaction is a product of $K_c$ values of the individual reactions.

When a reaction is reversed, the new $K_c$ = 1 / old $K_c$.

Start by analyzing the reactions provided the problem:

1. $2\ NO + Cl_2 \leftrightarrow 2\ NOCl$ $\quad K_c = 3.2 \times 10^3$
2. $2\ NO_2 \leftrightarrow 2\ NO + O_2$ $\quad K_c = 15.5$
3. $NOCl + \frac{1}{2}\ O_2 \leftrightarrow NO_2 + \frac{1}{2}\ Cl_2$ $\quad K_c = ?$

Reaction 3 can be created by combining reactions 1 and 2.

For reaction 1, notice that in reaction 3, NOCl is located on the left. To match reaction 3, reverse reaction 1:

$$2\ NOCl \leftrightarrow 2\ NO + Cl_2$$

$K_c$ of this reversed reaction is:

$$K_c = 1 / (3.2 \times 10^{-3}) = 312.5$$

Reaction 2 also needs to be reversed to match the position of NO in reaction 3:

$$2\ NO + O_2 \leftrightarrow 2\ NO_2$$

The new $K_c$ is:

$$K_c = 1 / 15.5 = 0.0645$$

*continued…*

Since $K$ of the resulting reaction is the product of $K_c$ from both reactions, add those reactions:

$2\ NOCl \leftrightarrow 2\ NO + Cl_2$

$2\ NO + O_2 \leftrightarrow 2\ NO_2$

---

$2\ NOCl + O_2 \leftrightarrow 2\ NO_2 + Cl_2$

$K_c = 312.5 \times 0.064 = 20.16$

Divide the equation by 2:

$NOCl + ½\ O_2 \leftrightarrow NO_2 + ½\ Cl_2$

The new $K_c$ is the square root of the initial $K_c$:

$K_c = \sqrt{20.16} = 4.49$

**83. C is correct.**

Increased pressure or increased concentration of reactants increases the *probability* of colliding with sufficient energy and orientation to overcome the energy of the activation barrier and proceed toward products.

**84. C is correct.**

The reaction rate is the speed at which reactants are consumed or a product is formed.

The rate of a chemical reaction at a constant temperature depends only on the concentrations of the substances that influence the rate. The reactants influence the rate of reaction, but occasionally, products can influence the reaction rate.

**85. A is correct.**

General formula for the equilibrium constant of a reaction:

$a\text{A} + b\text{B} \leftrightarrow c\text{C} + d\text{D}$

$K_{eq} = ([C]^c \times [D]^d) / ([A]^a \times [B]^b)$

For the reaction above:

$K_{eq} = [B] \times [C]^3 / [A]^2$

**86. B is correct.**

Rate law is calculated by comparing trials and determining how changes in the reactants' initial concentrations affect the reaction rate.

$$\text{rate} = k[A]^x \cdot [B]^y$$

where $k$ is the rate constant, and the exponents $x$ and $y$ are the partial reaction orders (i.e., determined experimentally). They are not equal to the stoichiometric coefficients.

**87. C is correct.**

Rates of formation/consumption of species in a reaction are proportional to their coefficients.

If the rate of a species is known, the rate of other species can be calculated using simple proportions:

$$\text{coefficient}_{\text{NOBr}} / \text{coefficient}_{\text{Br2}} = \text{rate}_{\text{formation NOBr}} / \text{rate}_{\text{consumption Br2}}$$

$$\text{rate}_{\text{consumption Br2}} = \text{rate}_{\text{formation NOBr}} / (\text{coefficient}_{\text{NOBr}} / \text{coefficient}_{\text{Br2}})$$

$$\text{rate}_{\text{consumption Br2}} = 4.50 \times 10^{-4}\ \text{mol L}^{-1}\ \text{s}^{-1} / (2 / 1)$$

$$\text{rate}_{\text{consumption Br2}} = 2.25 \times 10^{-4}\ \text{mol L}^{-1}\ \text{s}^{-1}$$

**88. C is correct.**

Activation energy is the barrier that must be overcome to transform reactant(s) into the product(s).

A reaction with lower activation energy (energy barrier) has an increased rate ($k$).

**89. A is correct.**

General formula for the equilibrium constant of a reaction:

$$aA + bB \leftrightarrow cC + dD$$

$$K_{eq} = ([C]^c \times [D]^d) / ([A]^a \times [B]^b)$$

For the reaction above: $K_{eq} = [C]^2 \cdot [D] / [A] \cdot [B]^2$

**90. D is correct.**

When changing the conditions of a reaction, Le Châtelier's principle states that the position of the equilibrium shifts to counteract the change.

If the reaction's concentration of water (i.e., product) is changed, equilibrium changes toward reactants.

**91. D is correct.**

For the general equation:

$a\text{A} + b\text{B} \leftrightarrow c\text{C} + d\text{D}$

Equilibrium constant is defined as:

$K_{eq} = ([\text{C}]^c \times [\text{D}]^d) / ([\text{A}]^a \times [\text{B}]^b)$

or

$K_{eq} = [\text{products}] / [\text{reactants}]$

If $K_{eq}$ is less than 1 (e.g., $4.3 \times 10^{-17}$), then the numerator (i.e., products) is smaller than the denominator (i.e., reactants), and fewer products have formed relative to the reactants.

Equilibrium lies to the left if the reaction favors reactants compared to products.

**92. B is correct.**

Primary function of catalysts is lowering the reaction's activation energy (energy barrier), thus increasing its rate ($k$).

Activation energy is the energy barrier for a reaction to proceed. The forward reaction is measured from the energy of the reactants to the highest energy level in the reaction.

*Transition state* signifies "*bond making*" and "*bond making*" events in the transformation of the reactant(s) into the product(s).

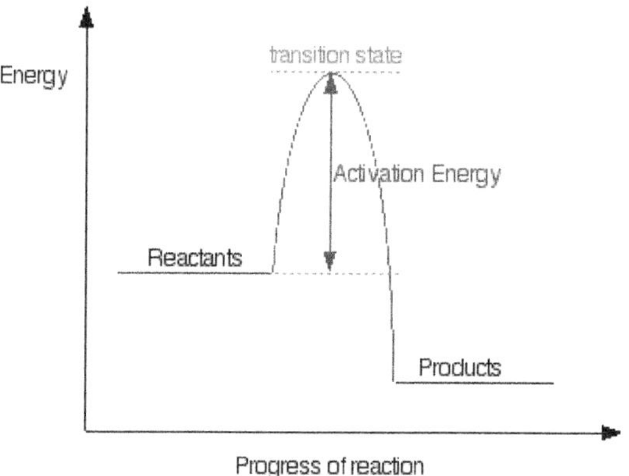


 

**100. D is correct.**

When two or more reactions are combined to create a new reaction, the $K_c$ of the resulting reaction is a product of $K_c$ values of the individual reactions.

When a reaction is reversed, new $K_c$ = 1 / old $K_c$.

$PCl_3 + Cl_2 \leftrightarrow PCl_5$ $\quad K_c = K_{c1}$

$2\ NO + Cl_2 \leftrightarrow 2\ NOCl$ $\quad K_c = K_{c2}$

Reverse the first reaction:

$PCl_5 \leftrightarrow PCl_3 + Cl_2$ $\quad K_c = 1 / K_{c1}$

Add the two reactions to create a third reaction:

$PCl_5 \leftrightarrow PCl_3 + Cl_2$ $\quad K_c = 1 / K_{c1}$

$2\ NO + Cl_2 \leftrightarrow 2\ NOCl$ $\quad K_c = K_{c2}$

---

$PCl_5 + 2\ NO \leftrightarrow PCl_3 + 2\ NOCl$

$K = [(1 / K_1) \times K_2] = K_2 / K_1$

===============================================================

**Practice Set 6: Questions 101–120**

===============================================================

**101. C is correct.**

*Mass action* expression is related to the equilibrium constant expression.

General formula for the *equilibrium constant* of a reaction:

$aA + bB \leftrightarrow cC + dD$

$K_{eq} = ([C]^c \times [D]^d) / ([A]^a \times [B]^b)$

To determine the mass action expression, only include species in *aqueous or gas phases*:

$K_c = [CrCl_3]^4 / [CCl_4]^3$

**102. A is correct.**

When changing the conditions of a reaction, Le Châtelier's principle states that the equilibrium position shifts to counteract the change. If the reaction's concentration is changed, the position of equilibrium changes.

Adding $KNO_2$ increases the amount of K+ and $NO_2^-$ (product). Therefore, the reaction shifts toward the reactants (i.e., to the left).

**103. D is correct.**

Rate expressions are determined experimentally from the rate data for a reaction. The calculation is not similar to that for the equilibrium constant.

The rate expression cannot be determined since rate data are not given for this reaction.

**104. D is correct.**

The rate law is calculated experimentally by comparing trials and determining how changes in the initial concentrations of the reactants affect the rate.

$\text{rate} = k[A]^x \cdot [B]^y$

where $k$ is the rate constant, and the exponents $x$ and $y$ are the partial reaction orders (i.e., determined experimentally). They are not equal to the stoichiometric coefficients.

To write a complete rate law, the order of each reactant is required.

*continued...*

 

Determine the order of reactant D by using 2 experiments where the concentrations of E are identical, and D concentrations are different: experiments 1 and 2.

Use data from these experiments to calculate the order of D:

$\text{Rate}_2 / \text{Rate}_1 = ([D]_2 / [D]_1)^{\text{order of D}}$

$(0.000500) / (0.000250) = (0.200 / 0.100)^{\text{order of D}}$

$2 = (2)^{\text{order of D}}$

Order of D = 1

Use the same procedure to calculate the order of E.

Data from experiments 1 and 3:

$\text{Rate}_3 / \text{Rate}_1 = ([E]_3 / [E]_1)^{\text{order of E}}$

$(0.001000) / (0.000250) = (0.500 / 0.250)^{\text{order of E}}$

$4 = (2)^{\text{order of E}}$

Order of E = 2

$\text{Rate} = k[D]\cdot[E]^2$

**105. D is correct.**

Rate of formation/consumption of species in a reaction is proportional to their coefficients.

If a species' rate is known, other species' rates can be calculated using simple proportions.

Because the formation rate of a species is known, there is no need to calculate the rate. (The extra information on the problem is just there to distract test-takers).

$\text{rate}_{\text{consumption H2}} / \text{rate}_{\text{formation NO2}} = \text{coefficient}_{\text{H2}} / \text{coefficient}_{\text{NO2}}$

$\text{rate}_{\text{consumption H2}} = (\text{coefficient}_{\text{H2}} / \text{coefficient}_{\text{NO2}}) \times \text{rate}_{\text{formation NO2}}$

$\text{rate}_{\text{consumption H2}} = (4 / 2) \times 2.8 \times 10^{-4}$ M/min

$\text{rate}_{\text{consumption H2}} = 5.6 \times 10^{-4}$ M/min

 

**106. C is correct.**

General formula for the equilibrium constant of a reaction:

$$aA + bB \leftrightarrow cC + dD$$

$$K_{eq} = ([C]^c \times [D]^d) / ([A]^a \times [B]^b)$$

For equilibrium constant calculation, only include species in aqueous or gas phases (which, in this problem, is all of them):

$$K_{eq} = [CO_2]^2 / [CO]^2 \times [O_2]$$

**107. B is correct.**

Reactions proceed faster at higher temperatures.

Catalysts lower the activation energy (i.e., energy barrier) for the reaction, and the reaction rate increases.

**108. D is correct.**

For the general equation:

$$aA + bB \leftrightarrow cC + dD$$

The equilibrium constant is defined as:

$$K_{eq} = ([C]^c \times [D]^d) / ([A]^a \times [B]^b)$$

or

$$K_{eq} = [\text{products}] / [\text{reactants}]$$

If $K_{eq}$ is less than 1 (e.g., $6.3 \times 10^{-14}$), then the numerator (i.e., products) is smaller than the denominator (i.e., reactants), and fewer products have formed relative to the reactants.

The equilibrium lies to the left if the reaction favors reactants compared to products.

**109. A is correct.**

Typically, the forward reaction's constant is denoted as $k_1$, and the reverse is denoted as $k_{-1}$.

The overall constant of step 1 would be $k_1 / k_{-1}$.

 

**110. D is correct.**

Catalyst lowers the activation energy ($E_a$) of a reaction.

The primary function of catalysts is to lower the reaction's activation energy, thus increasing its rate ($k$).

Although catalysts decrease the energy required to reach the rate-limiting transition state, they do *not* decrease the relative energy of products and reactants.

Catalysts provide an alternative pathway for the reaction to proceed to product formation. It lowers the *activation energy* (i.e., relative energy between reactants and transition state) and speeds the rate.

Catalysts do *not* affect the Gibbs free energy ($\Delta G$: stability of products vs. reactants) or the enthalpy ($\Delta H$: bond breaking in reactants or bond making in products).

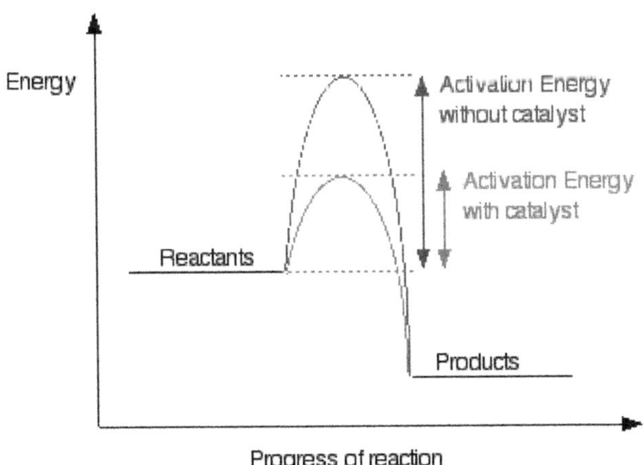


**111. D is correct.**

Le Châtelier's principle states that equilibrium shifts to counteract the change when changing the reaction conditions. If the reaction's concentration changes, the position of equilibrium changes.

Adding reactants or removing products shifts the equilibrium to the right.

**112. C is correct.**

Increased pressure or increased concentration of reactants increases the probability that the reactants collide with sufficient energy and orientation to overcome the energy of the activation barrier and proceed toward products.

**113. B is correct.**

Increasing the concentration of reactants (or increasing pressure) increases the reaction rate because there is an increased probability that two reactant molecules collide with sufficient energy to overcome the energy of activation and form products.

**114. B is correct.**

The number of collisions does not have to be large for a reaction. However, a minimum threshold of collisions must be reached before a reaction occurs. Reactions tend to happen more readily at higher temperatures because higher temperatures result in many collisions.

Some reactions only break bonds but do not form new bonds (e.g., radical cleavage of chlorine in the ozone decomposition process).

Some reactions only form bonds and do not break bonds (e.g., $Na^+ + Cl^- \rightarrow NaCl$).

All chemical bonds in the reactants do not need to break for a reaction to occur.

**115. C is correct.**

Le Châtelier's principle states that the equilibrium position shifts to counteract the change when changing reaction conditions. If the temperature, pressure, or volume changes, the equilibrium position will change.

A catalyst lowers the reaction's activation energy (energy barrier) and increases the reaction's rate (k). A catalyst does not affect the relative energy of the reactants and products and does not affect the equilibrium.

**116. D is correct**

All chemical reactions eventually reach equilibrium, the state at which the reactants and products are present in concentrations with no further tendency to change with time.

When the concentration in the reactant side of a reaction is reduced, some product transforms (i.e., the reaction is reversed) into reactant to compensate for the reduction of reactants. This results in a decrease in the concentration from the product side of the reaction.

**117. A is correct.**

Activated complexes (i.e., transition state for molecules undergoing bond breaking and bond making) decompose rapidly and have specific geometries that are highly reactive.

Activated complexes (as opposed to intermediates) cannot be isolated.

 

**118. A is correct.**

All chemical reactions eventually reach equilibrium, the state at which the reactants and products are present in concentrations with no further tendency to change with time.

Therefore, the rate in the forward and reverse directions is equal.

**119. A is correct.**

Note the following balanced reaction and calculation:

$2\ NH_3 \rightarrow N_2 + 3H_2$

$K = [\text{products}] / [\text{reactants}]$

$K = [N_2]\cdot[H_2]^3 / [NH_3]^2$

$K = (0.3)^3\cdot(0.2) / (0.1)^2$

**120. C is correct.**

Usually, to solve this kind of problem, the order of the reactants (X, Y, Z) is determined.

However, look at the answer choices; they pertain to the order of X. Therefore, calculate the order of X.

Find 2 experiments where the concentrations of other reactants (Y and Z) are constant: experiments 3 and 4.

Notice that when the concentration of X is reduced from 0.600 M to 0.200 M (i.e., a factor of 3), the rate decreases by a factor of 3 (from 15 to 5). Therefore, the reaction is first order with respect to X.

*Notes for active learning*

# 8 – Solution Chemistry: Detailed Explanations

---

**Practice Set 1: Questions 1–20**

---

**1. B is correct.**

Solubility is proportional to pressure. Use simple proportions to compare solubility in different pressures:

$P_1 / S_1 = P_2 / S_2$

$S_2 = P_2 / (P_1 / S_1)$

$S_2 = 4.5$ atm / (1.00 atm / 1.90 cc/100 mL)

$S_2 = 8.55$ cc / 100 mL

**2. C is correct.**

The correct interpretation of molarity is moles of solute per liter of solvent.

**3. A is correct.**

NaCl is a charged, ionic salt ($Na^+$ and $Cl^-$) and is highly water-soluble.

Hexanol can hydrogen bond with water due to the hydroxyl group, but the long hydrocarbon chain reduces its solubility, and the chains interact via hydrophobic interactions, forming micelles.

Aluminum hydroxide has poor solubility in water and requires the addition of Brønsted-Lowry acids to dissolve completely.

**4. A is correct.**

Spectator ions appear on each side of the ionic equation.

Ionic equation of the reaction:

$$Ba^{2+}\,(aq) + 2\,Cl^-\,(aq) + 2\,K^+\,(aq) + CrO_4^-\,(aq) \rightarrow BaCrO_4\,(s) + 2\,K^+\,(aq) + 2\,Cl^-\,(aq)$$

**5. D is correct.**

Higher pressure exerts more pressure on the solution, which allows more molecules to dissolve in the solvent.

Lower temperature increases gas solubility.

As the temperature increases, solvent and solute molecules move faster, making it more difficult for them to bond with the solute.

**6. A is correct.**

Solutes dissolve in the solvent to form a solution.

**7. B is correct.**

When the compound is dissolved in water, the attached hydrate/water crystals dissociate and become part of the water solvent.

Therefore, the only ions left are 1 $Co^{2+}$ and 2 $NO_3^-$.

**8. D is correct.**

The mass of a solution is calculated by taking the mass of the solution and container and subtracting the mass of the container.

**9. A is correct.**

An insoluble compound is incapable of being dissolved (especially with reference to water).

Hydroxide salts of Group I elements are soluble, while hydroxide salts of Group II elements (Ca, Sr, and Ba) are slightly soluble.

Salts containing nitrate ions ($NO_3^-$) are generally soluble.

Most sulfate salts are soluble. Important exceptions: $BaSO_4$, $PbSO_4$, $Ag_2SO_4$ and $SrSO_4$.

Hydroxide salts of transition metals and $Al^{3+}$ are insoluble.

Thus, $Fe(OH)_3$, $Al(OH)_3$ and $Co(OH)_2$ are not soluble.

**10. C is correct.**

% v/v solution = (volume of acetone / volume of solution) × 100%

% v/v solution = {25 mL / (25 mL + 75 mL)} × 100%

% v/v solution = 25%

**11. C is correct.**

The *van der Waals forces* in the hydrocarbons are relatively weak, while hydrogen bonds between $H_2O$ molecules are strong intermolecular bonds.

Bonds between polar $H_2O$ and nonpolar octane are weaker than those between two polar $H_2O$ molecules (i.e., hydrogen bonds).

**12. A is correct.**

To find $K_{sp}$, multiply the molarities (or concentrations) of the products (cC and dD).

For $K_{sp}$, only aqueous species are included in the calculation.

If any of the products have coefficients, raise the product to that coefficient power and multiply the concentration by that coefficient:

$$K_{sp} = [C]^c \cdot [D]^d$$

$$K_{sp} = [Cu^{2+}]^3 \cdot [PO_4^{3-}]^2$$

The reactant (aA) is solid and is not included in the $K_{sp}$ equation.

Solids are not included when calculating equilibrium constant expressions because their concentrations do not change the expression. Any change in their concentration is insignificant and is thus omitted.

**13. C is correct.**

*Like dissolves like* rule applies when a solvent is miscible by a solute with similar properties.

A polar solute (e.g., ethanol) is miscible with a polar solvent (e.g., water).

**14. C is correct.**

In a net ionic equation, substances that do not dissociate in aqueous solutions are written in their molecular form (not broken down into ions).

Gases, liquids, and solids are written in molecular form.

Some solutions do not dissociate into ions in water or do so in minimal amounts (e.g., weak acids such as HF, $CH_3COOH$). They are also written in molecular form.

**15. B is correct.**

*Supersaturated solution* contains more of the dissolved material than could be dissolved by the solvent under normal conditions. Increased heat allows for a solution to become supersaturated. The term also refers to the vapor of a compound with a higher partial pressure than the vapor pressure of that compound.

A supersaturated solution forms a precipitate if seed crystals are added as the solution reaches a lower energy state when solutes precipitate from the solution.

**16. D is correct.**

Mass % = mass of solute / mass of solution

Rearrange that equation to solve for mass of solution:

mass of solution = mass of solute / mass %

mass of solution = 122 g / 7.50%

mass of solution = 122 g / 0.075

mass of solution = 1,627 g

**17. B is correct.**

Determine moles of $NH_3$:

Moles of $NH_3$ = mass of $NH_3$ / molecular weight of $NH_3$

Moles of $NH_3$ = 15.0 g / (14.01 g/mol + 3 × 1.01 g/mol)

Moles of $NH_3$ = 15.0 g / (17.04 g/mol)

Moles of $NH_3$ = 0.88 mol

Determine the volume of the solution (solvent + solute):

Volume of solution = mass / density

Volume of solution = (250 g + 15 g) / 0.974 g/mL

Volume of solution = 272.1 mL

Convert volume to liters:

Volume = 272.1 mL × 0.001 L / mL

Volume = 0.2721 L

*continued...*

 

Divide moles by the volume to calculate molarity:

Molarity = moles / liter of solution

Molarity = 0.88 mol / 0.2721 L

Molarity = 3.23 M

**18. A is correct.**

Solutions with the highest concentration of ions have the highest boiling point.

Calculate the concentration of ions in each solution:

0.2 M $Al(NO_3)_3$ = 0. 2 M × 4 ions = 0.8 M

0. 2 M $MgCl_2$ = 0. 2 M × 3 ions = 0.6 M

0. 2 M glucose = 0. 2 M × 1 ion = 0.2 M (glucose does not dissociate into ions in solution)

0. 2 M $Na_2SO_4$ = 0. 2 M × 3 ions = 0.6 M

Water = 0 M

**19. A is correct.**

Calculate the number of ions in each option:

A: $Li_3PO_4 \rightarrow 3\ Li^+ + PO_4^{3-}$ (4 ions)

B: $Ca(NO_3)_2 \rightarrow Ca^{2+} + 2\ NO_3^-$ (3 ions)

C: $MgSO_4 \rightarrow Mg^{2+} + SO_4^{2-}$ (2 ions)

D: $(NH_4)_2SO_4 \rightarrow 2\ NH_4^+ + SO_4^{2-}$ (3 ions)

**20. D is correct.**

Electrolyte is a substance that produces an electrically conducting solution when dissolved in a polar solvent (e.g., water).

Dissolved electrolyte separates into positively-charged cations and negatively-charged anions.

Strong electrolytes dissociate entirely (or almost completely) because the resulting ions are stable in the solution.

 

==================================================================

**Practice Set 2: Questions 21–40**

==================================================================

**21. B is correct**

Note that carbonate salts help remove 'hardness' in water.

Generally, $SO_4$, $NO_3$ and Cl salts are soluble in water, while $CO_3$ salts are less soluble.

**22. A is correct.**

*Spectator ion* exists in the same form on the reactant and product sides of a chemical reaction.

Balanced equation:

$$Pb(NO_3)_2\ (aq) + H_2SO_4\ (aq) \rightarrow PbSO_4\ (s) + 2\ HNO_3\ (aq)$$

**23. C is correct.**

Polar solute is miscible with a polar solvent.

Ascorbic acid is polar and is, therefore, miscible in water.

**24. B is correct.**

Apply the formula for the molar concentration of a solution:

$$M_1V_1 = M_2V_2$$

Substitute the given volume and molar concentrations of HCl and solve for the final volume of HCl of the resulting dilution:

$$V_2 = [M_1V_1] / (M_2)$$

$$V_2 = [(2.00\ M\ HCl) \times (0.125\ L\ HCl)] / (0.400\ M\ HCl)$$

$$V_2 = 0.625\ L\ HCl$$

Solve for the volume of water needed to be added to the initial volume of HCl to obtain the final diluted volume of HCl:

$$V_{H_2O} = V_2 - V_1$$

$$V_{H_2O} = (0.625\ L\ HCl) - (0.125\ L\ HCl)$$

$$V_{H_2O} = 0.500\ L = 500\ mL$$

 

**25. A is correct.**

The *–ate* ending indicates the species with more oxygen than those ending in *–ite*.

However, it does not indicate a specific number of oxygen molecules.

**26. C is correct.**

Solubility is the property of a solid, liquid, or gaseous substance (i.e., solute); it dissolves in a solid, liquid, or gaseous solvent to form a solution (i.e., solute in the solvent).

Solubility of a substance depends on the physical and chemical properties of the solute and solvent and the temperature, pressure, and pH of the solution.

Gaseous solutes (e.g., oxygen) exhibit complex behavior with temperature. As the temperature rises, gases usually become less soluble in water but more soluble in organic solvents.

**27. A is correct.**

Saturated solution contains the maximum dissolved material in the solvent under normal conditions. Increased heat allows for a solution to become supersaturated.

Saturated solution refers to the vapor of a compound with a higher partial pressure than the vapor pressure of that compound.

Saturated solution forms a precipitate as more solute is added to the solution.

**28. D is correct.**

*Immiscible* refers to the property (of solutions) of when two or more substances (e.g., oil and water) are mixed and eventually separate into two layers.

*Miscible* is when two liquids are mixed, but they do not necessarily interact chemically.

In contrast to miscibility, *soluble* means the substance (solid, liquid, or gas) can *dissolve* in another solid, liquid, or gas. In other words, a substance *dissolves* when incorporated into another substance.

Also, in contrast to miscibility, solubility involves a *saturation point* at which a substance cannot dissolve further, and a mass, the *precipitate*, begins to form.

**29. D is correct.**

In this case, hydration means dissolving in water rather than reacting with water.

Therefore, the compound dissociates into its ions (without involving water in the chemical reaction).

**30. A is correct.**

Depending on the solubility of a solute, there are three possible results:

1) a dilute solution has less solute than the maximum amount that it can dissolve;

2) a saturated solution has the same amount as its solubility;

3) a precipitate forms if there is more solute than can be dissolved, so the excess solute separates from the solution (i.e., crystallization).

Precipitation lowers the concentration of the solute to the saturation level to increase the solution's stability.

Gas solubility is inversely proportional to temperature and proportional to pressure.

**31. D is correct.**

*Like dissolves like* means that polar substances tend to dissolve in polar solvents and nonpolar substances in nonpolar solvents.

Molecules that can form hydrogen bonds with water are soluble.

Salts are ionic compounds.

The anion and cation bond with the polar molecule of water and are soluble.

**32. B is correct.**

*Like dissolves like* means that polar substances tend to dissolve in polar solvents and nonpolar substances in nonpolar solvents.

Methanol ($CH_3OH$) is a polar molecule. Therefore, methanol is soluble in water because it can make a hydrogen bond with water.

**33. B is correct.**

To find $K_{sp}$, take molarities or concentrations of the products (cC and dD) and multiply them.

For $K_{sp}$, only aqueous species are included in the calculation.

If the products have coefficients in front, raise the product to that coefficient power and multiply the concentration by that coefficient:

$$K_{sp} = [C]^c \cdot [D]^d$$

$$K_{sp} = [Au^{3+}] \cdot [Cl^-]^3$$

*continued...*

 

Reactant (aA) is solid and is not included in the $K_{sp}$ equation.

Solids are not included when calculating equilibrium constant expressions because their concentrations do not change the expression. Any change in their concentration is insignificant and is thus omitted.

**34. B is correct.**

Break down the molecules into their constituent ions:

$$CaCO_3 + 2\ H^+ + 2\ NO_3^- \rightarrow Ca^{2+} + 2\ NO_3^- + CO_2 + H_2O$$

Remove species that appear on each side of the reaction:

$$CaCO_3 + 2\ H^+ \rightarrow Ca^{2+} + CO_2 + H_2O$$

**35. D is correct.**

According to the problem, 0.950 moles of nitrate ions are in a $Fe(NO_3)_3$ solution.

Because there are three nitrates ($NO_3$) ions for each $Fe(NO_3)_3$ molecule, the moles of $Fe(NO_3)_3$ is:

(1 mole / 3 mole) × 0.950 moles = 0.317 moles

Calculate the volume of solution:

Volume of solution = moles of solute / molarity

Volume of solution = 0.317 moles / 0.550 mol/L

Volume of solution = 0.576 L

Convert the volume into milliliters:

0.576 L × 1,000 mL/L = 576 mL

**36. A is correct.**

mEq represents the amount in milligrams of a solute equal to 1/1,000 of its gram-equivalent weight, considering the valency of the ion.

Millimolar (mM) = mEq / valence.

Divide the given concentration of $Ca^{2+}$ by 2:

$[Ca^{2+}]$ = [48 mEq $Ca^{2+}$] / 2 = 24 mM $Ca^{2+}$

$[Ca^{2+}]$ = 24 mM $Ca^{2+}$

*continued...*

Divide concentration of $Ca^{2+}$ by 1,000 to obtain the concentration of $Ca^{2+}$ in units of molarity:

$[Ca^{2+}] = 24\ \text{mM}\ Ca^{2+} \times [(1\ \text{M}) / (1{,}000\ \text{mM})]$

$[Ca^{2+}] = 0.024\ \text{M}\ Ca^{2+}$

**37. A is correct.**

Start by calculating the number of moles:

Moles of LiOH = mass of LiOH / molar mass of LiOH

Moles of LiOH = 36.0 g / (24.0 g/mol)

Moles of LiOH = 1.50 moles

Divide moles by volume to calculate molarity:

Molarity = moles / volume

Molarity = 1.50 moles / (975 mL × 0.001 L/mL)

Molarity = 1.54 M

**38. C is correct.**

Glucose content is 8.50 % (m/v).

Solute (glucose) is measured in grams, but the solution volume is measured in milliliters.

Therefore, the mass and volume of the solution are interchangeable (1 g = 1 mL).

% (m/v) of glucose = mass of glucose / volume of solution

volume of solution = mass of glucose / % (m/v) of glucose

volume of solution = 60 g / 8.50%

volume of solution = 706 mL

**39. D is correct.**

When AgCl dissociates, equal amounts of $Ag^+$ and $Cl^-$ are produced.

If the concentration of $Cl^-$ is $B$, this must be the concentration of $Ag^+$.

Therefore, $B$ can be the concentration of Ag.

*continued...*

 

The concentration of silver ion can be determined by dividing the $K_{sp}$ by the concentration of chloride ion:

$$K_{sp} = [Ag^+]\cdot[Cl^-].$$

Therefore, the concentration of Ag can be $A/B$ moles/liter.

**40. B is correct.**

Strong electrolytes dissociate entirely (or almost completely) in water.

Strong acids and bases dissociate almost completely, but weak ones only slightly.

===============================================================================

**Practice Set 3: Questions 41–60**

===============================================================================

**41. D is correct.**

Solute-solute and solvent-solvent attractions are essential in establishing bonding amongst themselves for solvent and solute.

Once mixed in a solution, the solute-solvent attraction becomes the major attraction force, but the other two forces (solute-solute and solvent-solvent attractions) are still present.

**42. B is correct.**

Since the empirical formula for magnesium iodide is $MgI_2$, two moles of dissolved $I^-$ result from each mole of dissolved $MgI_2$.

Therefore, if $[MgI_2] = 0.40$ M, then $[I^-] = 2(0.40\text{ M}) = 0.80$ M.

**43. B is correct.**

Ammonia forms hydrogen bonds, and $SO_2$ is polar.

**44. A is correct.**

Chlorite ion (chlorine dioxide anion) is $ClO_2^-$.

Chlorite is a compound that contains this group, with chlorine in an oxidation state of +3.

| Formula | $Cl^-$ | $ClO^-$ | $ClO_2^-$ | $ClO_3^-$ | $ClO_4^-$ |
|---|---|---|---|---|---|
| Anion name | chloride | hypochlorite | chlorite | chlorate | perchlorate |
| Oxidation state | −1 | +1 | +3 | +5 | +7 |

**45. D is correct.**

Water softeners cannot remove ions from the water without replacing them.

The process replaces the ions that cause scaling (precipitation) with non-reactive ions.

**46. B is correct.**

These compounds are rarely encountered in most chemistry problems and are part of the list of solubility rule exceptions.

Salts containing nitrate ions ($NO_3^-$) are generally soluble.

**47. A is correct.**

*Hydration* involves the interaction of water molecules with the solute. Water molecules exchange bonding relationships with the solute, breaking water-water bonds and forming water-solute bonds.

When an ion is hydrated, it is surrounded and bonded by water molecules.

The average number of water molecules bonding to an ion is its *hydration number*.

Hydration numbers can vary but often are 4 or 6.

**48. C is correct.**

Solvation describes the process whereby the solvent surrounds the solute molecules.

**49. D is correct.**

Miscible refers to the property (of solutions) when two or more substances (e.g., water and alcohol) are mixed without separating.

**50. C is correct.**

The *like dissolves like* rule applies when a solvent is miscible with a solute that has similar properties.

A nonpolar solute is immiscible with a polar solvent.

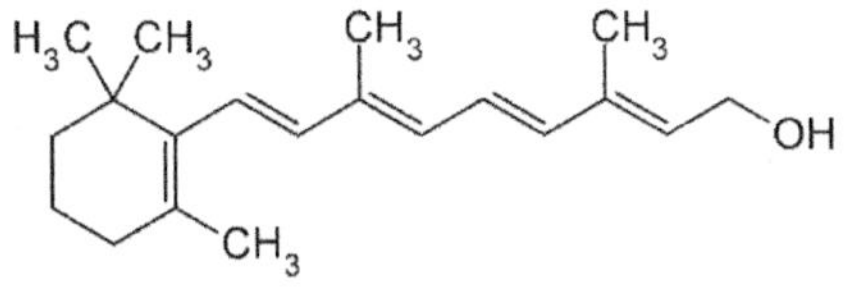

*Retinol (vitamin A)*

**51. B is correct.**

Only aqueous (*aq*) species are broken down into their ions for ionic equations.

Solids, liquids, and gases stay the same.

**52. C is correct.**

For $K_{sp}$, only aqueous species are included in the calculation.

The decomposition of $PbCl_2$:

$$PbCl_2 \rightarrow Pb^{2+} + 2\ Cl^-$$

$$K_{sp} = [Pb^{2+}]\cdot[Cl^-]^2$$

When $x$ moles of $PbCl_2$ fully dissociate, $x$ moles of Pb and $2x$ moles of $Cl^-$ are produced:

$$K_{sp} = (x)\cdot(2x)^2$$

$$K_{sp} = 4x^3$$

**53. B is correct.**

Strong electrolytes dissociate entirely (or almost completely) in water.

Strong acids and bases dissociate nearly completely (i.e., form stable anions), but weak acids and bases dissociate only slightly (i.e., form unstable anions).

**54. A is correct.**

*Spectator ion* exists in the same form on the reactant and product sides of a chemical reaction.

Balanced equation for potassium hydroxide and nitric acid:

$$KOH + HNO_3 \rightarrow KNO_3 + H_2O$$

Ionic equation:

$$K^+\ {}^-OH + H^+\ NO_3^- \rightarrow K^+\ NO_3^- + H_2O$$

**55. A is correct.**

When a solution is diluted, the solute (n) moles are constant.

However, the molarity and volume will change because n = MV:

$$n_1 = n_2$$

$$M_1V_1 = M_2V_2$$

$$V_2 = (M_1V_1) / M_2$$

$$V_2 = (0.20\ M \times 6.0\ L) / 14\ M$$

$$V_2 = 0.086\ L$$

Convert to milliliters:

$$0.086\ L \times (1{,}000\ mL / L) = 86\ mL$$

 

**56. B is correct.**

When a solution is diluted, the moles of solute are constant.

Use this formula to calculate the new molarity:

$M_1V_1 = M_2V_2$

160 mL × 4.50 M = 595 mL × $M_2$

$M_2$ = (160 mL / 595 mL) × 4.50 M

$M_2$ = 1.21 M

**57. B is correct.**

*Like dissolves like* means that polar substances tend to dissolve in polar solvents and nonpolar substances in nonpolar solvents.

Since benzene is nonpolar, look for a nonpolar substance.

Silver chloride is ionic, while $CH_2Cl_2$, $H_2S$ and $SO_2$ are polar.

**58. B is correct.**

Molarity can vary based on temperature because it involves the volume of solutions.

At different temperatures, the water volume varies slightly, affecting the molarity.

**59. D is correct.**

Concentration:

solute / volume

For example: 10 g / 1 liter = 10 g/liter

2 g / 1 liter = 2 g/liter

**60. C is correct.**

NaOH content is 5.0% (w/v). The solute (NaOH) is measured in grams, but the solution volume is measured in milliliters.

In this problem, the mass and volume of the solution are interchangeable (1 g = 1 mL).

mass of NaOH = % NaOH × volume of solution

mass of NaOH = 5.0% × 75.0 mL

mass of NaOH = 3.75 g

=====================================================================

**Practice Set 4: Questions 61–80**

=====================================================================

**61. D is correct.**

Adding NaCl increases [$Cl^-$] in solution (i.e., common ion effect), which increases the precipitation of $PbCl_2$ because the ion product increases.

This increase causes lead chloride to precipitate and the concentration of free chloride in solution to decrease.

**62. A is correct.**

AgCl has a stronger tendency to form than $PbCl_2$ because of its smaller $K_{sp}$.

Therefore, as AgCl forms, an equivalent amount of $PbCl_2$ dissolves.

**63. A is correct.**

For dissolution of $PbCl_2$:

$PbCl_2 \rightarrow Pb^{2+} + 2\ Cl^-$

$K_{sp} = [Pb^{2+}]\cdot[2\ Cl^-]^2 = 10^{-5}$

$K_{sp} = (x)\cdot(2x)^2 = 10^{-5}$

$4x^3 = 10^{-5}$

$x \approx 0.014$

$[Cl^-] = 2(0.014)$

$[Cl^-] = 0.028$

For dissolution of AgCl:

$AgCl \rightarrow Ag^+ + Cl^-$

$K_{sp} = [Ag^+]\cdot[Cl^-]$

$K_{sp} = (x)\cdot(x)$

$10^{-10} = x^2$

$x = 10^{-5}$

$[Cl^-] = 10^{-5}$

**64. A is correct.**

Intermolecular bonding (i.e., van der Waals) in alkanes is similar.

**65. C is correct.**

Ideally, dilute solutions are so dilute that solute molecules do not interact.

Therefore, the mole fraction of the solvent approaches one.

**66. D is correct.**

*Henry's law* states that, at a constant temperature, the amount of gas that dissolves in a volume of liquid is directly proportional to the partial pressure of that gas in equilibrium with that liquid.

*Tyndall effect* is light scattering by particles in a colloid (or particles in a very fine suspension).

*Colloidal suspension* contains microscopically dispersed insoluble particles (i.e., colloid) suspended throughout the liquid. The colloid particles are larger than those of the solution but not large enough to precipitate due to gravity.

**67. C is correct.**

Miscible refers to the property (of solutions) when two or more substances (e.g., water and alcohol) are mixed without separating.

**68. B is correct.**

Start by calculating moles of glucose:

Moles of glucose = mass of glucose / molar mass of glucose

Moles of glucose = 10.0 g / (180.0 g/mol)

Moles of glucose = 0.0555 mol

Then, divide moles by volume to calculate molarity:

Molarity of glucose = moles of glucose / volume of glucose

Molarity of glucose = 0.0555 mol / (100 mL × 0.001 L/mL)

Molarity = 0.555 M

**69. D is correct.**

An insoluble compound is incapable of being dissolved (especially with reference to water).

Hydroxide salts of Group I elements are soluble.

Hydroxide salts of Group II elements (Ca, Sr and Ba) are slightly soluble.

Hydroxide salts of transition metals and $Al^{3+}$ are insoluble. Thus, $Fe(OH)_3$, $Al(OH)_3$, $Co(OH)_2$ are not soluble.

Most sulfate salts are soluble. Important exceptions are $BaSO_4$, $PbSO_4$, $Ag_2SO_4$, and $SrSO_4$.

Salts containing $Cl^-$, $Br^-$ and $I^-$ are generally soluble. Exceptions are halide salts of $Ag^+$, $Pb^{2+}$ and $(Hg_2)^{2+}$. AgCl, $PbBr_2$, and $Hg_2Cl_2$ are insoluble.

Chromates are frequently insoluble. Examples: $PbCrO_4$, $BaCrO_4$

**70. C is correct.**

The overall reaction is:

$$Zn\ (s) + 2\ HCl\ (aq) \rightarrow ZnCl_2\ (aq) + H_2\ (g)$$

Because HCl and $ZnCl_2$ are ionic, they dissociate into ions in aqueous solutions:

$$Zn\ (s) + 2\ H^+\ (aq) + 2\ Cl^-\ (aq) \rightarrow Zn^{2+}\ (aq) + 2\ Cl^-\ (aq) + H_2\ (g)$$

The common ion ($Cl^-$) cancels to yield the net ionic reaction:

$$Zn\ (s) + 2\ H^+\ (aq) \rightarrow Zn^{2+}\ (aq) + H_2\ (g)$$

**71. D is correct.**

The *like dissolves like* rule applies when a solvent is miscible with a solute that has similar properties.

A polar solute (e.g., ketone, alcohol, or carboxylic acids) is miscible with a polar (water) solvent.

The three molecules are polar and are, therefore, soluble in water.

**72. A is correct.**

Calculating the solubility constant ($K_{sp}$) is similar to the equilibrium constant.

Concentration of each species is raised to the power of their coefficients and multiplied with each other.

*continued...*

 

For $K_{sp}$, only aqueous species are included in the calculation.

$$CaF_2\ (s) \rightarrow Ca^{2+}\ (aq) + 2F^-\ (aq)$$

Concentration of fluorine ions can be determined using the concentration of calcium ions:

$$[F^-] = 2 \times [Ca^{2+}]$$

$$[F^-] = 2 \times 0.00021\ M$$

$$[F^-] = 4.2 \times 10^{-4}\ M$$

Then, $K_{sp}$ can be determined:

$$K_{sp} = [Ca^{2+}]\cdot[F^-]^2$$

$$K_{sp} = (2.1 \times 10^{-4}\ M) \times (4.2 \times 10^{-4}\ M)^2$$

$$K_{sp} = 3.7 \times 10^{-11}$$

**73. B is correct.**

To determine the number of equivalents:

$$(4\ \text{moles / liter}) \times (3\ \text{equivalents / mole}) \times (1/3\ \text{liter})$$

$$= 4\ \text{equivalents}$$

**74. A is correct.**

Electrolyte is a substance that produces an electrically conducting solution when dissolved in a polar solvent (e.g., water).

Dissolved electrolyte separates into positively charged cations and negatively charged anions.

Electrolytes conduct electricity.

**75. A is correct.**

A frequently encountered insoluble salt in chemistry problems.

Most sulfate salts are soluble. Important exceptions are $BaSO_4$, $PbSO_4$, $Ag_2SO_4$, and $SrSO_4$.

**76. C is correct.**

When the ion product is equal to or greater than $K_{sp}$, precipitation of the salt occurs.

If the ion product value is less than $K_{sp}$, precipitation does not occur.

**77. C is correct.**

Gas solubility is inversely proportional to temperature, so a low temperature increases solubility.

The high pressure of $O_2$ above the solution increases the pressure of the solution (which improves solubility) and the amount of $O_2$ available for dissolving.

**78. C is correct.**

Molality is the number of solute moles dissolved in 1,000 grams of solvent.

Mass of the solution is 1,000 g + mass of solute.

Mass of solute ($CH_3OH$) = moles $CH_3OH$ × molecular mass of $CH_3OH$

Mass of solute ($CH_3OH$) = 8.60 moles × [12.01 g/mol + (4 × 1.01 g/mol) + 16 g/mol]

Mass of solute ($CH_3OH$) = 8.60 moles × (32.05 g/mol)

Mass of solute ($CH_3OH$) = 275.63 g

Total mass of solution: 1,000 g + 275.63 g = 1,275.63 g

Volume of solution = mass / density

Volume of solution = 1,275.63 g / 0.94 g/mL

Volume of solution = 1,357.05 mL

Divide moles by volume to calculate molarity:

Molarity = number of moles / volume

Molarity = 8.60 moles / 1,357.05 mL

Molarity = 6.34 M

**79. B is correct.**

By definition, a 15.0% aqueous solution of KI contains 15% KI and the remainder (100 % – 15% = 85%) is water.

If there is 100 g of KI solution, it has 15 g of KI and 85 g of water.

The answer choice of 15 g KI / 100 g water is incorrect because 100 g is the mass of the solution; the actual mass of water is 85 g.

 

**80. D is correct.**

Molarity equals moles of solute divided by liters of solution.

$1 \text{ cm}^3 = 1 \text{ mL}$

Molarity:

$(0.75 \text{ mol}) / (0.075 \text{ L}) = 10 \text{ M}$

==================================================================

**Practice Set 5: Questions 81–107**

==================================================================

**81. D is correct.**

Mass % of solute = (mass of solute / mass of seawater) × 100%

Mass % of solute = (1.35 g / 25.88 g) × 100%

Mass % of solute = 5.22%

**82. A is correct.**

Some solutions are colored (e.g., Kool-Aid powder dissolved in water).

The color of chemicals is a physical property of chemicals from (most commonly) the excitation of electrons due to the absorption of energy by the chemical. The observer sees not the absorbed color but the wavelength that is reflected.

Most simple inorganic (e.g., sodium chloride) and organic compounds (e.g., ethanol) are colorless.

Transition metal compounds are often colored because of transitions of electrons between *d*-orbitals of different energy.

Organic compounds tend to be colored when there is extensive conjugation (i.e., alternating double and single bonds), causing the energy gap between the HOMO (i.e., highest occupied molecular orbital) and LUMO (i.e., lowest unoccupied molecular orbital) to decrease. This brings the absorption band from the UV to the visible region.

Color is also due to the energy absorbed by the compound when an electron transitions from HOMO to LUMO.

A physical change is a change in physical properties, such as melting, transition to gas, changes to crystal form, color, volume, shape, size, and density.

**83. B is correct.**

Solubility product constant ($K_{sp}$) is a product of solubility of each ion in an ionic compound.

Start by writing the dissociation equation of AgCl in a solution:

$AgCl\ (s) \rightarrow Ag^+\ (aq) + Cl^-\ (aq)$

*continued…*

Solubility of AgCl is provided in the problem: $1.3 \times 10^{-4}$ mol/L. This is the maximum amount of AgCl that can be dissolved in water in standard conditions.

$1.3 \times 10^{-4}$ mol AgCl dissolved in one liter of water means $1.3 \times 10^{-4}$ mol of $Ag^+$ and $1.3 \times 10^{-4}$ mol of $Cl^-$ ions.

$K_{sp}$ is calculated using the same method as the equilibrium constant: ion concentration to the power of ion coefficient. Only aqueous species are included in the calculation.

For AgCl:

$$K_{sp} = [Ag] \cdot [Cl]$$

$$K_{sp} = (1.3 \times 10^{-4}) \cdot (1.3 \times 10^{-4})$$

$$K_{sp} = 1.7 \times 10^{-8}$$

**84. D is correct.**

Concentration:

solute / volume

For example: 10 g / 1 liter = 10 g/liter

5 g / 0.5 liter = 10 g/liter

**85. B is correct.**

Reaction is endothermic because the temperature of the solution drops as the reaction absorbs heat from the environment, and $\Delta H°$ is positive.

Solution is unsaturated at 1 molar, so dissolving more salt at standard conditions is spontaneous, and $\Delta G°$ is negative.

**86. D is correct.**

Heat of a solution is the enthalpy change (or energy absorbed as heat at constant pressure) when a solution forms.

For solution formation, solvent-solvent bonds and solute-solute bonds must be broken while solute-solvent bonds are formed.

Breaking bonds absorbs energy, while forming bonds releases energy.

If the heat of the solution is negative, energy is released.

**87. A is correct.**

The reaction is endothermic if the bonds formed have lower energy than the bonds broken.

From $\Delta G = \Delta H - T\Delta S$, the entropy of the system increases if the reaction is spontaneous.

**88. B is correct.**

Hydrogen bonding in $H_2O$ is stronger than van der Waals forces in the nonpolar hydrocarbon of benzene.

**89. D is correct.**

Balanced reaction:

$$BaCl_2 + K_2CrO_4 \rightarrow BaCrO_4 + 2\ KCl$$

Most Cl and K compounds are soluble; therefore, KCl is more likely to be soluble than $BaCrO_4$.

**90. C is correct.**

Sodium carbonate reacts with calcium and magnesium ions (responsible for the 'hard water' phenomena) and forms calcium carbonate and magnesium carbonate precipitates. This reduces the hardness of the water.

**91. C is correct.**

Calculate moles of $NaHCO_3$:

$$\text{Moles of } NaHCO_3 = \text{mass } NaHCO_3 / \text{molar mass } NaHCO_3$$

$$\text{Moles of } NaHCO_3 = 0.400 \text{ g} / 84.0 \text{ g/mol}$$

$$\text{Moles of } NaHCO_3 = 4.76 \times 10^{-3} \text{ mol}$$

In this reaction, $NaHCO_3$ and HCl have coefficients of 1.

Therefore, moles $NaHCO_3$ = moles HCl = $4.76 \times 10^{-3}$ mol.

Use this to calculate volume of HCl:

$$\text{volume of HCl} = \text{moles} / \text{molarity}$$

$$\text{volume of HCl} = 4.76 \times 10^{-3} \text{ mol} / 0.25 \text{ M}$$

$$\text{volume of HCl} = 0.019 \text{ L}$$

Convert volume to milliliters:

$$0.019 \text{ L} \times (1{,}000 \text{ mL/L}) = 19.0 \text{ mL}$$

**92. D is correct.**

Calculation of concentration in ppm (parts per million) is similar to percentage calculation; the difference is that the multiplication factor is $10^6$ ppm instead of 100%.

Concentration of contaminant = (mass of contaminant / mass of solution) × $10^6$ ppm

Concentration of contaminant = [($8.8 \times 10^{-3}$ g) / (5,246 g)] × $10^6$ ppm

Concentration of contaminant = 1.68 ppm

**93. A is correct.**

Bicarbonate ion is the conjugate base of carbonic acid: $H_2CO_3$

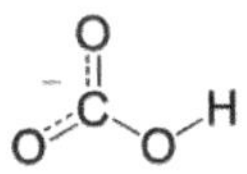

Bicarbonate ion carries a –1 formal charge and is the conjugate base of carbonic acid ($H_2CO_3$); it is the conjugate acid of the carbonate ion ($CO_2^{-3}$).

Equilibrium reactions are shown below:

$$CO_3^{2-} + 2\ H_2O \leftrightarrow HCO_3^- + H_2O + OH^- \leftrightarrow H_2CO_3 + 2\ OH^-$$

$$H_2CO_3 + 2\ H_2O \leftrightarrow HCO_3^- + H_3O^+ + H_2O \leftrightarrow CO_3^{2-} + 2\ H_3O^+$$

**94. D is correct.**

Solutes dissolve in the solvent to form a solution.

**95. D is correct.**

Hydration involves the interaction of water molecules with the solute. The water molecules exchange bonding relationships with the solute, breaking water-water bonds and forming water-solute bonds.

**96. C is correct.**

*Hydration* involves the interaction of water molecules with the solute. Water molecules exchange bonding relationships with the solute, breaking water-water bonds and forming water-solute bonds.

Hydrates are *not* new compounds but rather compound molecules surrounded by water crystals.

Colloidal suspension contains microscopically dispersed insoluble particles (i.e., colloid) suspended throughout the liquid. Colloid particles are larger than those of the solution but not large enough to precipitate due to gravity.

**97. A is correct.**

Reaction: 2 $AgNO_3$ (*aq*) + $K_2CrO_4$ (*aq*) → 2 $KNO_3$ (*aq*) + $AgCrO_4$ (*s*)

Salts containing nitrate ions ($NO_3^-$) are generally soluble.

**98. D is correct.**

The *like dissolves like* rule applies when a solvent is miscible with a solute that has similar properties.

A polar solute is miscible with a polar solvent; the rule applies to nonpolar solute / solvent.

The three molecules are nonpolar and, therefore, are miscible with the nonpolar liquid bromine. Attractive forces between each solute and liquid bromine solution are van der Waals.

**99. C is correct.**

Molecules can have covalent bonds in one situation and ionic bonds in another.

Hydrogen chloride (HCl) is a gas in which hydrogen and chlorine are covalently bound. However, if HCl is bubbled into the water, it ionizes completely to yield $H^+$ and $Cl^-$ of a hydrochloric acid solution.

Hydrochloric acid contains two nonmetals and is a covalently bonded molecule.

Hydrochloric acid is a strong acid and dissociates into ions when placed in water (i.e., strong electrolyte).

**100. D is correct.**

Adding solute to a liquid lowers the vapor pressure because some water molecules are bonding to the solute. This results in fewer water molecules available for vaporization, reducing the vapor pressure.

**101. A is correct.**

Calculating solubility constant ($K_{sp}$) is similar to the equilibrium constant.

The concentration of each species is raised to the power of their coefficients and multiplied with each other. For $K_{sp}$, only aqueous species are included in the calculation.

Barium hydroxide dissolves reversibly via the reaction:

$$Ba(OH)_2\ (s) \leftrightarrow Ba^{2+}\ (aq) + 2\ OH^-\ (aq)$$

Therefore, the equilibrium expression is:

$$K_{sp} = [Ba^{2+}]\cdot[^-OH]^2$$

 

**102. A is correct.**

Start by calculating the moles of $H_3PO_4$ required:

moles $H_3PO_4$ = molarity of solution × volume

moles $H_3PO_4$ = 0.2 M × (200 mL × 0.001 L/mL)

moles $H_3PO_4$ = 0.04 mole

Then, use this information to calculate mass:

Mass of $H_3PO_4$ = moles × molar mass

Mass of $H_3PO_4$ = 0.04 moles × [(3 × 1.01 g/mol) + 30.97 g/mol + (4 × 16 g/mol)]

Mass of $H_3PO_4$ = 0.04 moles × (3.03 g/mol + 30.97 g/mol + 64 g/mol)

Mass of $H_3PO_4$ = 3.9 g

**103. D is correct.**

Ionic product constant of water:

$K_w = [H_3O^+]·[^-OH]$

Rearrange the equation to solve for $[^-OH]$:

$[^-OH] = K_w / [H_3O^+]$

$[^-OH] = [1 \times 10^{-14}] / [1 \times 10^{-8}]$

$[^-OH] = 1 \times 10^{-6}$

**104. B is correct.**

Molarity of NaCl = moles of NaCl / volume of NaCl

Molarity of NaCl = 4.50 moles / 1.50 L

Molarity of NaCl = 3.0 M

**105. B is correct.**

Upon reaction with an acid, sulfite ($SO_3^{2-}$) compounds release $SO_2$ gas:

$2\ HNO_3\ (aq) + Na_2SO_3\ (aq) \rightarrow 2\ NaNO_3\ (aq) + H_2O\ (l) + SO_2\ (g)$

Because Na and $NO_3$ ions are present on each side, they are considered spectator ions and can be removed from the net ionic equation:

$2\ H^+\ (aq) + SO_3^{2-}\ (aq) \rightarrow H_2O\ (l) + SO_2\ (g)$

**106. D is correct.**

All the listed units are used in chemistry, but % (m/v) and % (m/m) are used more often (molarity and molality, respectively).

**107. D is correct.**

An electrolyte is a substance that produces an electrically conducting solution when dissolved in a polar solvent (e.g., water).

Dissolved electrolytes separate into positively-charged cations and negatively-charged anions.

Electrolytes conduct electricity.

 

*Notes for active learning*

*Notes for active learning*

# 9 – Acids and Bases: Detailed Explanations

==================================================================

**Practice Set 1: Questions 1–20**

==================================================================

**1. A is correct.**

Acids dissociate protons to form the conjugate base, while a conjugate base accepts a proton to form the acid.

**2. B is correct.**

pH = $-\log[H^+]$

pH = $-\log(0.10)$

pH = 1

**3. A is correct.**

Salt is a combination of acid and a base.

pH is determined by the acid and base that created the salt.

Combinations include:

Weak acid and strong base: salt is basic

Strong acid and weak base: salt is acidic

Strong acid and strong base: salt is neutral

Weak acid and weak base: could be anything (acidic, basic, or neutral)

Examples of strong acids and bases that commonly appear in chemistry problems:

Strong acids: HCl, HBr, HI, $H_2SO_4$, $HClO_4$ and $HNO_3$

Strong bases: NaOH, KOH, $Ca(OH)_2$ and $Ba(OH)_2$

The hydroxides of Group I and II metals are considered strong bases. Examples include LiOH (lithium hydroxide), NaOH (sodium hydroxide), KOH (potassium hydroxide), RbOH (rubidium hydroxide), CsOH (cesium hydroxide), $Ca(OH)_2$ (calcium hydroxide), $Sr(OH)_2$ (strontium hydroxide) and $Ba(OH)_2$ (barium hydroxide).

**4. C is correct.**

Brønsted-Lowry acid-base theory focuses on the ability to accept and donate protons ($H^+$).

Brønsted-Lowry acid is a term for a substance that donates a proton ($H^+$) in an acid-base reaction, while a Brønsted-Lowry base is a term for a substance that accepts a proton.

**5. C is correct.**

A solution's conductivity is correlated to the number of ions in a solution. The bulb shining brightly implies that the solution is a conductor, which means that the solution has a high concentration of ions.

**6. C is correct.**

An amphoteric compound reacts as an acid (i.e., donates protons) and a base (i.e., accepts protons).

One type of amphoteric species is amphiprotic molecules accepting or donating a proton ($H^+$). Examples of amphiprotic molecules include amino acids (i.e., an amine and carboxylic acid group) and self-ionizable compounds such as water.

**7. D is correct.**

Strong acids completely dissociate protons into the aqueous solution. The resulting anion is stable, which accounts for the ~100% ionization of the strong acid.

Weak acids do not entirely (or appreciably) dissociate protons into the aqueous solution. The resulting anion is unstable, which accounts for a small ionization of the weak acid.

Each of the listed strong acids forms an anion stabilized by resonance.

$HClO_4$ (perchloric acid) has a p$K_a$ of –10.

$H_2SO_4$ (sulfuric acid) is a diprotic acid with a p$K_a$ of –3 and 1.99.

$HNO_3$ (nitrous acid) has a p$K_a$ of –1.4.

**8. A is correct.**

To find pH of strong acid and strong base solutions (where $[H^+] > 10^{-6}$), use the equation:

$$pH = -\log[H_3O^+]$$

Given that $[H^+] < 10^{-6}$, the pH = 4.

Note that this approach only applies to strong acids and strong bases.

 

**9. B is correct.**

The self-ionization (autoionization) of water is an ionization reaction in pure water or an aqueous solution in which a water molecule, $H_2O$, deprotonates (loses the nucleus of one of its hydrogen atoms) to become a hydroxide ion ($^-OH$).

**10. B is correct.**

An electrolyte is a substance that dissociates into cations (i.e., positive ions) and anions (i.e., negative ions) when placed in solution.

$NH_3$ is a weak acid with a p$K_a$ of 38. Therefore, it does not readily dissociate into $H^+$ and $^-NH_2$.

$HCO_3^-$ (bicarbonate) has a p$K_a$ of about 10.3 and is considered a weak acid because it does not dissociate completely. Therefore, it does not readily dissociate into $H^+$ and $CO_3^{2-}$.

HCN (nitrile) is a weak acid with a p$K_a$ of 9.3.

Therefore, it does not readily dissociate into $H^+$ and $^-CN$ (i.e., cyanide).

**11. B is correct.**

If the $[H_3O^+] = [^-OH]$, the solution has a pH of 7 and is neutral.

**12. D is correct.**

The Arrhenius acid-base theory states that acids produce $H^+$ ions (protons) in $H_2O$ solution, and bases produce $OH^-$ ions (hydroxide) in $H_2O$ solution.

**13. C is correct.**

Brønsted-Lowry acid-base theory focuses on the ability to accept and donate protons ($H^+$).

Brønsted-Lowry acid is a term for a substance that donates a proton in an acid-base reaction, while a Brønsted-Lowry base is a term for a substance that accepts a proton.

The definition is expressed in terms of an equilibrium expression

acid + base $\leftrightarrow$ conjugate base + conjugate acid.

**14. B is correct.**

According to the Brønsted-Lowry acid-base theory:

Acid (reactant) dissociates a proton to become the conjugate base (product).

Base (reactant) gains a proton to become the conjugate acid (product).

*continued...*

The definition is expressed in terms of an equilibrium expression:

acid + base ↔ conjugate base + conjugate acid

**15. C is correct.**

pH scale ranges from 1 to 14, with 7 being neutral.

Acidic solutions have a pH below 7, while basic solutions have a pH above 7.

The pH scale is a log scale where 7 has 50% deprotonated and 50% protonated (neutral) species (1:1 ratio).

At pH 6, there are 1 deprotonated : 10 protonated species. The ratio is 1:10.

At pH 5, there are 1 deprotonated : 100 protonated species. The ratio is 1:100.

A pH change of 1 unit changes the ratio by 10×.

Lower pH (< 7) results in more protonated species (e.g., cation), while an increase in pH (> 7) results in more deprotonated species (e.g., anion).

**16. A is correct.**

pI is the symbol for isoelectric point: the pH where a protein ion has zero net charge.

To calculate pI of amino acids with 2 p$K_a$ values, take the average of the p$K_a$'s:

$$pI = (pK_{a1} + pK_{a2}) / 2$$

$$pI = (2.2 + 4.2) / 2$$

$$pI = 3.2$$

**17. D is correct.**

Acidic solutions contain hydronium ions ($H_3O^+$). These ions are in the aqueous form because they are dissolved in water.

Although chemists often write $H^+$ (*aq*), referring to a single hydrogen nucleus (a proton), it exists as the hydronium ion ($H_3O^+$).

**18. C is correct.**

Equivalence point is when chemically equivalent quantities of acid and base have been mixed.

The moles of acid are equivalent to the moles of base.

*continued...*

 

Endpoint (related to, but not the same as the equivalence point) refers to the point at which the indicator changes color in a colorimetric titration. The endpoint can be found by an indicator, such as phenolphthalein (i.e., it turns colorless in acidic solutions and pink in basic solutions).

Buffer is an aqueous solution from a weak acid and its conjugate base or vice versa.

Buffered solutions resist changes in pH and are often used to keep the pH at a nearly constant value in many chemical applications. It readily absorbs or releases protons ($H^+$) and $^-OH$.

When an acid is added to the solution, the buffer releases $^-OH$ and accepts $H^+$ ions from acid.

When a base is added, the buffer accepts $^-OH$ ions from the base and releases protons ($H^+$).

Using the Henderson-Hasselbalch equation:

$$pH = pK_a + \log([A^-] / [HA]),$$

where [HA] = concentration of the weak acid, in units of molarity; $[A^-]$ = concentration of the conjugate base, in units of molarity.

$$pK_a = -\log(K_a)$$

where $K_a$ = acid dissociation constant.

From the question, the concentration of the conjugate base equals the concentration of the weak acid. This equates to the following expression:

$$[HA] = [A^-]$$

Rearranging: $[A^-] / [HA]$, like in the Henderson-Hasselbalch equation:

$$[A^-] / [HA] = 1$$

Substituting the value into the Henderson-Hasselbalch equation:

$$pH = pK_a + \log(1)$$

$$pH = pK_a + 0$$

$$pH = pK_a$$

When pH = $pK_a$, the titration is in the buffering region.

**19. D is correct.**

Ionic product constant of water:

$$K_w = [H_3O^+]\cdot[^-OH]$$

$$[^-OH] = K_w / [H_3O^+]$$

$$[^-OH] = [1 \times 10^{-14}] / [7.5 \times 10^{-9}]$$

$$[^-OH] = 1.3 \times 10^{-6}$$

**20. C is correct.**

Consider the $K_a$ presented in the question – which species is more acidic or basic than $NH_3$?

$NH_4^+$ is correct because it is $NH_3$ after absorbing one proton.

Concentration of $H^+$ in water is proportional to $K_a$. $NH_4^+$ is the conjugate acid of $NH_3$.

A: $H^+$ is acidic.

B: $NH_2^-$ is $NH_3$ with one less proton. If a base loses a proton, it would be an even stronger base with a higher proton affinity, so it is not weaker than $NH_3$.

D: Water is neutral.

==================================================================

**Practice Set 2: Questions 21–40**

==================================================================

**21. C is correct.**

Acid as a reactant produces a conjugate base, while a base reactant produces a conjugate acid.

Conjugate base of a chemical species is that species after $H^+$ has dissociated.

Therefore, the conjugate base of $HSO_4^-$ is $SO_4^{2-}$.

Conjugate base of $H_3O^+$ is $H_2O$.

**22. A is correct.**

Start by calculating the moles of $Ca(OH)_2$:

Moles of $Ca(OH)_2$ = molarity $Ca(OH)_2$ × volume of $Ca(OH)_2$

Moles of $Ca(OH)_2$ = 0.1 M × (30 mL × 0.001 L/mL)

Moles of $Ca(OH)_2$ = 0.003 mol

Use coefficients from the reaction equation to determine moles of $HNO_3$:

Moles of $HNO_3$ = (coefficient of $HNO_3$) / [coefficient $Ca(OH)_2$ × moles of $Ca(OH)_2$]

Moles of $HNO_3$ = (2 / 1) × 0.003 mol

Moles of $HNO_3$ = 0.006 mol

Divide moles by molarity to calculate volume:

Volume of $HNO_3$ = moles of $HNO_3$ / molarity of $HNO_3$

Volume of $HNO_3$ = 0.006 mol / 0.2 M

Volume of $HNO_3$ = 0.03 L

Convert volume to milliliters:

0.03 L × 1000 mL / L = 30 mL

**23. D is correct.**

Acidic solutions have a pH less than 7 due to a higher concentration of $H^+$ ions than $^-OH$ ions.

Basic solutions have a pH > 7 due to a higher concentration of $^-OH$ ions relative to $H^+$ ions.

**24. D is correct.**

An amphoteric compound reacts as an acid (i.e., donates protons) and a base (i.e., accepts protons).

Amphoteric molecules include amino acids (i.e., an amine and carboxylic acid group) and self-ionizable compounds such as water.

**25. B is correct.**

Strong acids (i.e., reactants) proceed towards products.

$$K_a = [\text{products}] / [\text{reactants}]$$

Molecule with the largest $K_a$ is the strongest acid.

$$pK_a = -\log K_a$$

Molecule with the smallest $pK_a$ is the strongest acid.

Strong acids dissociate a proton to produce the weakest conjugate base (i.e., most stable anion).

Weak acids dissociate a proton to produce the strongest conjugate base (i.e., least stable anion).

**26. B is correct.**

Balanced reaction:

$$H_3PO_4 + 3\ LIOH = Li_3PO_4 + 3\ H_2O$$

In the neutralization of acids and bases, the result is salt and water.

Phosphoric acid and lithium hydroxide react, so lithium phosphate and water are the resulting compounds.

**27. B is correct.**

Stability of the compound determines base strength. If the compound is unstable in its present state, it seeks a bonding partner (e.g., H+ or another atom) by donating its electrons for the new bond formation.

The 8 strong bases are: LiOH (lithium hydroxide), NaOH (sodium hydroxide), KOH (potassium hydroxide), $Ca(OH)_2$ (calcium hydroxide), RbOH (rubidium hydroxide), $Sr(OH)_2$, (strontium hydroxide), CsOH (cesium hydroxide) and $Ba(OH)_2$ (barium hydroxide).

**28. D is correct.**

Formula for pH:

$$pH = -\log[H^+]$$

Rearrange to solve for $[H^+]$:

$$[H^+] = 10^{-pH}$$

$$[H^+] = 10^{-2} \text{ M}$$

$$[H^+] = 0.01 \text{ M}$$

**29. C is correct.**

| Ionization | Dissociation |
|---|---|
| Process produces new charged particles | Separation of charged particles existing in a compound |
| Involves polar covalent compounds or metals | Involves ionic compounds |
| Involves covalent bonds between atoms | Involves ionic bonds in compounds |
| Produces charged particles | Produces charged or electrically neutral particles |
| Irreversible | Reversible |
| Example: $HCl \rightarrow H^+ + Cl^-$ $Mg \rightarrow Mg^{2+} + 2\,e^-$ | Example: $PbBr_2 \rightarrow Pb^{2+} + 2\,Br^-$ |

**30. C is correct.**

A base is a chemical substance with a pH greater than 7 and feels slippery because it dissolves the fatty acids and oils from the skin, reducing the friction between skin cells.

Under acidic conditions, litmus paper is red, and under basic conditions, it is blue.

Many bitter-tasting foods are alkaline because bitter compounds often contain amine groups, weak bases.

Acids are known to have a sour taste (e.g., lemon juice) because the sour taste receptors on the tongue detect the dissolved hydrogen ($H^+$) ions.

 

**31. A is correct.**

The 7 strong acids are HCl (hydrochloric acid), $HNO_3$ (nitric acid), $H_2SO_4$ (sulfuric acid), HBr (hydrobromic acid), HI (hydroiodic acid), $HClO_3$ (chloric acid), and $HClO_4$ (perchloric acid).

The 8 strong bases are: LiOH (lithium hydroxide), NaOH (sodium hydroxide), KOH (potassium hydroxide), $Ca(OH)_2$ (calcium hydroxide), RbOH (rubidium hydroxide), $Sr(OH)_2$, (strontium hydroxide), CsOH (cesium hydroxide) and $Ba(OH)_2$ (barium hydroxide).

**32. C is correct.**

Arrhenius acid-base theory states that acids produce $H^+$ ions in $H_2O$ solution, and bases produce $^-OH$ ions in $H_2O$ solution.

Brønsted-Lowry acid-base theory focuses on the ability to accept and donate protons ($H^+$).

Brønsted-Lowry acid is a term for a substance that donates a proton in an acid-base reaction, while a Brønsted-Lowry base is a term for a substance that accepts a proton.

Lewis acids are electron pair acceptors, whereas Lewis bases are electron pair donors.

**33. A is correct.**

By Brønsted-Lowry acid-base theory:

Acid (reactant) dissociates a proton to become the conjugate base (product).

Base (reactant) gains a proton to become the conjugate acid (product).

The definition is expressed in terms of an equilibrium expression:

acid + base ↔ conjugate base + conjugate acid

**34. D is correct.**

With polyprotic acids (i.e., more than one $H^+$ present), the $pK_a$ indicates the pH at which the $H^+$ is deprotonated. If the pH goes above the first $pK_a$, one proton dissociates, and so on.

In this example, the pH is above the first and second $pK_a$, so two acid groups are deprotonated while the third acidic proton is unaffected.

 

**35. A is correct.**

It is important to identify the acid that is active in the reaction. The parent acid is defined as the *most protonated* form of the buffer. The number of dissociating protons an acid can donate depends on the charge of its conjugate base.

$Ba_2P_2O_7$ is given as one of the products in the reaction. Because barium is a group 2B metal, it has a stable oxidation state +2. Because two barium cations are present in the product, the charge of $P_2O_7$ ion (the conjugate base in the reaction) must be –4.

Therefore, the fully protonated form of this conjugate must be $H_4P_2O_7$, a tetraprotic acid, because it has 4 protons that can dissociate.

**36. C is correct.**

The greater the concentration of $H_3O^+$, the more acidic the solution is.

**37. B is correct.**

A triprotic acid has three protons that can dissociate.

**38. B is correct.**

KCl and NaI are salts. Two salts only react if one of the products precipitates.

In this example, the products (KI and NaCl) are soluble in water, so they do not react.

**39. B is correct.**

Sodium acetate is a basic compound because acetate is the conjugate base of acetic acid, a weak acid ("the conjugate base of a weak acid acts as a base in the water").

Adding a base to a solution, even a buffered solution, increases the pH.

**40. D is correct.**

An acid anhydride is a compound with two acyl groups bonded to the same oxygen atom.

Anhydride means *without water* and is formed via dehydration (i.e., removal of $H_2O$) reaction.

==================================================================

**Practice Set 3: Questions 41–60**

==================================================================

**41. A is correct.**

Brønsted-Lowry acid-base theory focuses on the ability to accept and donate protons ($H^+$).

Brønsted-Lowry acid is a term for a substance that donates a proton ($H^+$) in an acid-base reaction, while a Brønsted-Lowry base is a term for a substance that accepts a proton.

**42. A is correct.**

Strong acids (i.e., reactants) proceed towards products.

$$K_a = [\text{products}] / [\text{reactants}]$$

The molecule with the largest $K_a$ is the strongest acid.

$$pK_a = -\log K_a$$

The molecule with the smallest $pK_a$ is the strongest acid.

Strong acids dissociate a proton to produce the weakest conjugate base (i.e., most stable anion).

Weak acids dissociate a proton to produce the strongest conjugate base (i.e., least stable anion).

**43. D is correct.**

Learn the ions involved in boiler scale formations: $CO_3^{2-}$ and the metal ions.

**44. B is correct.**

Buffer is an aqueous solution that consists of a weak acid and its conjugate base or vice versa.

Buffered solutions resist changes in pH and are often used to keep the pH at a nearly constant value in many chemical applications. It readily absorbs or releases protons ($H^+$) and $^-OH$.

When an acid is added to the solution, the buffer releases $^-OH$ and accepts $H^+$ ions from the acid.

A pair of weak acids/bases and their conjugate is needed to create a buffer solution or salt that contains an ion from the weak acid/base.

Since the problem indicates that sulfoxylic acid ($H_2SO_2$), which has a $pK_a$ 7.97, needs to be in the mixture, the other component would be an $HSO_2^-$ ion (bisulfoxylate) of $NaHSO_2$.

 

**45. C is correct.**

Acidic salt is a salt that still contains $H^+$ in its anion. It is formed when a polyprotic acid is partially neutralized, leaving at least 1 $H^+$.

For example:

$H_3PO_4 + 2\ KOH \rightarrow K_2HPO_4 + 2\ H_2O$: (partial neutralization, $K_2HPO_4$ is acidic salt)

While:

$H_3PO_4 + 3\ KOH \rightarrow K_3PO_4 + 3\ H_2O$: (complete neutralization, $K_3PO_4$ is not acidic salt)

**46. D is correct.**

An acid anhydride is a compound with two acyl groups bonded to the same oxygen atom. Anhydride means *without water* and is formed via dehydration (i.e., removal of $H_2O$) reaction.

**47. A is correct.**

By the Brønsted-Lowry definition, an acid donates protons while a base accepts protons.

On the product side of the reaction, $H_2O$ acts as a base (i.e., the conjugate base of $H_3O^+$), and HCl acts as an acid (i.e., the conjugate acid of $Cl^-$).

**48. D is correct.**

The ratio of the conjugate base to the acid must be determined from the pH of the solution and the p$K_a$ of the acidic component in the reaction.

In the reaction, $H_2PO_4^-$ acts as the acid, and $HPO_4^{2-}$ acts as the base, so the p$K_a$ of $H_2PO_4^-$ should be used in the equation.

Substitute the given values into the Henderson-Hasselbalch equation:

$pH = pK_a + \log[\text{salt} / \text{acid}]$

$7.35 = 6.87 + \log[\text{salt} / \text{acid}]$

Since $H_2PO_4^-$ is acting as the acid, subtract 6.87 from each side:

$0.48 = \log[\text{salt} / \text{acid}]$

The log base is 10, so the inverse log gives:

$10^{0.48} = (\text{salt} / \text{acid})$

$(\text{salt} / \text{acid}) = 3.02$

Ratio between the conjugate base or salt and the acid is 3.02 / 1.

**49. B is correct.**

An electrolyte is a substance that dissociates into cations (i.e., positive ions) and anions (i.e., negative ions) when placed in solution.

The light bulb is dimly lit, indicating that the solution contains only a low concentration (i.e., partial ionization) of the ions.

An electrolyte produces an electrically conducting solution when dissolved in a polar solvent (e.g., water). The dissolved ions disperse uniformly through the solvent.

If an electrical potential (i.e., voltage) is applied to such a solution, the cations of the solution migrate towards the electrode (i.e., an abundance of electrons). In contrast, the anions migrate towards the electrode (i.e., a deficit of electrons).

**50. B is correct.**

An acid is a chemical substance with a pH of less than 7, producing $H^+$ ions in water. An acid can be neutralized by a base (i.e., a substance with a pH above 7) to form a salt.

Acids are known to have a sour taste (e.g., lemon juice) because the sour taste receptors on the tongue detect the dissolved hydrogen ($H^+$) ions.

However, acids are not known to have a slippery feel; this is characteristic of bases. Bases feel slippery because they dissolve the fatty acids and oils from the skin and reduce the friction between the skin cells.

**51. C is correct.**

The Arrhenius acid-base theory states that acids produce $H^+$ ions (protons) in $H_2O$ solution, and bases produce $^-OH$ ions (hydroxide) in $H_2O$ solution.

The Brønsted-Lowry acid-base theory focuses on the ability to accept and donate protons ($H^+$).

A Brønsted-Lowry acid is a term for a substance that donates a proton in an acid-base reaction, while a Brønsted-Lowry base is a term for a substance that accepts a proton.

**52. A is correct.**

By Brønsted-Lowry acid-base theory:

Acid (reactant) dissociates a proton to become the conjugate base (product).

Base (reactant) gains a proton to become the conjugate acid (product).

The definition is expressed in terms of an equilibrium expression:

$$\text{acid} + \text{base} \leftrightarrow \text{conjugate base} + \text{conjugate acid}$$

 

**53. A is correct.**

By Brønsted-Lowry acid-base theory:

Acid (reactant) dissociates a proton to become the conjugate base (product).

Base (reactant) gains a proton to become the conjugate acid (product).

The definition is expressed in terms of an equilibrium expression:

acid + base ↔ conjugate base + conjugate acid.

HCl dissociates completely and, therefore, is a strong acid.

Strong acids produce weak (i.e., stable) conjugate bases.

**54. D is correct.**

A diprotic acid has two protons that can dissociate.

**55. A is correct.**

Protons ($H^+$) migrate between amino acid and solvent, depending on the pH of the solvent and p$K_a$ of functional groups on the amino acid.

Carboxylic acid groups can donate protons, while amine groups can receive protons.

For the carboxylic acid group:

If the pH of solution < p$K_a$ : the group is protonated and neutral

If the pH of solution > p$K_a$ : the group is deprotonated and negative

For the amine group:

If the pH of solution < p$K_a$ : the group is protonated and positive

If the pH of solution > p$K_a$ : the group is deprotonated and neutral

**56. C is correct.**

Acidic solutions contain hydronium ions ($H_3O^+$). These ions are in the aqueous form because they are dissolved in water.

Although chemists often write $H^+$ (*aq*), referring to a single hydrogen nucleus (a proton), it exists as the hydronium ion ($H_3O^+$).

**57. A is correct.**

With a $K_a$ of $10^{-5}$, the pH of a 1 M solution of the carboxylic acid, $CH_3CH_2CH2CO_2H$, would be 5 and is a weak acid.

Only $CH_3CH_2CH_2CO_2H$ is a weak acid because it yields a (relatively) unstable anion.

**58. C is correct.**

Condition 1: pH = 4

$$[H_3O^+] = 10^{-pH} = 10^{-4}$$

Condition 2: pH = 7

$$[H_3O^+] = 10^{-pH} = 10^{-7}$$

Ratio of $[H_3O^+]$ in condition 1 and condition 2:

$$10^{-4} : 10^{-7} = 1{,}000 : 1$$

Solution with a pH of 4 has 1,000 times greater $[H^+]$ than a solution with a pH of 7.

Note: $[H_3O^+]$ is equivalent to $[H^+]$

**59. B is correct.**

pH of a buffer is calculated using the Henderson-Hasselbalch equation:

$$pH = pK_a + \log([\text{conjugate base}] / [\text{conjugate acid}])$$

When [acid] = [base], the fraction is 1.

Log 1 = 0,

$$pH = pK_a + 0$$

If $K_a$ of the acid is $4.6 \times 10^{-4}$ (between $10^{-4}$ and $10^{-3}$), $pK_a$ (therefore pH) is between 3 and 4.

**60. C is correct.**

$K_w = [H^+]\cdot[^-OH]$ is the definition of the ionization constant for water.

==================================================================

**Practice Set 4: Questions 61–80**

==================================================================

**61. A is correct.**

Weak acids react with a strong base and are converted to their weak conjugate base, creating an overall basic solution.

Strong acids do react with a strong base, creating a neutral solution.

Weak acids partially dissociate when dissolved in water; unlike strong acids, they do not readily form ions.

Strong acids are much more corrosive than weak acids.

**62. C is correct.**

$K_w$ is the water ionization constant (or *water autoprotolysis constant*).

It can be determined experimentally and equals $1.011 \times 10^{-14}$ at 25 °C ($1.00 \times 10^{-14}$ is used).

**63. D is correct.**

All options are correct descriptions of the reaction, but the correct choice is the most descriptive.

**64. A is correct.**

Brønsted-Lowry acid-base theory focuses on the ability to accept and donate protons ($H^+$).

Brønsted-Lowry acid is a term for a substance that donates a proton in an acid-base reaction, while a Brønsted-Lowry base is a term for a substance that accepts a proton.

**65. A is correct.**

Buffered solutions resist changes in pH and are often used to keep the pH at a nearly constant value in many chemical applications. It readily absorbs or releases protons ($H^+$) and $^-OH$.

$H_2SO_4$ is a strong acid. Weak acids and their salts are good buffers.

A buffer is an aqueous solution that consists of a weak acid and its conjugate base or vice versa.

When an acid is added to the solution, the buffer releases $^-OH$ and accepts $H^+$ ions from the acid.

When a base is added, the buffer accepts $^-OH$ ions from the base and releases protons ($H^+$).

**66. D is correct.**

The main use of litmus paper is to test whether a solution is acidic or basic.

Litmus paper can test for water-soluble gases that affect acidity or alkalinity; the gas dissolves in the water, and the resulting solution colors the litmus paper. For example, alkaline ammonia gas causes the litmus paper to change from red to blue.

*Blue litmus paper* turns red under acidic conditions, and *red litmus paper* turns blue under basic or alkaline conditions, with the color change occurring over the pH range 4.5-8.3 at 25 °C (77 °F).

Neutral litmus paper is purple.

Litmus can be prepared as an aqueous solution that functions similarly. Under acidic conditions, the solution is red; under basic conditions, the solution is blue.

The properties of turning litmus paper blue, bitter taste, slippery feel, and neutralizing acids are true of bases.

An acidic solution has opposite qualities.

It has a pH lower than 7 and turns litmus paper red.

It neutralizes bases, tastes sour, and does not feel slippery.

**67. A is correct.**

An electrolyte is a substance that dissociates into cations (i.e., positive ions) and anions (i.e., negative ions) when placed in solution.

Electrolytes produce an electrically conducting solution when dissolved in a polar solvent (e.g., water). The dissolved ions disperse uniformly through the solvent.

If an electrical potential (i.e., voltage) is applied to such a solution, the cations of the solution migrate towards the electrode (i.e., an abundance of electrons). In contrast, the anions migrate towards the electrode (i.e., a deficit of electrons).

An acid is a substance that ionizes when dissolved in suitable ionizing solvents such as water.

If a high proportion of the solute dissociates to form free ions, it is a strong electrolyte.

If most of the solute does not dissociate, it is a weak electrolyte.

The more free ions present, the better the solution conducts electricity.

 

**68. A is correct.**

The 7 strong acids are HCl (hydrochloric acid), $HNO_3$ (nitric acid), $H_2SO_4$ (sulfuric acid), HBr (hydrobromic acid), HI (hydroiodic acid), $HClO_3$ (chloric acid), and $HClO_4$ (perchloric acid).

**69. B is correct.**

The activity series determines if a metal displaces another metal in the solution.

Reaction occurs if the added metal is above (i.e., activity series) the metal currently bonded with the anion.

An activity series ranks substances in their order of relative reactivity.

For example, magnesium metal can displace hydrogen ions from solution, so it is more reactive than elemental hydrogen:

$$Mg\ (s) + 2\ H^+\ (aq) \rightarrow H_2\ (g) + Mg^{2+}\ (aq)$$

Zinc can displace hydrogen ions from solution, so zinc is more reactive than elemental hydrogen:

$$Zn\ (s) + 2\ H^+\ (aq) \rightarrow H_2\ (g) + Zn^{2+}\ (aq)$$

Magnesium metal can displace zinc ions from solution:

$$Mg\ (s) + Zn^{2+}\ (aq) \rightarrow Zn\ (s) + Mg^{2+}\ (aq)$$

Metal activity series with the most active (i.e., most strongly reducing) metals appear at the top and the least active metals near the bottom.

Li: $2\ Li\ (s) + 2\ H_2O\ (l) \rightarrow LiOH\ (aq) + H_2\ (g)$

K: $2\ K\ (s) + 2\ H_2O\ (l) \rightarrow 2\ KOH\ (aq) + H_2\ (g)$

Ca: $Ca\ (s) + 2\ H_2O\ (l) \rightarrow Ca(OH)_2\ (s) + H_2\ (g)$

Na: $2\ Na\ (s) + 2\ H_2O\ (l) \rightarrow 2\ NaOH\ (aq) + H_2\ (g)$

The above can displace $H_2$ from water, steam, or acids

Mg: $Mg\ (s) + 2\ H_2O\ (g) \rightarrow Mg(OH)_2\ (s) + H_2\ (g)$

Al: $2\ Al\ (s) + 6\ H_2O\ (g) \rightarrow 2\ Al(OH)_3\ (s) + 3\ H_2\ (g)$

Mn: $Mn\ (s) + 2\ H_2O\ (g) \rightarrow Mn(OH)_2\ (s) + H_2\ (g)$

Zn: $Zn\ (s) + 2\ H_2O\ (g) \rightarrow Zn(OH)_2\ (s) + H_2\ (g)$

Fe: $Fe\ (s) + 2\ H_2O\ (g) \rightarrow Fe(OH)_2\ (s) + H_2\ (g)$

*continued...*

The above can displace $H_2$ from steam or acids

Ni: $Ni\ (s) + 2\ H^+\ (aq) \rightarrow Ni^{2+}\ (aq) + H_2\ (g)$

Sn: $Sn\ (s) + 2\ H^+\ (aq) \rightarrow Sn^{2+}\ (aq) + H_2\ (g)$

Pb: $Pb\ (s) + 2\ H^+\ (aq) \rightarrow Pb^{2+}\ (aq) + H_2\ (g)$

The above can displace $H_2$ from acids only

$H_2 > Cu > Ag > Pt > Au$

The above cannot displace $H_2$

**70. D is correct.**

Arrhenius acid-base theory states that acids produce $H^+$ ions (protons) in $H_2O$ solution, and bases produce $OH^-$ ions (hydroxide) in $H_2O$ solution.

$Al(OH)_3\ (s)$ is insoluble in water; therefore, it cannot function as an Arrhenius base.

**71. A is correct.**

By Brønsted-Lowry acid-base theory:

Acid (reactant) dissociates a proton to become the conjugate base (product).

Base (reactant) gains a proton to become the conjugate acid (product).

The definition is expressed in terms of an equilibrium expression:

acid + base ↔ conjugate base + conjugate acid

**72. D is correct.**

Henderson-Hasselbach equation:

$$pH = pK_a + \log(A^- / HA)$$

Buffer is an aqueous solution that consists of weak acid and its conjugate base or vice versa.

Buffered solutions resist changes in pH and are often used to keep the pH at a nearly constant value in many chemical applications. It readily absorbs or releases protons ($H^+$) and $^-OH$.

When an acid is added to the solution, the buffer releases $^-OH$ and accepts $H^+$ ions from acid.

**73. C is correct.**

Weakest acid has the smallest $K_a$ (or largest $pK_a$).

Weakest acid has the strongest (i.e., least stable) conjugate base.

**74. D is correct.**

Balanced reaction:

$$2\ H_3PO_4 + 3\ Ba(OH)_2 \rightarrow Ba_3(PO_4)_2 + 6\ H_2O$$

There are 2 moles of $H_3PO_4$ in a balanced reaction.

However, acids are categorized by the number of $H^+$ per mole of acid.

For example:

HCl is a monoprotic acid (has one $H^+$ to dissociate).

$H_2SO_4$ is a diprotic acid (has two $H^+$ to dissociate).

$H_3PO_4$ is a triprotic acid (has three $H^+$ to dissociate).

**75. B is correct.**

$Na^+$ forms a strong base (NaOH), and S forms a weak acid ($H_2S$); it undergoes a hydrolysis reaction in water:

$$Na_2S\ (aq) + 2\ H_2O\ (l) \rightarrow 2\ NaOH\ (aq) + H_2S$$

Ionic equation for individual ions:

$$2\ Na^+ + S^{2-} + 2\ H_2O \rightarrow 2\ Na^+ + 2\ OH^- + H_2S$$

Removing $Na^+$ ions from each side of the reaction:

$$S^{2-} + 2\ H_2O \rightarrow 2\ OH^- + H_2S$$

Removing $H_2O$ from each side of the reaction:

$$S^{2-} + H_2O \rightarrow OH^- + HS^-$$

**76. A is correct.**

$HNO_3$ is a strong acid, which means that $[HNO_3] = [H_3O^+] = 0.0765$.

$$pH = -\log[H_3O^+]$$

$$pH = -\log(0.0765)$$

$$pH = 1.1$$

**77. C is correct.**

Strong acids dissociate a proton to produce weakest conjugate base (i.e., most stable anion).

Weak acids dissociate a proton to produce strongest conjugate base (i.e., least stable anion).

Acetic acid ($CH_3COOH$) has a p$K_a$ of about 4.8 and is considered weak because it does not dissociate completely. Therefore, it does not readily dissociate into $H^+$ and $CH_3COO^-$.

Each of the listed strong acids forms an anion stabilized by resonance.

HBr (hydrobromic acid) has a p$K_a$ of –9.

$HNO_3$ (nitrous acid) has a p$K_a$ of –1.4.

$H_2SO_4$ (sulfuric acid) is a diprotic acid with a p$K_a$ of –3 and 1.99.

HCl (hydrochloric acid) has a p$K_a$ of –6.3.

**78. A is correct.**

It is important to recognize the chromate ions:

Dichromate: $Cr_2O_7^{2-}$

Chromium (II): $Cr^{2+}$

Chromic/Chromate: $CrO_4^-$

**79. B is correct.**

Brønsted-Lowry acid-base theory focuses on the ability to accept and donate protons ($H^+$).

Brønsted-Lowry acid is a substance that donates a proton in an acid-base reaction.

Brønsted-Lowry base is a substance that accepts a proton.

**80. C is correct.**

In Arrhenius theory, acids dissociate in an aqueous solution to produce $H^+$ (hydrogen ions).

In Arrhenius theory, bases dissociate in an aqueous solution to produce $^-OH$ (hydroxide ions).

 

---

**Practice Set 5: Questions 81–100**

---

**81. B is correct.**

The acid requires two equivalents of a base to be fully titrated and, therefore, is a diprotic acid.

Using the fully protonated sulfuric acid ($H_2SO_4$) as an example:

At point A, the acid is 50% fully protonated and 50% singly deprotonated (50% $H_2SO_4$: 50% $HSO_4^-$).

At point B, the acid exists in the singly deprotonated form (100% $HSO_4^-$).

At point C, the acid exists as 50% singly deprotonated $HSO_4^-$ and 50% doubly deprotonated $SO_4^{2-}$.

At point D, the acid exists as 100% $SO_4^{2-}$.

Point A is p$K_{a1}$, and point C is p$K_{a2}$ (i.e., strongest buffering regions).

Point B and D are *equivalence points* (i.e., weakest buffering region)

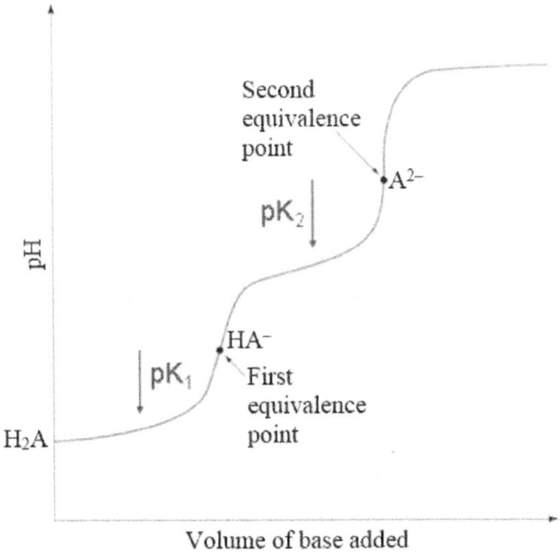


*Titration curve: addition of a strong base to diprotic acid ($H_2A$)*

**82. C is correct.**

Point C on the graph for the question is p$K_{a2}$. At this point, the acid exists as 50% singly deprotonated (e.g., $HSO_4^-$) and 50% doubly deprotonated (e.g., $SO_4^{2-}$).

Henderson-Hasselbalch equation:

$$pH = pK_{a2} + \log[\text{salt / acid}]$$

$$pH = pK_{a2} + \log[50\% / 50\%]$$

$$pH = pK_{a2} + \log[1]$$

$$pH = pK_{a2} + 0$$

$$pH = pK_{a2}$$

**83. A is correct.**

Point A is p$K_{a1}$. At this point, the acid exists as 50% fully protonated (e.g., $H_2SO_4$) and 50% singly deprotonated (e.g., $HSO_4^-$).

Henderson-Hasselbalch equation:

$$pH = pK_{a1} + \log[\text{salt / acid}]$$

$$pH = pK_{a1} + \log[50\% / 50\%]$$

$$pH = pK_{a1} + \log[1]$$

$$pH = pK_{a1} + 0$$

$$pH = pK_{a1}$$

**84. B is correct.**

At point B on the graph (i.e., equivalence point), the acid exists in the singly deprotonated form only (100% $HSO_4^-$ for the example of $H_2SO_4$).

**85. B is correct.**

Buffer region is the flattest region on the curve (the region that resists pH increases with added base). This diprotic acid has two buffering regions: p$K_{a1}$ (around point A) and p$K_{a2}$ (around point C).

Buffer is an aqueous solution that consists of weak acid and its conjugate base or vice versa.

*continued...*

 

Buffered solutions resist changes in pH and are often used to keep the pH at a nearly constant value in many chemical applications. It readily absorbs or releases protons ($H^+$) and $^-OH$.

When acid is added to the solution, the buffer releases $^-OH$ and accepts $H^+$ ions from the acid.

When base is added, the buffer accepts $^-OH$ ions from the base and releases protons ($H^+$).

**86. D is correct.**

Sodium hydroxide turns into soap (i.e., saponification) from the reaction with the fatty acid esters and oils on the fingertips (i.e., skin).

Fatty acid esters react with NaOH by releasing free fatty acids, which act as soap surrounding grease with their nonpolar (i.e., hydrophobic) ends, while their polar (i.e., hydrophilic) ends orient towards water molecules. This decreases friction and accounts for the slippery feel of NaOH interacting with the skin.

**87. C is correct.**

I: water is produced in a neutralization reaction.

II: this reaction is a common example of neutralization, but it is not always true.

For example, weak bases react with strong acids:

$$AH + B \rightleftharpoons A^- + BH^+$$

Alternatively, strong bases react with weak acids:

$$AH + H_2O \rightleftharpoons H_3O^+ + A^-$$

III: neutralization occurs when acid donates a proton to the base.

**88. D is correct.**

Strong acids *dissociate entirely*, or almost completely, when in water. They dissociate into a positively charged hydrogen ion ($H^+$) and another negatively charged ion.

An example is hydrochloric acid (HCl), which dissociates into $H^+$ and $Cl^-$ ions.

*Polarity* refers to the distribution of electrons in a bond. If a molecule is polar, one side has a partial positive charge and a partial negative charge.

The more polar the bond, the easier it is for a molecule to dissociate into ions; therefore, the acid is more strongly acidic.

**89. D is correct.**

Brønsted-Lowry acid-base theory focuses on the ability to accept and donate protons ($H^+$).

Brønsted-Lowry acid is a term for a substance that donates a proton in an acid-base reaction, while a Brønsted-Lowry base is a term for a substance that accepts a proton.

**90. B is correct.**

Formula to calculate pH of an acidic buffer:

$[H^+] = ([acid] / [salt]) \times K_a$

Calculate $H^+$ from pH:

$[H^+] = 10^{-pH}$

$[H^+] = 10^{-4}$

Calculate $Ka$ from $pK_a$:

$K_a = 10^{-pKa}$

$K_a = 10^{-3}$

Substitute values into the buffer equation:

$[H^+] = ([acid] / [salt]) \times K_a$

$10^{-4} = ([acid] / [salt]) \times 10^{-3}$

$10^{-1} = [acid] / [salt]$

$[salt] / [acid] = 10$

Therefore, the ratio of salt to acid is 10:1.

**91. D is correct.**

Electrolyte is a substance that dissociates into cations (i.e., positive ions) and anions (i.e., negative ions) when placed in solution. An electrolyte produces an electrically conducting solution when dissolved in a polar solvent (e.g., water).

Dissolved ions disperse uniformly through the solvent. If an electrical potential (i.e., voltage) is applied to such a solution, the cations of the solution migrate towards the electrode (i.e., an abundance of electrons).

In contrast, the anions migrate towards the electrode (i.e., a deficit of electrons).

*continued...*

If a high proportion of the solute dissociates to form free ions, it is a strong electrolyte.

If most of the solute does not dissociate, it is a weak electrolyte.

The more free ions present, the better the solution conducts electricity.

**92. D is correct.**

*Litmus* (i.e., dyes extracted from lichens) tests whether a solution is acidic or basic.

*Wet litmus paper* can be used to test for water-soluble acidic or alkaline gases that dissolve in the water and produce color changes in the litmus paper.

*Neutral litmus paper* is purple, with color changes occurring over the pH range of 4.5 to 8.3.

*Red litmus paper* turns blue when exposed to alkaline ammonia gas (basic conditions).

*Blue litmus paper* turns red under acidic conditions.

**93. C is correct.**

The more acidic molecule has a lower p$K_a$.

**94. A is correct.**

Acetic acid ($CH_3COOH = HC_2H_3O_2$) is commonly known as vinegar. It has a p$K_a$ of about 4.8 and is considered a weak acid because it does not dissociate completely.

HI (hydroiodic acid) has a p$K_a$ of –10.

$HClO_4$ (perchloric acid) has a p$K_a$ of –10.

HCl (hydrochloric acid) has a p$K_a$ of –6.3.

$HNO_3$ (nitrous acid) has a p$K_a$ of –1.4.

**95. C is correct.**

Acid anhydride is a compound with two acyl groups bonded to the same oxygen atom.

$H_3CCO_3CCH_3$

Anhydride means *without water* and is formed via dehydration (i.e., removal of $H_2O$) reaction.

**96. B is correct.**

Acids and bases react to form $H_2O$ and salts: NaOH + HCl $\rightleftharpoons$ $Na^+Cl^-$ (a salt) + $H_2O$ (water)

Brønsted-Lowry acid-base theory focuses on the ability to accept and donate protons ($H^+$).

Brønsted-Lowry acid is a substance that donates a proton in an acid-base reaction.

Brønsted-Lowry base is a substance that accepts a proton.

Arrhenius acid-base theory states that acids produce $H^+$ ions in $H_2O$ solution.

Arrhenius acid-base theory states that bases produce $^-OH$ ions in $H_2O$ solution.

Lewis acids are electron-pair acceptors, whereas Lewis bases are electron-pair donors.

**97. D is correct.**

Bronsted-Lowry acid-base theory focuses on the ability to accept and donate protons.

The definition is expressed in terms of an equilibrium expression:

acid + base ↔ conjugate base + conjugate acid.

With an acid, HA, the equation can be written symbolically as:

$HA + B \leftrightarrow A^- + HB^+$

**98. C is correct.**

By Brønsted-Lowry acid-base theory:

Acid (reactant) dissociates a proton to become the conjugate base (product).

Base (reactant) gains a proton to become the conjugate acid (product).

The definition is expressed in terms of an equilibrium expression:

acid + base ↔ conjugate base + conjugate acid

**99. D is correct.**

The addition of hydroxide (i.e., NaOH) decreases the solubility of magnesium hydroxide due to the common ion effect.

Therefore, the amount of undissociated $Mg(OH)_2$ increases.

**100. A is correct.**

The weakest acid has the highest p$K_a$ (and lowest $K_a$).

===========================================================================

**Practice Set 6: Questions 101–120**

===========================================================================

**101. C is correct.**

Strong acids dissociate entirely (or almost completely) because they form a stable anion.

Hydrogen cyanide (HCN) has a p$K_a$ of 9.3, and hydrogen sulfide ($H_2S$) has a p$K_a$ of 7.0.

In this example, the strong acids include HI, HBr, HCl, $H_3PO_4$, and $H_2SO_4$.

**102. D is correct.**

Bromothymol blue is a pH indicator often used for solutions with neutral pH near 7 (e.g., managing the pH of pools and fish tanks). Bromothymol blue acts as a weak acid in a solution that can be protonated or deprotonated. It appears yellow when protonated (lower pH), blue when deprotonated (higher pH), and bluish-green in neutral solution.

Methyl red has a p$K_a$ of 5.1 and is a pH indicator dye that changes color in acidic solutions: it turns red in pH under 4.4, orange between 4.4 and 6.2, and yellow in pH over 6.2.

Phenolphthalein is used as an indicator for acid-base titrations. A weak acid can dissociate protons ($H^+$ ions) in solutions. The phenolphthalein molecule is colorless, and the phenolphthalein ion is pink. It turns colorless in acidic solutions and pink in basic solutions. With basic conditions, the phenolphthalein (neutral) ⇌ ions (pink) equilibrium shifts to the right, leading to more ionization as $H^+$ ions are removed.

**103. A is correct.**

Anhydride means *without water* and is formed via dehydration (i.e., removal of $H_2O$) reaction. Therefore, a basic anhydride is a base without water.

**104. C is correct.**

The problem is asking for $K_a$ of an acid, indicating that the acid in question is weak.

In aqueous solutions, a hypothetical weak acid HX will partly dissociate and create this equilibrium:

$$HX \leftrightarrow H^+ + X^-$$

with an acid equilibrium constant, or $K_a = [H^+]·[X^-] / [HX]$

*continued...*

To calculate $K_a$, the concentration of species is needed.

From the given pH, the concentration of $H^+$ ions can be calculated:

$$[H^+] = 10^{-pH}$$

$$[H^+] = 10^{-7}\ M$$

Number of $H^+$ and $X^-$ ions are equal; the concentration of $X^-$ ions is $10^{-7}$ M.

Those ions came from the dissociated acid molecules. According to the problem, only 24% of the acid is dissociated. Therefore, the rest of acid molecules (100% – 24% = 76%) did not dissociate. Use the simple proportion of percentages to calculate the concentration of HX:

$$[HX] = (76\% / 24\%) \times 1 \times 10^{-7}\ M$$

$$[HX] = 3.17 \times 10^{-7}\ M$$

Use concentration values to calculate $K_a$:

$$K_a = [H^+]\cdot[X^-] / [HX]$$

$$K_a = [(1 \times 10^{-7}) \times (1 \times 10^{-7})] / (3.17 \times 10^{-7})$$

$$K_a = 3.16 \times 10^{-8}$$

Calculate $pK_a$:

$$pK_a = -\log K_a$$

$$pK_a = -\log (3.16 \times 10^{-8})$$

$$pK_a = 7.5$$

**105. B is correct.**

With a neutralization reaction, the cations and anions on reactants switch pairs, resulting in salt and water.

**106. D is correct.**

Strongest acid has the largest $K_a$ value (or the lowest $pK_a$).

**107. C is correct.**

If the $[H_3O^+]$ is greater than $1 \times 10^{-7}$, it is an acidic solution.

It is a basic solution if the $[H_3O^+]$ is less than $1 \times 10^{-7}$.

If the $[H_3O^+]$ equals $1 \times 10^{-7}$, the solution has a pH of 7 and is neutral.

 

**108. A is correct.**

Electrolyte is a substance that dissociates into cations (i.e., positive ions) and anions (i.e., negative ions) when placed in solution. An electrolyte produces an electrically conducting solution when dissolved in a polar solvent (e.g., water). The dissolved ions disperse uniformly through the solvent.

If an electrical potential (i.e., the voltage generated by the battery) is applied, the cations of the solution migrate towards the electrode (i.e., an abundance of electrons).

In contrast, the anions migrate towards the electrode (i.e., a deficit of electrons).

**109. D is correct.**

$H_3PO_4 \rightarrow H_2PO_4^- \rightarrow HPO_4^{2-}$

In Arrhenius theory, acids dissociate in an aqueous solution to produce $H^+$ (hydrogen ions).

In Arrhenius theory, bases dissociate in an aqueous solution to produce $^-OH$ (hydroxide ions).

In Brønsted–Lowry theory, acids and bases are defined by how they react.

The definition is expressed in terms of an equilibrium expression:

acid + base ↔ conjugate base + conjugate acid.

With an acid, HA, the equation can be written symbolically as:

$HA + B \leftrightarrow A^- + HB^+$

**110. D is correct.**

Arrhenius acid-base theory states that acids produce $H^+$ ions in $H_2O$ solution, and bases produce $OH^-$ ions in $H_2O$ solution.

Arrhenius acid-base theory states that neutralization happens when acid-base reactions produce water and salt and must occur in an aqueous solution.

There are additional ways to classify acids and bases.

Brønsted-Lowry acid-base theory focuses on the ability to accept and donate protons.

Lewis acid-base theory focuses on the ability to accept and donate electrons.

**111. A is correct.**

Brønsted-Lowry acid-base theory focuses on the ability to accept and donate protons ($H^+$).

Brønsted-Lowry acid is a term for a substance that donates a proton in an acid-base reaction, while a Brønsted-Lowry base is a term for a substance that accepts a proton.

Arrhenius acid-base theory focuses on the ability to produce $H^+$ and $^-OH$ ions.

Arrhenius acid–base theory states that neutralization happens when acid–base reactions produce water and salt and must take place in an aqueous solution.

Lewis acid–base theory focuses on the ability to accept and donate electrons.

**112. B is correct.**

By Brønsted-Lowry acid–base theory:

Acid (reactant) dissociates a proton to become the conjugate base (product).

Base (reactant) gains a proton to become the conjugate acid (product).

The definition is expressed in terms of an equilibrium expression:

acid + base ↔ conjugate base + conjugate acid

**113. C is correct.**

Water is neutral and has equal hydroxide ($^-OH$) and hydronium ($H_3O^+$) ion concentrations.

**114. D is correct.**

The pI (isoelectric point) for an amino acid is defined as the pH for which an ionizable molecule has a net charge of zero.

In general, the net charge on the molecule is affected by pH, as it can become more positively or negatively charged due to the gain or loss of protons ($H^+$), respectively.

Amphoteric compounds react as acid (i.e., donates protons) and base (i.e., accepts protons).

Zwitterion is a molecule with a positive (cation) and negative (anion) region within the same molecule.

Zwitterions are amphoteric. Amino acids (amino and carboxyl end) are an example of amphoteric molecules.

 

**115. B is correct.**

*Spectator ions* exist as a reactant and product in a chemical equation. A net ionic equation ignores the spectator ions that were part of the original equation.

Zwitterion is a molecule with a positive (cation) and negative (anion) region within the same molecule. Zwitterion is amphoteric.

Amino acids (amino and carboxyl end) are an example of amphoteric molecules.

**116. A is correct.**

Salts that result from the reaction of strong acids with strong bases are neutral.

For example:

$HCl + NaOH \leftrightarrow NaCl + H_2O$

strong acid, strong base, neutral

**117. D is correct.**

Polyprotic means two $H^+$ that can dissociate.

**118. B is correct.**

Acid dissociates a proton to form the conjugate base, while a conjugate base accepts a proton to form the acid.

**119. C is correct.**

The products of a neutralization reaction (e.g., salt and water) are not corrosive.

NaOH and HCl are very corrosive.

However, they form NaCl, or common table salt, which is not corrosive after a neutralization reaction.

**120. C is correct.**

Start by calculating the moles of $CaCO_3$:

Moles of $CaCO_3$ = mass of $CaCO_3$ / molecular mass of $CaCO_3$

Moles of $CaCO_3$ = 0.5 g / 100.09 g/mol

Moles of $CaCO_3$ = 0.005 mol

*continued...*

Use coefficients in the reaction equation to find moles of $HNO_3$:

Moles of $HNO_3$ = (coefficient of $HNO_3$ / coefficient of $CaCO_3$) × moles of $CaCO_3$

Moles of $HNO_3$ = (2/1) × 0.005 mol

Moles of $HNO_3$ = 0.01 mol

Divide moles by volume to calculate molarity:

Molarity of $HNO_3$ = moles of $HNO_3$ / volume of $HNO_3$

Molarity of $HNO_3$ = 0.01 mol / (25 mL × 0.001 L/mL)

Molarity of $HNO_3$ = 0.4 M

 

---

**Practice Set 7: Questions 121–140**

---

**121. D is correct.**

Strong acids dissociate a proton to produce weakest conjugate base (i.e., most stable anion).

Weak acids dissociate a proton to produce strongest conjugate base (i.e., least stable anion).

Hydrofluoric acid (HF) has a p$K_a$ of about 3.8 and is considered weak because it does not dissociate completely. The $F^-$ anion is the least stable halogen anion (due to its small valence shell). $^-OH$ is the conjugate base of $H_2O$.

$NaNH_2$ (sodium amide) is a strong base with a p$K_a$ of 38.

$HNO_3$ (nitrous acid) has a p$K_a$ of –1.4.

HI (hydroiodic acid) has a p$K_a$ of –10.

**122. C is correct.**

Methyl red has a p$K_a$ of 5.1 and is a pH indicator dye that changes color in acidic solutions: it turns red at pH under 4.4, orange at pH 4.4-6.2, and yellow over 6.2.

Phenolphthalein is used as an indicator for acid–base titrations. It is a weak acid which can dissociate protons ($H^+$ ions) in solutions.

Phenolphthalein molecule is colorless, and the phenolphthalein ion is pink. It turns colorless in acidic solutions and pink in basic solutions.

With basic conditions, the phenolphthalein (neutral) ⇌ ions (pink) equilibrium shifts to the right, leading to more ionization as $H^+$ ions are removed.

Bromothymol blue is a pH indicator often used for solutions with neutral pH near 7 (e.g., managing the pH of pools and fish tanks).

Bromothymol blue acts as a weak acid in a solution that can be protonated or deprotonated. It appears yellow when protonated (lower pH), blue when deprotonated (higher pH), and bluish-green in neutral solution.

**123. C is correct.**

Balanced reaction:

$$H_2CO_3 + 2\ KOH\ (aq) \rightarrow K_2CO_3\ (aq) + 2\ H_2O\ (l)$$

With a neutralization reaction, cations and anions on reactants switch pairs, resulting in salt and water.

**124. D is correct.**

Strong acids completely dissociate protons into the aqueous solution. The resulting anion is stable, which accounts for the ~100% ionization of the acid.

**125. B is correct.**

A solution is acidic if the $[H_3O^+]$ exceeds $1 \times 10^{-7}$.

If the $[H_3O^+]$ is less than $1 \times 10^{-7}$, it is a basic solution.

If the $[H_3O^+]$ equals $1 \times 10^{-7}$, it has a pH of 7, and the solution is neutral.

**126. A is correct.**

Electrolytes dissociate into ions when dissolved in water.

$CH_4$, as with most hydrocarbon compounds, is not an electrolyte and will not dissociate.

**127. B is correct.**

Weak acids form unstable conjugate bases, and therefore, the equilibrium lies on the side of the acid.

**128. B is correct.**

Arrhenius acid–base theory states that acids produce $H^+$ ions (protons) in $H_2O$ solution, and bases produce $^-OH$ ions (hydroxide) in $H_2O$ solution.

Brønsted-Lowry acid–base theory focuses on the ability to accept and donate protons ($H^+$).

Brønsted-Lowry acid is a substance that donates a proton in an acid–base reaction, while a Brønsted-Lowry base is a substance that accepts a proton.

**129. B is correct.**

Arrhenius acid–base theory states that acids produce $H^+$ ions (protons) in $H_2O$ solution, and bases produce $^-OH$ ions (hydroxide) in $H_2O$ solution.

 

**130. D is correct.**

Brønsted-Lowry acid–base theory focuses on the ability to accept and donate protons ($H^+$).

Brønsted-Lowry acid is a substance that donates a proton in an acid–base reaction, while a Brønsted-Lowry base is a substance that accepts a proton.

**131. D is correct.**

By Brønsted-Lowry acid–base theory:

Acid (reactant) dissociates a proton to become the conjugate base (product).

Base (reactant) gains a proton to become the conjugate acid (product).

The definition is expressed in terms of an equilibrium expression:

acid + base ↔ conjugate base + conjugate acid.

**132. A is correct.**

Because $HNO_3$ is a strong acid, $[HNO_3] = [H_3O^+] = 0.045$ M.

$pH = -\log[H_3O^+]$

$pH = -\log(0.045)$

$pH = 1.35$

**133. D is correct.**

pI (isoelectric point) for an amino acid is defined as the pH for which an ionizable molecule has a net charge of zero. In general, the net charge on the molecule is affected by pH, as it can become more positively or negatively charged due to the gain or loss of protons ($H^+$), respectively.

**134. C is correct.**

Brønsted-Lowry acid–base theory focuses on the ability to accept and donate protons ($H^+$).

Brønsted-Lowry acid is a substance that donates a proton in an acid–base reaction, while a Brønsted-Lowry base is a substance that accepts a proton.

Arrhenius acid–base theory states that acids produce $H^+$ ions in $H_2O$ solution and bases produce $^-OH$ ions in $H_2O$ solution.

**135. D is correct.**

All reactions either increase the production of $H^+$ or $^-OH$ or decrease $H^+$ or $^-OH$ production and resist changes in pH (a feature of a buffer system).

Buffer is an aqueous solution that consists of a weak acid and its conjugate base or vice versa.

Buffered solutions resist changes in pH and are often used to keep the pH at a nearly constant value in many chemical applications. It readily absorbs or releases protons ($H^+$) and $^-OH$.

When acid is added to the solution, the buffer releases $^-OH$ and accepts $H^+$ ions from the acid.

When base is added, the buffer accepts $^-OH$ ions from the base and releases protons ($H^+$).

**136. A is correct.**

A standard solution contains a precisely known concentration of an element or substance, usually determined to 3-4 significant digits.

Standard solutions are often used to determine the concentration of other substances, such as solutions in titrations.

**137. B is correct.**

*Amphoteric* means that the compound can act as an acid *or* base.

**138. D is correct.**

*Inflection point* is halfway between the beginning of the curve (before any titrant has been added) and the equivalence point (where the titrant has neutralized the starting material).

At inflection point, concentrations of the two species (the acid and conjugate base) are equal.

Henderson-Hasselbalch equation:

$$pH = pK_a + \log[\text{conjugate base}] / [\text{acid}]$$

Buffer is an aqueous solution from a weak acid and its conjugate base or vice versa.

Buffered solutions resist changes in pH and are often used to keep the pH at a nearly constant value in many chemical applications. It readily absorbs or releases protons ($H^+$) and $^-OH$.

When acid is added to the solution, the buffer releases $^-OH$ and accepts $H^+$ ions from the acid.

**139. A is correct.**

Three possible ways to balance this reaction:

2 LiOH (*aq*) + $H_2SO_4$ (*aq*) = 2 $H_2O$ (*l*) + $Li_2SO_4$ (*aq*) – salt, so this is probably the correct balance

2 LiOH (*aq*) + $H_2SO_4$ (*aq*) = 2 $H_2O$ + $Li_2SO_4$ (*aq*)

LiOH (*aq*) + $H_2SO_4$ (*aq*) = $H_2O$ (*l*) + $LiHSO_4$ (*aq*) – *hydrogen sulfate (bisulfate)*

**140. B is correct.**

Protons ($H^+$) migrate between amino acid and solvent, depending on the pH of the solvent and $pK_a$ of functional groups on the amino acid.

Carboxylic acid groups can donate protons, while amine groups can receive protons.

For the carboxylic acid group:

If the pH of solution < $pK_a$ : group is protonated and neutral

If the pH of solution > $pK_a$ : group is deprotonated and negative

For the amine group:

If the pH of solution < $pK_a$ : group is protonated and positive

If the pH of solution > $pK_a$ : group is deprotonated and neutral

*Notes for active learning*

# 10 – Electrochemistry: Detailed Explanations

==================================================================

**Practice Set 1: Questions 1–20**

==================================================================

**1. D is correct.**

In electrochemical (i.e., galvanic) cells, oxidation occurs at the anode, and reduction occurs at the cathode.

$Br^-$ is oxidized at the anode, not the cathode.

**2. C is correct.**

As ionization energy increases, it is more difficult for electrons to be released by a substance. This is because the release of electrons would increase the charge of the substance, which is the definition of oxidation.

Substances with higher ionization energy are more likely to be reduced. If a substance is reduced in a redox reaction, it is an oxidizing agent because it facilitates the oxidation of the other reactant.

Therefore, with the increase of ionization energy, a substance is more likely to be reduced or be a stronger oxidizing agent.

**3. A is correct.**

Half-reaction:

$$H_2S \rightarrow S_8$$

Balancing half-reaction in acidic conditions:

*Step 1:* Balance atoms except for H and O

$$8\ H_2S \rightarrow S_8$$

*Step 2:* To balance oxygen, add $H_2O$ to the side with fewer oxygen atoms

There is no oxygen at all, so skip this step.

*Step 3:* To balance hydrogen, add $H^+$:

$$8\ H_2S \rightarrow S_8 + 16\ H^+$$

*continued...*

 

*Balance charges by adding electrons to the side with a greater/more positive total charge*

Total charge on left side: 0

Total charge on right side: 16(+1) = +16

Add 16 electrons to right side:

$8\ H_2S \rightarrow S_8 + 16\ H^+ + 16\ e^-$

**4. B is correct.**

Electrolytic cell is a nonspontaneous electrochemical cell requiring electrical energy (e.g., battery) to initiate the reaction.

Anode is positive, and the cathode is the negative electrode.

For electrolytic and galvanic cells, oxidation occurs at the anode, while reduction occurs at the cathode.

Therefore, Co metal is produced at the cathode because it is a reduction product (from an oxidation number of +3 on the left to 0 on the right). Co will not be produced at the anode.

**5. C is correct.**

The anode in galvanic cells attracts anions.

Anions in solutions flow toward the anode, while cations flow toward the cathode.

Oxidation (i.e., loss of electrons) occurs at the anode.

Positive ions are formed while negative ions are consumed at the anode.

Therefore, negative ions flow toward the anode to equalize the charge.

**6. A is correct.**

By convention, the reference standard for potential is always hydrogen reduction.

**7. C is correct.**

Calculate the oxidation numbers of species involved and look for the oxidized species (increase in oxidation number).

Cd's oxidation number increases from 0 on the left side of the reaction to +2 on the right.

**8. D is correct.**

Oxidation number of Fe increases from 0 on the left to +3 on the right.

 

**9. D is correct.**

Salt bridges contain cations (positive ions) and anions (negative ions).

Anions flow towards the oxidation half-cell because the oxidation product is positively charged, and the anions are required to balance the charges within the cell.

The opposite is true for cations; they flow towards the reduction half-cell.

**10. B is correct.**

A redox reaction, or oxidation-reduction reaction, involves the transfer of electrons between two reacting substances. An oxidation reaction refers explicitly to a substance losing electrons, and a reduction reaction refers to a substance gaining electrons.

Oxidation and reduction reactions alone are half-reactions because they occur together as a whole reaction.

The key to this question is *transferring* electrons between *two* species, referring to the whole redox reaction in its entirety, not just one half-reaction.

An electrochemical reaction occurs during the passage of electric current and involves redox reactions. However, it is not the correct answer to the question posed.

**11. A is correct.**

Cell reaction is:

$Co\ (s) + Cu^{2+}\ (aq) \rightarrow Co^{2+}\ (aq) + Cu\ (s)$

Separate it into half-reactions:

$Cu^{2+}\ (aq) \rightarrow Cu\ (s)$

$Co\ (s) \rightarrow Co^{2+}\ (aq)$

Potentials provided are written in this format:

$Cu^{2+}\ (aq)\ |\ Cu\ (s)$

+0.34 V

It means that for the reduction reaction:

$Cu^{2+}\ (aq) \rightarrow Cu\ (s)$, the potential is +0.34 V

Reverse reaction or oxidation reaction is:

$Cu\ (s) \rightarrow Cu^{2+}\ (aq)$ has opposing potential value: –0.34 V

*continued...*

Obtain the potential values for each half-reaction.

Reverse sign for potential values of oxidation reactions:

$Cu^{2+}$ (*aq*) → Cu (*s*) = +0.34 V (reduction)

Co (*s*) → $Co^{2+}$ (*aq*) = +0.28 V (oxidation)

Determine standard cell potential:

Standard cell potential = sum of half-reaction potential

Standard cell potential = 0.34 V + 0.28 V

Standard cell potential = 0.62 V

**12. C is correct.**

Other methods listed involve two or more energy conversions between light and electricity.

It is an electrical device that converts light energy directly into electricity by the photovoltaic effect (i.e., chemical and physical processes).

The operation of a photovoltaic (PV) cell has the following requirements:

1) Light is absorbed, which excites electrons.

2) Separation of charge carries opposite types.

3) The separated charges are transferred to an external circuit.

In contrast, solar panels supply heat by absorbing sunlight.

A photoelectrolytic / photoelectrochemical cell is a photovoltaic cell or a device that splits water directly into hydrogen and oxygen using only solar illumination.

**13. A is correct.**

Each half-cell contains an electrode; two half-cells are required to complete a reaction.

**14. B is correct.**

A: electrolysis can be performed on any metal, not only iron.

C: electrolysis will not boil water nor raise the ship.

D: it is probably unlikely that the gases would stay in the compartments.

 

**15. C is correct.**

Balanced reaction:

$$Zn\ (s) + CuSO_4\ (aq) \rightarrow Cu\ (s) + ZnSO_4\ (aq)$$

Zn is a stronger reducing agent (more likely oxidized) than Cu. This can be determined by each element's standard electrode potential (E°).

**16. D is correct.**

Positive cell potential indicates that the reaction is spontaneous, favoring products.

Cells that generate electricity spontaneously are considered galvanic cells.

**17. B is correct.**

A joule is a unit of energy.

**18. C is correct.**

Electrolysis of aqueous sodium chloride yields hydrogen and chlorine, with aqueous sodium hydroxide remaining in the solution.

Sodium hydroxide is a strong base, which means the solution will be basic.

**19. C is correct.**

Balanced equation:

$$Ag^+ + e^- \rightarrow Ag\ (s)$$

Formula to calculate deposit mass:

mass of deposit = (atomic mass × current × time) / 96,500 C

mass of deposit = [107.86 g × 3.50 A × (12 min × 60 s/min)] / 96,500 C

mass of deposit = 2.82 g

**20. D is correct.**

Anode is the electrode where oxidation occurs.

Salt bridge provides electrical contact between the half-cells.

Cathode is the electrode where *reduction* occurs.

Spontaneous electrochemical cells are called *galvanic cells*.

---

**Practice Set 2: Questions 21–40**

---

**21. A is correct.**

Ionization energy (IE) is required to release one electron from an element.

Higher IE means the element is more stable.

Elements with high IE usually only need one or two more electrons to achieve stable configuration (e.g., complete valence shell – 8 electrons, such as the noble gases, or 2 electrons, as in hydrogen). These elements are more likely to gain an electron and reach stability than to lose an electron.

When an atom gains electrons, its oxidation number goes down and is reduced. Therefore, elements with high IE are easily reduced.

In oxidation-reduction reactions, species that undergo reduction are oxidizing agents because their presence allows the other reactant to be oxidized. Because elements with high IE are easily reduced, they are strong oxidizing agents.

Reducing agents are species that undergo oxidation (i.e., lose electrons).

Elements with high IE do not undergo oxidation readily and, therefore, are weak reducing agents.

**22. D is correct.**

Balancing a half-reaction in basic conditions:

First few steps are identical to balancing reactions in acidic conditions.

*Step 1: Balance atoms except for H and O*

$C_8H_{10} \rightarrow C_8H_4O_4^{2-}$ C is balanced

*Step 2: To balance oxygen, add $H_2O$ to the side with fewer oxygen atoms*

$C_8H_{10} + 4\ H_2O \rightarrow C_8H_4O_4^{2-}$

*Step 3: To balance hydrogen, add $H^+$ to the opposing side of $H_2O$ added in the previous step*

$C_8H_{10} + 4\ H_2O \rightarrow C_8H_4O_4^{2-} + 14\ H^+$

This next step is the unique additional step for basic conditions.

*continued...*

 

*Step 4: Add equal amounts of $OH^-$ on both sides. The number of $OH^-$ should match the number of $H^+$ ions. Combine $H^+$ and $OH^-$ on the same side to form $H_2O$. If there are $H_2O$ molecules on each side, subtract accordingly to end up with $H_2O$ on one side only.*

There are 14 $H^+$ ions on the right, so add 14 $OH^-$ ions on both sides:

$$C_8H_{10} + 4\ H_2O + 14\ OH^- \rightarrow C_8H_4O_4^{2-} + 14\ H^+ + 14\ OH^-$$

Combine $H^+$ and $OH^-$ ions to form $H_2O$:

$$C_8H_{10} + 4\ H_2O + 14\ OH^- \rightarrow C_8H_4O_4^{2-} + 14\ H_2O$$

$H_2O$ molecules are on both sides, which cancel, and some $H_2O$ remain on one side:

$$C_8H_{10} + 14\ OH^- \rightarrow C_8H_4O_4^{2-} + 10\ H_2O$$

*Step 5: Balance charges by adding electrons to the side with a greater/more positive total charge*

Total charge on left side: $14(-1) = -14$

Total charge on the right side: $-2$

Add 12 electrons to the right side:

$$C_8H_{10} + 14\ OH^- \rightarrow C_8H_4O_4^{2-} + 10\ H_2O + 12\ e^-$$

**23. A is correct.**

In electrochemical (i.e., galvanic) cells, *oxidation* occurs at *anode*, and *reduction* at *cathode.*

Therefore, $CO_2$ is produced at the anode because it is an oxidation product (carbon's oxidation number increases from 0 on the left to +2 on the right).

**24. A is correct.**

$Cu^{2+}$ is reduced (i.e., gains electrons), while $Sn^{2+}$ is oxidized (i.e., loses electrons).

Salt bridges contain cations (positive ions) and anions (negative ions).

Anions flow towards the oxidation half-cell because the oxidation product is positively charged, and the anions are required to balance the charges within the cell. The opposite is true for cations; they flow towards the reduction half-cell.

*continued...*

 

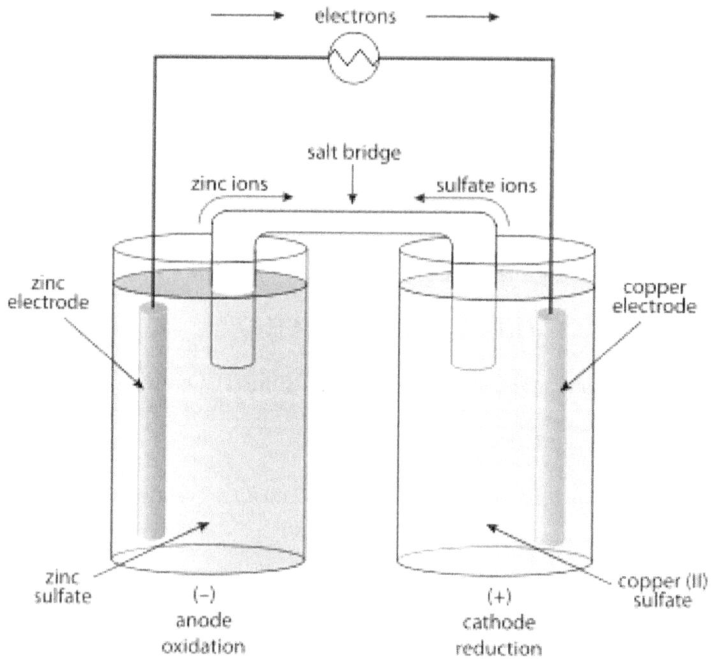


*A schematic example of a Zn–Cu galvanic cell*

**25. B is correct.**

In all cells, reduction occurs at the cathode, while oxidation occurs at the anode.

**26. C is correct.**

ZnO is being reduced; the oxidation number of Zn decreases from +2 on the left to 0 on right.

**27. D is correct.**

Consider the reactions:

$$Ni\ (s) + Ag^{+}\ (aq) \rightarrow Ag\ (s) + Ni^{2+}\ (aq)$$

Because it is spontaneous, Ni is more likely to be oxidized than Ag. Ni is oxidized, while Ag is reduced in this reaction.

Arrange the reactions so the metal that was oxidized in a reaction is reduced in the following reaction:

$$Ni\ (s) + Ag^{+}\ (aq) \rightarrow Ag\ (s) + Ni^{2+}\ (aq)$$

$$Cd\ (s) + Ni^{2+}\ (aq) \rightarrow Ni\ (s) + Cd^{2+}\ (aq)$$

$$Al\ (s) + Cd^{2+}\ (aq) \rightarrow Cd\ (s) + Al^{3+}\ (aq)$$

*continued...*

 

Lastly, in this reaction, Ag ($s$) + $H^+$ ($aq$) → no reaction should be first because Ag was not oxidized, so it should be before a reaction where Ag is reduced.

Ag ($s$) + $H^+$ ($aq$) → no reaction

Ni ($s$) + $Ag^+$ ($aq$) → Ag ($s$) + $Ni^{2+}$ ($aq$)

Cd ($s$) + $Ni^{2+}$ ($aq$) → Ni ($s$) + $Cd^{2+}$ ($aq$)

Al ($s$) + $Cd^{2+}$ ($aq$) → Cd ($s$) + $Al^{3+}$ ($aq$)

Metal being oxidized in the last reaction has the highest tendency to be oxidized.

**28. A is correct.**

The purpose of the salt bridge is to balance the charges between the two chambers/half-cells.

Oxidation creates cations at the anode, while reduction reduces cations at the cathode.

Ions in the bridge travel to those chambers to balance the charges.

**29. B is correct.**

Since G° = –nFE, when E° is positive, G is negative.

**30. C is correct.**

Electrons travel through the wires that connect the cells instead of the salt bridge. The function of the salt bridge is to provide ions to balance charges at the cathode and anode.

**31. A is correct.**

Electrochemistry is the branch of physical chemistry that studies chemical reactions at the interface of an ionic conductor (i.e., the electrolyte) and an electrode.

Electric charges move between the electrolyte and the electrode through redox reactions, and chemical energy is converted to electrical energy.

**32. B is correct.**

Light energy from the sun causes the electron to move towards the silicon wafer, starting the electric generation process.

**33. B is correct.**

Oxidation number (or *oxidation state*) indicates an atom's degree of oxidation (i.e., loss of electrons).

If an atom is electron-poor, it has lost electrons and would have a positive oxidation number.

If an atom is electron-rich, it has gained electrons and would have a negative oxidation number.

**34. B is correct.**

Reduction potential measures a substance's ability to acquire electrons (i.e., undergo reduction).

The more positive the reduction potential, the more likely the substance will be reduced. Reduction potential is generally measured in volts.

**35. D is correct.**

Batteries run down and must be recharged, while fuel cells do not run down because they can be refueled.

**36. A is correct.**

*Disproportionation reaction* is when a species undergoes oxidation and reduction in the same reaction.

The first step in balancing this reaction is to write the substance undergoing disproportionation twice on the reactant side.

**37. A is correct.**

The terms spontaneous *electrochemical*, *galvanic*, and *voltaic* are synonymous.

An example of a voltaic cell is an alkaline battery – it generates electricity spontaneously.

Electrons flow from the anode (oxidation half-cell) to the cathode (reduction half-cell).

**38. B is correct.**

Reaction at the anode:

$$2\ H_2O \rightarrow O_2 + 4\ H^+ + 4\ e^-$$

Oxygen gas is released, and $H^+$ ions are added to the solution, which causes the solution to become acidic (i.e., lowers the pH).

 

**39. D is correct.**

Nonspontaneous electrochemical cells are electrolysis cells.

In electrolysis, the metal being reduced is produced at the cathode.

**40. B is correct.**

Electrolysis is nonspontaneous because it needs an electric current from an external source to occur.

==================================================================

**Practice Set 3: Questions 41–60**

==================================================================

**41. C is correct.**

Electronegativity indicates the tendency of an atom to attract electrons.

An atom that attracts electrons strongly would be more likely to pull electrons from another atom. As a result, this atom would be a strong oxidizer because it would cause other atoms to be oxidized when the electrons are pulled toward the atom with high electronegativity.

A strong oxidizing agent is a weak reducing agent – they are opposing attributes.

**42. A is correct.**

$MnO_2$ is being reduced into $Mn_2O_3$.

Oxidation number increases from +4 on the left to +3 on the right.

**43. C is correct.**

Electrolytic cell: needs electrical energy input

Battery: spontaneously produces electrical energy. A dry cell is a type of battery.

Half-cell: does not generate energy by itself.

**44. B is correct.**

Anode in galvanic cells attracts anions.

Anions in solutions flow toward the anode, while cations flow toward the cathode.

Oxidation (i.e., loss of electrons) occurs at the anode.

Positive ions are formed, while negative ions are consumed at the anode.

Therefore, negative ions flow toward the anode to equalize the charge.

**45. B is correct.**

In electrochemical (i.e., galvanic) cells, oxidation occurs at the anode, and reduction occurs at the cathode.

Al metal is produced at the cathode because it is a reduction product (from an oxidation of +3 on the left to 0 on the right).

 

**46. B is correct.**

Anions in solutions flow toward the anode, while cations flow toward the cathode.

Oxidation is the loss of electrons. Sodium is a group I element and, therefore, has a single valence electron. During oxidation, the Na becomes $Na^+$ with a complete octet.

**47. A is correct.**

A salt bridge contains cations (positive ions) and anions (negative ions). Anions flow towards the oxidation half-cell because the oxidation product is positively charged, and the anions are required to balance the charges within the cell.

The opposite is true for cations; they flow towards the reduction half-cell. Anions in the salt bridge should flow from Cd to Zn half-cell.

**48. B is correct.**

This can be determined using the electrochemical series:

| **Equilibrium** | **E°** |
|---|---|
| $Li^+$ (*aq*) + $e^-$ ↔ Li (*s*) | –3.03 volts |
| $K^+$ (*aq*) + $e^-$ ↔ K (*s*) | –2.92 |
| *$Ca^{2+}$ (*aq*) + 2 $e^-$ ↔ Ca (*s*) | –2.87 |
| *$Na^+$ (*aq*) + $e^-$ ↔ Na (s) | –2.71 |
| $Mg^{2+}$ (*aq*) + 2 $e^-$ ↔ Mg (s) | –2.37 |
| $Al^{3+}$ (*aq*) + 3 $e^-$ ↔ Al (*s*) | –1.66 |
| $Zn^{2+}$ (*aq*) + 2 $e^-$ ↔ Zn (*s*) | –0.76 |
| $Fe^{2+}$ (*aq*) + 2 $e^-$ ↔ Fe (*s*) | –0.44 |
| $Pb^{2+}$ (*aq*) + 2 $e^-$ ↔ Pb (*s*) | –0.13 |
| 2 $H^+$ (*aq*) + 2 $e^-$ ↔ $H_2$ (*g*) | 0.0 |
| $Cu^{2+}$ (*aq*) + 2 $e^-$ ↔ Cu (*s*) | +0.34 |
| $Ag^+$ (*aq*) + $e^-$ ↔ Ag (*s*) | +0.80 |
| $Au^{3+}$ (*aq*) + 3 $e^-$ ↔ Au (*s*) | +1.50 |

For a substance to act as an oxidizing agent for another substance, the oxidizing agent must be located below the substance being oxidized in the series. Cu is being oxidized, which means only $Ag^+$ or $Au^{3+}$ can oxidize Cu.

**49. D is correct.**

Anions in solutions flow toward the anode, while cations flow toward the cathode.

Oxidation (i.e., loss of electrons) occurs at the anode.

Positive ions are formed while negative ions are consumed at the anode.

Therefore, negative ions flow toward the anode to equalize the charge.

**50. B is correct.**

Electrochemistry is the branch of physical chemistry that studies chemical reactions at the interface of an ionic conductor (i.e., the electrolyte) and an electrode.

Electric charges move between the electrolyte and the electrode through a series of redox reactions, and chemical energy is converted to electrical energy.

**51. A is correct.**

E° tends to be negative and G positive because electrolytic cells are nonspontaneous.

Electrons must be forced into the system for the reaction to proceed.

**52. C is correct.**

Ion must be oxidized to create chlorine gas ($Cl_2$) from chloride ion ($Cl^-$).

In electrolysis, oxidation occurs at the anode, which is the positive electrode.

The positively-charged electrode would absorb electrons from the ion and transfer them to the cathode for reduction.

**53. B is correct.**

In electrolytic cells:

Anode is a positively charged electrode where oxidation occurs.

Cathode is a negatively charged electrode where reduction occurs.

**54. C is correct.**

Nonspontaneous electrochemical cells are electrolysis cells.

In electrolysis, the metal that is being oxidized continuously dissolves at the anode.

 

**55. D is correct.**

Dry-cell batteries are the common disposable household batteries. They are based on zinc and manganese dioxide cells.

**56. C is correct.**

*Disproportionation* is a redox reaction in which a species is *simultaneously reduced and oxidized* to form two products.

Unbalanced reaction: $HNO_2 \rightarrow NO + HNO_3$

Balanced reaction: $3\ HNO_2 \rightarrow 2\ NO + HNO_3 + H_2O$

$2\ H_2O \rightarrow 2\ H_2 + O_2$: decomposition

$H_2SO_3 \rightarrow H_2O + SO_2$: decomposition

$Mg + H_2SO_4 \rightarrow MgSO_4 + H_2$: single replacement

**57. D is correct.**

Electrolysis is the same process in reverse for the chemical process inside a battery.

Electrolysis is often used to separate elements.

**58. A is correct.**

Fuel cell automobiles are fueled by hydrogen; the only emission is water (the statement above is the reverse of the true statement).

**59. A is correct.**

Impure copper is oxidized so that it would happen at the anode. Then, pure copper would plate out (i.e., be reduced) on the cathode.

**60. B is correct.**

Electrolysis is a chemical reaction resulting from electrical energy passing through a liquid electrolyte.

Electrolysis utilizes direct electric current (DC) to drive an otherwise non-spontaneous chemical reaction. The voltage needed for electrolysis is called the decomposition potential.

==========================================================================

**Practice Set 4: Questions 61–80**

==========================================================================

**61. D is correct.**

Calculate the oxidation numbers of species involved and identify the oxidized species (increased oxidation number).

Zn's oxidation number increases from 0 on the left side of reaction I to +2 on the right.

**62. C is correct.**

The lack of an aqueous solution makes it a dry cell. An example of a dry cell is an alkaline battery.

**63. B is correct.**

Half-reaction: $C_2H_6O \rightarrow HC_2H_3O_2$

Balancing half-reaction in acidic conditions:

*Step 1: Balance atoms except for H and O*

$C_2H_6O \rightarrow HC_2H_3O_2$ (C is balanced)

*Step 2: To balance oxygen, add $H_2O$ to the side with fewer oxygen atoms*

$C_2H_6O + H_2O \rightarrow HC_2H_3O_2$

*Step 3: To balance hydrogen, add $H^+$ to the opposing side of $H_2O$ added in the previous step*

$C_2H_6O + H_2O \rightarrow HC_2H_3O_2 + 4\ H^+$

*Step 4: Balance charges by adding electrons to the side with a greater/more positive total charge*

Total charge on the left side: 0

Total charge on right side: 4(+1) = +4

Add 4 electrons to the right side:

$C_2H_6O + H_2O \rightarrow HC_2H_3O_2 + 4\ H^+ + 4\ e^-$

**64. B is correct.**

In electrochemical (i.e., galvanic) cells, oxidation occurs at the anode and reduction at the cathode.

$CCl_4$ is produced at the anode because it is an oxidation product (carbon's oxidation number increases from 0 on the left to +4 on the right).

**65. C is correct.**

Battery: spontaneously produces electrical energy. A dry cell is a type of battery.

Half-cell: does not generate energy by itself

Electrolytic cell: needs electrical energy input

**66. D is correct.**

Anions in solutions flow toward the anode, while cations flow toward the cathode.

Oxidation (i.e., loss of electrons) occurs at the anode.

Positive ions are formed, while negative ions are consumed at the anode.

Therefore, negative ions flow toward the anode to equalize the charge.

**67. A is correct.**

A reducing agent is an oxidized reactant (i.e., loses electrons).

Therefore, the best-reducing agent is most easily oxidized.

Reversing each of the half-reactions shows that the oxidation of Cr (*s*) has a potential of +0.75 V, which is greater than the potential (+0.13 V) for the oxidation of $Sn^{2+}$ (*aq*).

Thus, Cr (*s*) is a stronger reducing agent.

**68. B is correct.**

Salt bridges contain cations (positive ions) and anions (negative ions).

Anions flow towards the oxidation half-cell because the oxidation product is positively charged, and the anions are required to balance the charges within the cell.

The opposite is true for cations; they flow toward the reduction half-cell.

Anions in the salt bridge should flow from Cd to Zn half-cell.

**69. B is correct.**

When zinc is added to HCl, the reaction is:

$Zn + HCl \rightarrow ZnCl_2 + H_2O$,

which means that Zn is oxidized into $Zn^{2+}$.

The number provided in the problem is reduction potential, so to obtain the oxidation potential, flip the reaction:

$Zn\ (s) \rightarrow Zn^{2+} + 2\ e^-$

$E°$ is inverted and becomes +0.76 V.

Because $E°$ is higher than hydrogen's value (which is set to 0), the reaction occurs.

**70. C is correct.**

Calculate the mass of metal deposited in cathode:

*Step 1: Calculate total charge using current and time*

$Q = \text{current} \times \text{time}$

$Q = 1\ A \times (10\ \text{minutes} \times 60\ \text{s/minute})$

$Q = 600\ A{\cdot}s = 600\ C$

*Step 2: Calculate moles of electrons that have the same amount of charge*

$\text{moles } e^- = Q / 96{,}500\ C/mol$

$\text{moles } e^- = 600\ C / 96{,}500\ C/mol$

$\text{moles } e^- = 6.22 \times 10^{-3}\ mol$

*Step 3: Calculate moles of metal deposit*

Half-reaction of zinc ion reduction:

$Zn^{2+}\ (aq) + 2\ e^- \rightarrow Zn\ (s)$

$\text{moles of Zn} = (\text{coefficient Zn} / \text{coefficient } e^-) \times \text{moles } e^-$

$\text{moles of Zn} = (½) \times 6.22 \times 10^{-3}\ mol$

$\text{moles of Zn} = 3.11 \times 10^{-3}\ mol$

*Step 4: Calculate mass of metal deposit*

$\text{mass Zn} = \text{moles Zn} \times \text{molecular mass of Zn}$

$\text{mass Zn} = 3.11 \times 10^{-3}\ mol \times (65\ g/mol)$

$\text{mass Zn} = 0.20\ g$

**71. A is correct.**

Oxidation number of Mg increases from 0 to +2, which means that Mg loses electrons (i.e., is oxidized).

Conversely, the oxidation number of Cu decreases, gaining electrons (i.e., is reduced).

**72. D is correct.**

Electrolysis uses a direct electric current to provide electricity to a nonspontaneous redox process to drive the reaction. The direct electric current must be passed through an ionic substance or solution that contains electrolytes.

Electrolysis is often used to separate elements.

**73. D is correct.**

Oxidation is the loss of electrons, while reduction is the gain of electrons.

**74. D is correct.**

E° for the cell is always positive.

**75. B is correct.**

Galvanic cells are spontaneous and generate electrical energy.

**76. D is correct.**

In electrolytic cells:

Anode is a positively charged electrode where oxidation occurs.

Cathode is a negatively charged electrode where reduction occurs.

**77. B is correct.**

Balanced equation:

$Cu^{2+} + 2\ e^- \rightarrow Cu\ (s)$

Formula to calculate deposit mass:

mass of deposit = atomic mass × moles of electron

*continued...*

Since Cu has 2 electrons per atom, multiply the moles by 2:

mass of deposit = atomic mass × (2 × moles of electron)

mass of deposit = atomic mass × (2 × current × time) / 96,500 C

4.00 g = 63.55 g × (2 × 2.50 A × time) / 96,500 C

time = 4,880 s

time = (4,880 s × 1 min/60s × 1 hr/60 min) = 1.36 hr

**78. C is correct.**

Alkaline and dry cells are standard disposable batteries and are non-rechargeable.

Fuel cells require fuel that is going to be consumed and are non-rechargeable.

**79. A is correct.**

Electrolytic cells do not occur spontaneously; the reaction only occurs with the addition of external electrical energy.

**80. D is correct.**

A redox reaction, or oxidation-reduction reaction, involves the transfer of electrons between two reacting substances.

An oxidation reaction refers explicitly to the substance losing electrons, and a reduction reaction refers to the substance gaining electrons.

Oxidation and reduction reactions alone are half-reactions because they form a whole reaction.

Therefore, half-reaction can represent a separate oxidation or reduction process.

# 11 – Atomic and Electronic Structure: Detailed Explanations

========================================================================

**Practice Set 1: Questions 1–20**

========================================================================

**1. A is correct.**

Though alpha particles have low penetrating power and high ionizing power, they are not harmless.

All forms of radiation have risks and cannot be thought of as entirely harmless.

**2. C is correct.**

A beta particle (β) is a high-energy, high-speed electron ($\beta^-$) or positron ($\beta^+$) emitted in the radioactive decay of an atomic nucleus.

Electron emission ($\beta^-$ decay) occurs in an unstable atomic nucleus with an excess of neutrons, whereby a neutron is converted into a proton, an electron, and an electron antineutrino.

Positron emission ($\beta^+$ decay) occurs in an unstable atomic nucleus with an excess of protons, whereby a proton is converted into a neutron, a positron, and an electron neutrino.

**3. D is correct.**

The de Broglie wavelength is given as:

$$\lambda = h / p$$

where $h$ is Planck's constant and $p$ is momentum

$$p = mv$$

$$\lambda_1 = h / mv$$

$$\lambda_2 = h / m(2v)$$

$$\lambda_2 = \tfrac{1}{2}h / mv$$

$$\lambda_2 = \tfrac{1}{2} \lambda_1$$

λ decreases by factor of 2

 

**4. B is correct.**

The Bohr model places electrons around the nucleus of the atom at discrete energy levels.

The Balmer series line spectra agreed with the Bohr model because the energy of the observed photons in each spectrum matched the transition energy of electrons within these discrete predicted states.

**5. C is correct.**

A radioactive element is an element that spontaneously emits radiation in the form of one or a combination of the following: alpha radiation, beta radiation, gamma radiation.

**6. D is correct.**

This is an example of an electron capture nuclear reaction.

When this happens, the atomic number decreases by one, but the mass number stays the same.

$$^{100}_{44}\text{Ru} + {}^{0}_{-1}\text{e}^- \rightarrow {}^{100}_{43}\text{Tc}$$

Ru: 100 = mass number (# protons + # neutrons)

Ru: 44 = atomic number (# protons)

From the periodic table, Tc is the element with 1 less proton than Ru.

**7. B is correct.**

A *nucleon* is a particle that makes up the nucleus of an atom.

The two known nucleons are *protons* and *neutrons*.

**8. D is correct.**

An *alpha particle* is composed of two neutrons and two protons and is identical to the nucleus of a $^4$He atom.

Total mass of two alpha particles:

$2 \times (2 \text{ neutrons} + 2 \text{ protons}) = 8$

Mass of a $^9$Be atom:

5 neutrons + 4 protons = 9

Mass of a $^9$Be atom > total mass of two alpha particles

The mass of a $^9$Be atom is greater than the mass of two alpha particles, so its mass is greater than twice the mass of a $^4$He atom.

**9. C is correct.**

The superscript is the mass number (atomic weight), which is both neutrons and protons.

The subscript is the atomic number, which is the number of protons.

Therefore, the number of neutrons is equal to the superscript minus the subscript.

181 – 86 = 95, which is the greatest number of neutrons among the choices.

**10. A is correct.**

The number of neutrons and protons must be equal after the reaction.

Thus, the sum of the atomic number before and mass number before should be equal after the reaction.

Mass number (superscript):

$$(1 + 235) - (131 + 3) = 102$$

*continued…*

Atomic number (subscript):

$$92 - (53) = 39$$

${}^{102}_{39}Y$ balances the reaction.

**11. B is correct.**

Balmer equation is given by:

$$\lambda = B[(n^2) / (n^2 - 2^2)]$$

$$\lambda = (3.6 \times 10^{-7})\cdot[(12^2) / (12^2 - 2^2)]$$

$$\lambda = 3.7 \times 10^{-7}\ m$$

where $c$ is the speed of light:

$$c = \lambda f$$

Convert wavelength to frequency:

$$f = c / \lambda$$

$$f = (3 \times 10^8\ m/s) / (3.7 \times 10^{-7}\ m)$$

$$f = 8.1 \times 10^{14}\ s^{-1} = 8.1 \times 10^{14}\ Hz$$

**12. D is correct.**

*Gamma-rays* are the most penetrating form of radiation because they are the highest energy and least ionizing.

A gamma-ray passes through a given amount of material without imparting as much of its energy into removing electrons from atoms and ionizing them as other forms of radiation do.

Gamma rays retain more of their energy passing through matter and can penetrate further.

**13. A is correct.**

Use the *Rydberg Formula*:

$$E = hf$$

$$f = c / \lambda$$

$$E = (hc)\cdot(1 / \lambda)$$

$1 / \lambda = R(1 / n_1^2 - 1 / n_2^2)$, where $n_1 = 1$ and $n_2 = 2$

$$E = hcR[(1 / n_1^2) - (1 / n_2^2)]$$

$$E = (4.14 \times 10^{-15}\ \text{eV·s})\cdot(3 \times 10^8\ \text{m/s})\cdot(1.097 \times 10^7\ \text{m}^{-1})\cdot[(1 / 1^2) - (1 / 2^2)]$$

$$E = 13.6[1 - (1 / 4)]$$

$$E = 10.2\ \text{eV}$$

The positive energy indicates that a photon was absorbed and not emitted.

**14. C is correct.**

In $\beta^-$ (beta minus) decay, the atomic number (subscript) increases by 1, but the atomic mass stays constant.

$${}^{87}_{37}\text{Rb} \rightarrow {}^{87}_{38}\text{Sr} + {}^{0}_{-1}\text{e} + {}^{0}_{0}\nu$$

Sr is the element with 1 more proton (subscript) than Rb. ${}^{0}_{0}\nu$ represents an electron antineutrino.

**15. C is correct.**

The *nucleus* of an atom is bound by the strong nuclear force from the nucleons within it.

The strong nuclear force must overcome the Coulomb repulsion of the protons (due to their like charges).

Neutrons help stabilize and bind the nucleus by contributing to the strong nuclear force so that it is greater than the Coulomb repulsion experienced by the protons.

 

**16. A is correct.**

*Geiger-Muller counters* operate using a Geiger-Muller tube, consisting of a high voltage shell and small rod in the center, filled with a low-pressure inert gas (e.g., argon).

When exposed to radiation (specifically particle radiation), the radiation particles ionize atoms of the argon allowing for a brief charge to be conducted between the high voltage rod and outer shell.

The electric pulse is then displayed visually or via audio to indicate radioactivity.

**17. B is correct.**

The *half-life* calculation:

$$A = A_0(½)^{t/h}$$

where $A_0$ = original amount, t = time elapse and $h$ = half-life

If three half-life pass, $t = 3h$

$$A = (1)·(½)^{3h/h}$$

$$A = (1)·(½)^3$$

$$A = 0.125 = 12.5\%$$

**18. D is correct.**

In the Lyman series, electron transitions always go from $n \geq 2$ to $n = 1$.

**19. B is correct.**

Most of the volume of an atom is occupied by empty space.

**20. A is correct.**

Alpha particles are positively charged due to their protons, while beta minus particles are negatively charged (i.e., electrons).

When exposed to a magnetic field, alpha particles and electrons deflect in opposite directions due to their opposite charges and thus experience opposite forces due to the magnetic field.

## Practice Set 2: Questions 21–40

**21. D is correct.**

Chemical reactions store and release energy in their chemical bonds but do not convert mass into energy.

However, nuclear reactions convert a small amount of mass into energy, which can be measured and calculated via the equation:

$$E = \Delta mc^2$$

**22. C is correct.**

The Curie is a non-SI unit of radioactivity equivalent to $3.7 \times 10^{10}$ decays (disintegrations) per second. It is named after the early radioactivity researchers Marie and Pierre Curie.

**23. B is correct.**

The *atomic numbers*: ${}^{235}_{92}\text{U} \rightarrow {}^{141}_{56}\text{Ba} + {}^{92}_{36}\text{Kr}$

The subscripts on each side of the expression sum to 92, so adding a proton (${}^{1}_{1}\text{H}$) to the right side would not balance.

The superscripts sum to 235 on the left and sum to 233 on the right.

Add two neutrons (${}^{1}_{0}\text{n} + {}^{1}_{0}\text{n}$) to the right side to balance both sides of the equation.

**24. D is correct.**

The *atomic number* is the subscript and represents the number of protons, and the superscript is the mass number and is the sum of protons and neutrons.

The number of neutrons can be found by taking the difference between the mass number and atomic number.

In this example, there are 16 protons and 18 neutrons.

 

**25. A is correct.**

The *half-life* calculation: $A = A_0(½)^{t/h}$

where $A_0$ = original amount, t = time elapse and $h$ = half-life

Consider: (2 days)·(24 hours / 1 day) = 48 hours

$$A = (1)\cdot(½)^{(48/12)}$$

$$A = 0.5^4$$

$$A = 0.0625 = 1/16$$

**26. C is correct.**

An *alpha particle* consists of two protons and two neutrons and is identical to a helium nucleus so that it can be written as $^{4}_{2}He$

For a nuclear reaction to be written correctly, it must be balanced, and the sum of superscripts and subscripts must be equal on both sides of the reaction.

The superscripts add to 238, and the subscripts add to 92 on both sides. Therefore, it is the only balanced answer.

**27. D is correct.**

The question is asking for the λ of the emitted photon so use the Rydberg Formula:

$$1 / \lambda = R(1 / n_1^2 - 1 / n_2^2)$$

$$\lambda = 1 / [R(1 / n_1^2 - 1 / n_2^2)]$$

Use $n_1 = 5$ and $n_2 = 20$ because we are solving for λ of an emitted (not absorbed) photon.

$$\lambda = 1 / [(1.097 \times 10^7\ m^{-1})\cdot(1 / 5^2 - 1 / 20^2)]$$

$$\lambda = 2.43\ \mu m$$

**28. D is correct.**

The superscript represents the mass number when writing a nuclear reaction, while the subscript represents the atomic number.

A nuclear reaction is balanced when the sum of superscripts (mass number) and subscripts (atomic number) is equal on both sides of the reaction.

**29. C is correct.**

A *blackbody* is an ideal system that absorbs 100% of all light incident upon it and reflects none, and it also emits 100% of the radiation it generates.

Therefore, it has perfect absorption and emissivity.

**30. D is correct.**

*Carbon dating* relies upon a steady creation of $^{14}C$ and knowledge of the rate of creation at various points in time to determine the approximate age of objects.

If a future archeologist is unaware of nuclear bomb testing and the higher levels of $^{14}C$ created, then the dates they calculate for an object would be too young.

This is because a higher amount of $^{14}C$ would be present in samples and make them seem as if they had not had time to decay and thus appear younger.

**31. A is correct.**

*Gamma radiation* is an electromagnetic wave and is not a particle.

Thus, when gamma radiation is emitted, the atomic number and mass number remain the same.

**32. C is correct.**

In $\beta^-$ decay, a neutron is converted to a proton, and an electron and electron antineutrino are emitted.

In $\beta^+$ decay, a proton is converted to a neutron, and a positron and an electron neutrino are emitted.

$$^{14}_{6}C \rightarrow {}^{14}_{7}N + e^+ + \nu_e$$

**33. D is correct.**

In a balanced nuclear equation, the number of nucleons must be conserved.

The sum of mass numbers and atomic numbers must be equal on both sides of the equation.

The product (i.e., daughter nuclei) should be on the right side of the equation.

**34. B is correct.**

The *atomic number* is the number of protons within the nucleus that characterizes the element's nuclear and chemical properties.

**35. C is correct.**

The *Balmer series* is the name of the emission spectrum of hydrogen when electrons transition from a higher state to the n = 2 state.

Within the Balmer series, there are four visible spectral lines with colors ranging from red to violet (i.e., ROY G BIV)

**36. D is correct.**

Electrons were discovered through early experiments with electricity, specifically in high voltage vacuum tubes (cathode ray tubes).

Beams of electrons were observed traveling through these tubes when high voltage was applied between the anode and cathode, and the electrons struck fluorescent material at the back of the tube.

**37. B is correct.**

Beams a and c deflect when an electric field is applied, indicating they have a net charge and therefore must be particles.

Beam b is undisturbed by the applied electric field, indicating it has no net charge and must be a high-energy electromagnetic wave since all other forms of radioactivity (alpha and beta radiation) are charged particles.

**38. C is correct.**

Beam a is composed of negatively charged particles, while beam c is composed of positively charged particles; therefore, both beams are deflected by the electric field.

Beam b is also composed of particles; however, they are neutral because the electric field does not deflect them.

An example of this kind of radiation would be a gamma-ray, which consists of neutral photons.

**39. C is correct.**

A *helium nucleus* is positively charged, deflected away from the top plate, and attracted toward the negative plate.

 

**40. B is correct.**

5.37 eV is the energy required to excite the electron from the ground state to the zero-energy state.

Calculate the *wavelength of a photon* with this energy:

$E = hf$

$f = c / \lambda$

$E = hc / \lambda$

$\lambda = hc / E$

$\lambda = (4.14 \times 10^{-15}\ \text{eV·s})\cdot(3 \times 10^{8}\ \text{m/s}) / (5.37\ \text{eV})$

$\lambda = 2.3 \times 10^{-7}\ \text{m}$

---

**Practice Set 3: Questions 41–60**

---

**41. D is correct.**

Elements with atomic numbers 84 and higher are radioactive because the strong nuclear force binding the nucleus cannot overcome the Coulomb repulsion from the high number of protons within the atom.

Thus, these nuclei are unstable and emit alpha radiation to decrease the number of protons within the nucleus.

**42. D is correct.**

Beta particles, like all forms of ionizing radiation, cannot be considered harmless.

**43. C is correct.**

*Gamma rays* are the highest-energy electromagnetic radiation and require extremely dense materials to shield effectively.

Several centimeters of lead is required to reduce gamma rays considerably.

**44. B is correct.**

*Planck's constant* quantizes the amount of energy that can be absorbed or emitted.

Therefore, it sets a discrete lowest amount of energy for energy transfer.

**45. D is correct.**

The decay rate of any radioactive isotope or element is constant and independent of temperature, pressure, or surface area.

**46. B is correct.**

Larger nuclei (atomic number above 83) tend to decay because the attractive force of the nucleons (strong nuclear force) has a limited range, and the nucleus is larger than this range. Therefore, these nuclei tend to emit alpha particles to decrease the size of the nucleus.

Smaller nuclei are not large enough to encounter this problem, but some isotopes have an irregular ratio of neutrons to protons and become unstable.

$^{14}$Carbon has 8 neutrons and 6 protons, and its neutron to proton ratio is too large. Therefore, it is unstable and radioactive.

**47. C is correct.**

Positron emission occurs during $\beta^+$ decay.

In $\beta^+$ decay, a proton converts to a neutron and emits a positron and electron neutrino.

The decay can be expressed as:

$$^{44}_{21}\text{Sc} \rightarrow {}^{44}_{20}\text{Ca} + e^+ + \nu_e$$

**48. C is correct.**

A *scintillation counter* operates by detecting light flashes from the scintillator material.

When radiation strikes the scintillator crystal (often NaI), a light flash is emitted and detected by a photomultiplier tube, which then passes an electronic signal to audio or visual identification equipment.

**49. B is correct.**

When a Geiger counter clicks, it indicates that it has detected the radiation from one nucleus decaying.

The click could be from an alpha, beta, or even gamma-ray source but cannot be determined without other information.

**50. A is correct.**

*Pauli Exclusion Principle* states that no two electrons can have the same quantum numbers in an atom.

Thus, every electron in an atom has a unique set of quantum numbers, and a particular set belongs to only one electron.

**51. D is correct.**

The atomic number indicates the number of protons within the nucleus of an atom.

**52. C is correct.**

*Gamma rays* are high-energy electromagnetic radiation rays and have no charge or mass.

Due to their high energy and speed, they have a high penetrating power.

**53. B is correct.**

Calculate *mass defect*:

$m_1$ = (2 protons)·(1.0072764669 amu) + (2 neutrons)·(1.0086649156 amu)

$m_1$ = 4.031882765 amu

$\Delta m$ = 4.031882765 amu – 4.002602 amu

$\Delta m$ = 0.029280765 amu

Convert to kg:

$\Delta m$ = (0.029280765 amu / 1)·(1.6606 × $10^{-27}$ kg / 1 amu)

$\Delta m$ = 4.86236 × $10^{-29}$ kg

Find *binding energy*:

$E = \Delta mc^2$

$E$ = (4.86236 × $10^{-29}$ kg)·(3 × $10^8$ m/s)$^2$

$E$ = 4.38 × $10^{-12}$ J ≈ 4.4 × $10^{-12}$ J

**54. D is correct.**

Boron has an atomic number of five and thus has five protons and five electrons.

The 1*s* orbital is filled with two electrons; then, the 2*s* orbital is filled with two electrons.

Only one electron remains, and it is in the 2*p* orbital, leaving it partially filled. The electron configuration of boron: $1s^22s^22p$

**55. C is correct.**

Einstein's theory of special relativity that expresses the equivalence of mass and energy.

$E = mc^2$

$m = E / c^2$

$m$ = (3.85 × $10^{26}$ J) / (3 × $10^8$ m/s)$^2$

$m$ = 4.3 × $10^9$ kg

**56. D is correct.**

When any form of ionizing radiation interacts with the body, the high-energy particles or electromagnetic waves cause atoms within the tissue to ionize. This process creates unstable ions and free radicals that are damaging and pose serious health risks.

**57. B is correct.**

When a gamma-ray is emitted, the atom must lose energy due to the conservation of energy. Thus, the atom has less energy than before.

**58. A is correct.**

The lower-case symbol "n" signifies a neutron.

An *atomic number* of zero indicates that there are no protons.

A *mass number* of one indicates that it contains a neutron.

**59. C is correct.**

The *uncertainty principle* states that the position and momentum of a particle cannot be simultaneously measured over a set precision. Additionally, energy and time cannot be simultaneously known over a set precision.

Mathematically this is stated as:

$$\Delta x \Delta p > \hbar / 2$$

$$\Delta E \Delta t > \hbar / 2$$

where $x$ = position, $p$ = momentum, $\hbar$ = reduced Planck's constant, $E$ = energy, $t$ = time

**60. D is correct.**

$$A = A_0(½)^{t / h}$$

where $A_o$ = original amount, $t$ = time elapse and $h$ = half-life

$$0.03 = (1)\cdot(½)^{(t / 20{,}000 \text{ years})}$$

$$\ln (0.03) = (t / 20{,}000 \text{ years})\cdot\ln (½)$$

$$t = 101{,}179 \text{ years}$$

============================================================

**Practice Set 4: Questions 61–80**

============================================================

**61. B is correct.**

A *positron* is the antiparticle of an electron; the same mass, but an opposite charge of +1.

**62. C is correct.**

*Beta radiation* is characterized as $\beta^-$ or $\beta^+$. In $\beta^-$ decay, an electron is released, while in $\beta^+$ decay, a positron is released.

Beta radiation is the only type of radiation that emits a negatively charged particle (electron) deflects towards the positive electrode.

**63. D is correct.**

The *principal quantum number* "n" describes the size of the orbital, and large values of n indicate a larger orbital shell size.

Consequently, the radial distance between the nucleus and the outer bounds of the orbital shell will increase with increasing values of n.

**64. A is correct.**

The *electron* was discovered in 1897 by J. J. Thompson.

The *proton* was discovered c. 1920 by Ernest Rutherford.

The *neutron* was discovered in 1932 by James Chadwick.

**65. C is correct.**

*Alpha particles* consist of 2 protons and 2 neutrons, hence the +2 charge and mass of 4 amu.

They have a high mass; thus, they have a low speed and a low penetrating power.

**66. A is correct.**

$^{56}Fe$ has the highest binding energy because its nucleus is "at the midpoint" of nuclear size.

Thus, the strong nuclear force and the electromagnetic repulsion are most balanced, and the nucleus is at the lowest energy configuration.

Note: all nuclei are most stable at the lowest energy configuration.

**67. B is correct.**

*Nuclear fusion* is the process whereby lighter nuclei fuse to form a heavier element with a greater atomic number, equal to the sum of the atomic numbers of the combining nuclei.

**68. D is correct.**

The reactants are $^{15}_{7}\text{N} + ^{1}_{1}\text{p}$

The superscripts sum to 16 while the subscripts sum to 8.

$^{12}\text{C} + ^{4}\text{He}$ is the only possible set of products because the superscript and subscript add to 16 and 8, respectively.

$$^{15}_{7}\text{N} + ^{1}_{1}\text{p} \rightarrow \ ^{12}_{6}\text{C} + ^{4}_{2}\text{He}$$

The products $^{14}\text{B} + ^{2}\text{Li}$ sum to the correct superscript and subscript, but lithium has three protons, so the representation $^{2}\text{Li}$ is not possible.

**69. B is correct.**

Smoke detectors use a small amount of $^{241}$americium to detect the smoke.

**70. C is correct.**

During $\beta^-$ decay, a neutron is converted into a proton, increasing the atomic number (# protons) by 1, while the atomic mass (sum of protons and neutrons) remains constant.

**71. D is correct.**

Because nuclide I has a half-life of about a day, only a small fraction of nuclide I will be present after a few days, and its emitted radioactivity will be greatly reduced.

Nuclide II has a half-life of about a week.

After a few days, some will have decayed, but a larger fraction (greater than 50%) will be present, so its radioactivity is not considerably reduced.

Thus, nuclide II contributes most to the radioactivity of the sample after a few days.

**72. A is correct.**

The *photoelectric effect* is when increasing the intensity of light upon a metal increases the electron ejection rate but does not increase their kinetic energy.

**73. D is correct.**

The *d* shell has 5 orbitals with two electrons per orbital for a total of 10 electrons.

**74. B is correct.**

All elements with atomic numbers greater than 83 will be radioactive.

*Bismuth* has an atomic number of 83, so all successive elements after bismuth are radioactive.

**75. A is correct.**

A $\gamma$ ray has 0 protons and 0 neutrons. It is an electromagnetic wave.

**76. D is correct.**

*Balmer equation*:

$$\lambda = B(n^2 / n^2 - 2^2)$$

The wavelength is shorter when n becomes larger, such that the term in parentheses approaches 1.

Thus, when $n = \infty$ (the term in parentheses) is minimized (goes to 1), the shortest $\lambda$ is produced.

$$\lambda = (3.645 \times 10^{-7}\ \text{m})\cdot[\infty^2 / (\infty^2 - 2^2)]$$

$$\lambda = 3.645 \times 10^{-7}\ \text{m}$$

**77. C is correct.**

The sum of protons and neutrons in the nucleus is the *mass number*.

**78. B is correct.**

*Beta decay* is more powerful than alpha decay but less powerful than gamma rays.

*Alpha decay* is the weakest form of radiation and can be blocked by a sheet of paper.

Beta-decay is more powerful and can penetrate the skin, paper, or even a light layer of clothing, such as a T-shirt. Thicker materials, such as leather, are needed to provide beta radiation shielding.

*Gamma radiation* is the most powerful and requires thick lead barriers or immersion in water to shield effectively.

**79. D is correct.**

Alpha particles are composed of two protons and two neutrons, which is identical to a helium nucleus ($^{4}_{2}He$).

The mass number is the combined number of protons and neutrons; thus, the mass number of an alpha particle is 4 (2 protons + 2 neutrons = 4 mass number).

**80. C is correct.**

$^{0}_{-1}e^{-}$ is the notation for an electron identical to a beta particle from $\beta^{-}$ decay.

---

**Practice Set 5: Questions 81–100**

---

**81. C is correct.**

The reactants are ${}^{3}_{2}\text{He} + {}^{3}_{2}\text{He}$, so the superscripts must sum to 6 while the subscripts must sum to 4.

**82. D is correct.**

The Sievert is the SI unit for a dose of ionizing radiation and is equal to 100 rems.

**83. A is correct.**

Positron emission only occurs in $\beta^+$ decay and does not occur in $\alpha$ decay.

**84. D is correct.**

Heavier elements contain more neutrons than protons to increase the strength of the nuclear strong force binding the nucleus together.

The increase in nuclear strong force magnitude is necessary because heavier elements contain more protons, and the Coulomb repulsion force within the nucleus is higher within these elements.

**85. A is correct.**

In classical wave theory, the three main predictions for the photoelectric effect are:

1) Intensity of light should be proportional to the KE of the photoelectrons (i.e., the photocurrent).

2) Photoelectric effect should occur for any light, regardless of frequency.

3) There should be a delay between radiation contact and the initial release of electrons.

**86. B is correct.**

*Half-life* formula:

$$A = A_0(\tfrac{1}{2})^{t/h}$$

where $A_0$ = original amount, t = time elapse, $h$ = half-life

$$A = (10\text{ g})\cdot(\tfrac{1}{2})^{(6\text{ days} / 2\text{ days})}$$

$$A = 1.25\text{ grams}$$

**87. D is correct.**

*Nuclear fusion* is the process whereby two lighter elements fuse to form a heavier element.

Nuclear fusion produces non-radioactive elements, releases a larger amount of energy (as heat), and is the energy source of stars (including the Sun).

**88. C is correct.**

*Alpha radiation* consists of alpha particles that have two neutrons and two protons.

Alpha particles are the most massive form of radiation and have the highest charge of +2 thus; they possess the least penetrating ability.

**89. A is correct.**

The *Bohr model* separates electrons into discrete energy levels, with the levels (shells) increasing distance from the nucleus.

However, the energy difference between the adjacent shells decreases as they get further from the nucleus.

**90. C is correct.**

The rem is the notation for the Roentgen equivalent in humans and measures the biological damage of ionizing radiation. One rem = .01 Sieverts.

**91. B is correct.**

An X-ray tube produces X-rays by accelerating electrons through a vacuum tube using a potential difference.

If the voltage (potential difference) is doubled, the electric potential energy is doubled:

$PE_{electric} = qV$

$2PE_{electric} = q(2V)$

The electric potential energy is equal to the kinetic energy given to the electrons, which is equal to the energy of the X-rays produced:

$PE_{electric} = KE_{electrons} = E_{X\text{-}ray}$

$2PE_{electric} = 2KE_{electrons} = 2E_{X\text{-}ray}$

$E = hf$

$f = c / \lambda$

*continued…*

 

$$E = h(c / \lambda)$$

The energy of an electromagnetic wave is inversely proportional to the wavelength, so by doubling the energy, the X-ray wavelength will be reduced by half.

**92. C is correct.**

*Gamma rays* are electromagnetic radiation; thus, they are not composed of particles.

*Alpha* and *beta decay* change the atomic number through particle ejection or neutron/proton conversion.

**93. A is correct.**

In alpha decay, the nucleus loses 2 protons and 2 neutrons in the form of an alpha particle, which is identical to a helium nucleus ($^{4}_{2}He$):

$$^{235}_{92}U \rightarrow {}^{4}_{2}He + {}^{231}_{90}Th$$

**94. B is correct.**

$^{36}_{17}Cl$ has 19 neutrons and 17 protons; thus, it has an excess of neutrons. It will probably not undergo $\beta^+$ decay because this would convert a proton to a neutron and increase the neutron-to-proton ratio.

Because it is a smaller nucleus (an atomic number less than 83), it will probably not undergo alpha decay because this form of decay is most often found in larger nuclei that exceed the bounds of the strong nuclear force. It will likely undergo $\beta^-$ decay because this converts an excess neutron to a proton and decreases the neutron-to-proton ratio.

**95. D is correct.**

A positron is the antiparticle of an electron and has the same mass but a charge of +1e.

**96. D is correct.**

*Radioactive elements* (or *isotopes*) decay at a constant rate known as the *half-life*, the time needed for half the original sample to decay. All decay is spontaneous and forms a stable element or isotope.

Bismuth has the atomic number 83, and elements or isotopes with an atomic number higher than 83 are radioactive. Thus, all elements heavier than bismuth are radioactive.

Radioactivity is part of the natural environment in cosmic rays and trace quantities of radioactive elements or isotopes.

**97. D is correct.**

Atomic mass is specified by the sum of the protons and neutrons within the nucleus of an atom.

*Isotopes* have more or fewer neutrons than the characteristic element; thus, they have different masses.

**98. D is correct.**

All statements are correct.

*Nucleons* are protons and neutrons, and their mass is different outside the nucleus versus within it. Nucleons in the nucleus change mass slightly due to some mass being converted to bond energy.

When nuclei are broken apart, the energy released is from the mass of the nucleons and is converted into bond energy.

**99. C is correct.**

$$E = mc^2$$

Calculate the mass needed to produce $10^{12}$ J of energy:

$$10^{12}\ \text{J} = m(3 \times 10^8\ \text{m/s})^2$$

$$m = 10^{-5}\ \text{kg}$$

$$m = 10^{-2}\ \text{g}$$

**100. D is correct.**

In the *photoelectric effect*, the frequency of the incident photon must be greater than a specified minimum value to eject an electron.

*Frequency* is directly proportional to photon energy:

$$E = hf$$

Thus, a minimum frequency corresponds to the minimum energy needed for the metal to eject an electron.

Wavelength is inversely proportional to photon energy:

$$E = hc / \lambda$$

The photon wavelength must be less than a minimum value for the minimum energy to be attained.

---

**Practice Set 6: Questions 101–120**

---

**101. A is correct.**

A *beta particle* is an electron ejected from the nucleus with a charge of –1e and an atomic mass of 0 amu.

Beta particles have medium penetrating power, and they are more penetrating than alpha particles but less penetrating than gamma rays.

**102. B is correct.**

*Control rods* in a nuclear reactor are composed of neutron-absorbing material and regulate the flux of neutrons within the reactor, therefore regulating the fission chain reaction.

**103. D is correct.**

*Alpha particles* consist of 2 neutrons and 2 protons and are identical to a $^4$He nucleus.

**104. A is correct.**

The *spin quantum number* –½ or +½ describes the angular momentum of an electron.

Each orbital of an atom can hold two electrons with an opposite spin of +/–½.

**105. C is correct.**

Ernest Rutherford discovered the nucleus when he was using a radioactive alpha emitter to shoot alpha particles at gold foil to measure the angle of deflection.

**106. B is correct.**

The number of nucleons in a nuclear reaction is always conserved, so II must be incorrect.

**107. D is correct.**

Beta-decay occurs as either $\beta^+$ (positron emission) or $\beta^-$ (electron emission).

$\beta^+$ decay: $\quad {}^{A}_{Z}X \rightarrow {}^{A}_{Z-1}Y + {}^{0}_{+1}e + {}^{0}_{0}\nu_e$

$\beta^-$ decay: $\quad {}^{A}_{Z}X \rightarrow {}^{A}_{Z+1}Y + {}^{0}_{-1}e + {}^{0}_{0}\nu_e$

Thus, in beta decay, the nucleus gains or loses a proton.

**108. B is correct.**

*Positrons* are the antiparticles to electrons with a positive electric charge of + e.

Positron Symbol: ${}^{0}_{+1}e$ or $e^+$

Electron Symbol: ${}^{0}_{-1}e$ or $e^-$

**109. D is correct.**

An alpha particle consists of two protons and two neutrons.

Missing an alpha particle results in the atomic number decreasing by 2 and the mass number decreases by 4.

**110. C is correct.**

The half-life equation can be used to determine radioactivity after a set amount of time has elapsed.

$$A = A_0(½)^{t/h}$$

where $A_o$ = original amount, $t$ = time elapse and $h$ = half-life

$$25 \text{ mCi} = (400 \text{ mCi})\cdot(½)^{(t / 14.3 \text{ days})}$$

$$\ln (0.0625) = (t / 14.3 \text{ days})\cdot\ln (½)$$

$$t = 57.2 \text{ days}$$

**111. B is correct.**

Beta particles are high-energy, high-speed electrons or positrons.

They can be written as:

$\beta^-$ decay (electron emission): ${}^{0}_{-1}e$ or $e^-$

$\beta^+$ decay (positron emission): ${}^{0}_{+1}e$ or $e^+$

**112. A is correct.**

A neon discharge works by applying a voltage to a tube of neon gas such that the electrons gain energy and are promoted from their ground state to an excited state (higher energy orbital).

As an electron goes back to its ground state, it releases the energy it gained in the form of a photon. This photon has a characteristic wavelength corresponding to the elemental gas within the tube; for neon, this wavelength corresponds to a red color.

 

**113. C is correct.**

Increasing the brightness (intensity) increases the number of photons emitted by the light source but does not change the energy or speed of the emitted photons.

Photon energy is dependent upon frequency, and photon speed is always constant within a specific medium.

**114. B is correct.**

Electron capture occurs when the nucleus absorbs an electron and emits a neutrino.

Additionally, in this process, a proton is converted to a neutron.

**115. D is correct.**

The *mass number* of an atom is the sum of protons and neutrons within the nucleus and is, therefore, the sum of the nucleons within the atom.

**116. C is correct.**

Similar chemical properties are due to similar valence electron configurations.

Elements are grouped in vertical columns to display these similarities; thus, the element below neon has similar chemical properties with the next larger atomic number.

Argon has an atomic number of 18, corresponding to 18 protons and 18 electrons.

**117. A is correct.**

Isotopes of iron and nickel ("the iron group") have the maximum binding energy per nucleon of the nucleus.

This group has the maximum binding energy per nucleon because the number of nucleons within these elements maximizes the strong nuclear force without maximizing Coulomb repulsion within the nucleus.

Lighter elements do not have as many nucleons and therefore have less energy from the strong nuclear force.

In comparison, heavier elements have enough protons that Coulomb repulsion lowers their binding energy.

**118. A is correct.**

The intensity of electromagnetic radiation with respect to distance from the point source:

$$I = S / 4\pi r^2$$

where $I$ = intensity, $S$ = point source strength, and $r$ = radial distance from the point source.

The intensity of the radiation from the point source is inversely proportional to the square of the distance from the point source.

**119. C is correct.**

Half-life formula:

$$A = A_0(½)^{t\ /\ h}$$

where $A_0$ = original amount, t = time elapse and $h$ = half-life

$$A = (1)·(½)^{(60\text{ hours} / 15\text{ hours})}$$

$$A = 0.5^4$$

$$A = 0.0625 = 6.25\%$$

**120. B is correct.**

Ionizing radiation can be high-energy charged particles (alpha, beta) or high-energy electromagnetic waves (X-rays, gamma rays).

All are termed ionizing because they possess enough energy to remove electrons from atoms or molecules and ionize them.

 

---

**Practice Set 7: Questions 121–140**

---

**121. D is correct.**

*Positron emission* occurs during $\beta^+$ decay. In $\beta^+$ decay:

$\beta^+$ decay:

$${}^{A}_{Z}X \rightarrow {}^{A}_{Z-1}Y + {}^{0}_{+1}e + {}^{0}_{0}\nu_e$$

The mass number remains the same because a proton is converted to a neutron, but the atomic number decreases by one.

The nuclear mass does not change because there is an equal number of nucleons within the nucleus before and after the decay.

**122. C is correct.**

Nuclear reaction:

$${}^{55}_{28}Ni \rightarrow {}^{55}_{27}Co + e^+ + \nu_e$$

where Co = product, $e^+$ = positron and $\nu_e$ = electron neutrino

In positron emission ($\beta^+$ decay), a proton in the nucleus converts to a neutron while releasing a positron and an electron neutrino.

The atomic number decreases by one, but the mass number stays constant.

**123. A is correct.**

*Gamma rays* are high-energy electromagnetic waves, and they have the highest energy of all radiation (e.g., alpha, beta, and electromagnetic spectrum).

**124. A is correct.**

In alpha decay, two protons and two neutrons are released as alpha particles.

Thus, the new atomic number is Z–2.

Then when the beta minus decay occurs, a neutron is converted to a proton, giving the atomic number Z–1.

 

**125. C is correct.**

Alpha radiation is the emission of a particle with two neutrons and two protons identical to a helium nucleus.

The alpha particles have a +2 charge due to the two protons and thus are deflected towards the negative electrode of two electrically charged plates.

**126. D is correct.**

In *nuclear fusion*, lighter elements fuse to form heavier elements.

During this process, some of the mass of the three helium nuclei is converted to energy by:

$$E = mc^2$$

Thus, the net mass of the three helium nuclei is slightly greater than that of the carbon nucleus due to mass conversion into energy.

**127. B is correct.**

One of the obvious errors of the planetary model is that electrons cannot orbit the nucleus without experiencing acceleration (due to a change in direction).

As such, the electron would lose its energy as photons (accelerating charges produce electromagnetic radiation) and eventually collapse into the nucleus.

**128. C is correct.**

*Gamma radiation* would be the best type of radiation for medical imaging because it penetrates the furthest.

**129. D is correct.**

Alpha particles have the greatest charge of +2 and consist of two protons and two neutrons.

Beta particles (electrons or positrons) have +/–1, and gamma radiation has no charge as an electromagnetic wave.

**130. B is correct.**

*Catalysts* are substances that lower the activation energy of chemical reactions and bond formation without being consumed within the reaction themselves.

Nuclear reactions cannot have their rate increased by a catalyst because the reaction takes place at a nuclear scale and does not involve forming bonds.

 

**131. D is correct.**

Mass is conserved in a chemical reaction; no particles are created or destroyed.

The atoms are rearranged to form products from reactants.

In nuclear reactions, the mass difference is due to mass converting into energy.

**132. C is correct.**

Positron emission occurs during $\beta^+$.

The decay equation: $^{13}_{7}N \rightarrow {}^{13}_{6}C + e^+ + \nu_e$

A proton converts into a neutron and a positron is ejected with an electron neutrino. The daughter nuclide is $^{13}C$.

**133. D is correct.**

*Radon gas* is a natural decay product of uranium that accounts for the greatest source of yearly radiation exposure in humans.

Radon is more hazardous to smokers due to the combined carcinogenic effects of smoking and radiation exposure.

**134. D is correct.**

The strong nuclear force is the strongest of the four fundamental forces.

However, it only acts within a small range of distance (about the diameter of the nucleus).

**135. D is correct.**

Natural line broadening is the extension of a spectral line over a range of frequencies.

This occurs partly due to the uncertainty principle, which relates the time in which an atom is excited to the energy of its emitted photon.

$$\Delta E \Delta t > \hbar / 2$$

where $\hbar$ is reduced Planck's constant

Because energy is related to frequency by:

$$E = hf$$

where $h$ is Planck's constant

The range of frequencies observed (broadening) is due to the uncertainty in energy outlined by the uncertainty principle.

 

**145. D is correct.**

Use Wien's displacement law:

$\lambda_{max} = b / T$

$T = (2.9 \times 10^{-3}\ K{\cdot}m) / (580 \times 10^{-9}\ m)$

$T = 5{,}000\ K$

**137. A is correct.**

Gamma radiation is a high-energy electromagnetic wave, and as such, it has no charge and will not deflect within an electric field.

**138. C is correct.**

Nuclei with atomic numbers over 83 are inherently unstable and thus radioactive.

This limit in size is because the strong nuclear force has a very short range, and as the nucleus gets larger, the strong nuclear force cannot overcome the Coulomb repulsion from the protons within the nucleus.

**139. D is correct.**

A mass of an element or compound can be measured by its molar mass.

The molar mass relates the mass of the element or compound to a discrete number of subunits (atoms for elements, molecules for compounds).

**140. B is correct.**

The *rem* is short for the Roentgen equivalent in man and is designed to measure the biological damage of ionizing radiation.

It does not measure the number of particles absorbed or emitted, nor is it the maximum occupational safety exposure limit for radiation.

 

*Notes for active learning*

*Notes for active learning*

# 12 – Quantum Mechanics: Detailed Explanations

===============================================================

**Practice Set 1: Questions 1–20**

===============================================================

**1. D is correct.**

The energy of each incident photon is transferred to an electron, which must then overcome the material's work function to be ejected.

Therefore:

$hc / \lambda$ = Work function

$\lambda = (6.626 \times 10^{-34}\ \text{J·s} \cdot 3.00 \times 10^{8}\ \text{m/s}) / (1.90\ \text{eV} \cdot 1.60 \times 10^{-19}\ \text{J/eV})$

$\lambda = 6.53 \times 10^{-7}\ \text{m}$

$\lambda = 653\ \text{nm}$

**2. B is correct.**

The power varies with the number of photons per second and the energy of the photons; therefore, I is incorrect.

The energy of the photons varies with the frequency of the light; therefore, III is incorrect.

The intensity of a laser beam depends on the number of photons and the energy of each photon.

The energy of the photons varies linearly with the frequency.

If the photon frequency doubles, the energy doubles.

Since the number of photons is unchanged, the intensity doubles.

**3. C is correct.**

The energy of each incident photon is transferred to an electron, which must then overcome the material's work function to be ejected.

Therefore, the maximum kinetic energy remaining in any electron is:

E = Energy in 240 nm photon – work function

*continued…*

$2.5 \text{ eV} = hc / \lambda - \text{work function}$

Work function = $(6.626 \times 10^{-34} \text{ J·s} \cdot 3.00 \times 10^{8} \text{ m/s}) / (240 \times 10^{-9} \text{ m}) - 2.58 \text{ eV}$

Work function = $(8.28 \times 10^{-19} \text{ J}) / (1.60 \times 10^{-19} \text{ J/eV}) - 2.58 \text{ eV}$

Work function = 2.6 eV

**4. C is correct.**

The minimum energy of the electron-positron pair is its rest mass ($E = 2m_{electron}\, c^2$).

The photon that creates the pair must have more than the minimum energy.

The energy of the photon ($hv$) must therefore be:

$hv > 2m_{electron}\, c^2$

$v > 2m_{electron}\, c^2 / h$

$v > 2 \cdot (9.11 \times 10^{-31} \text{ kg}) \cdot (3.00 \times 10^{8} \text{ m/s})^2 / (6.626 \times 10^{-34} \text{ J·s})$

$v > 2.47 \times 10^{20} \text{ Hz}$

**5. D is correct.**

The energy of each incident photon is transferred to an electron, which must then overcome the material's work function to be ejected.

Therefore:

$hc / \lambda > \text{Work function}$

$\lambda < (6.626 \times 10^{-34} \text{ J·s} \cdot 3.00 \times 10^{8} \text{ m/s}) / (2.20 \text{ eV} \cdot 1.60 \times 10^{-19} \text{ J/eV})$

$\lambda < 564 \text{ nm}$

**6. C is correct.**

Anderson used a cloud chamber to observe the effects of cosmic rays.

A cloud chamber is a device filled with saturated water or alcohol vapor and held in a magnetic field.

As the cosmic rays interact with the molecules in the vapor, they form high-energy charged particles.

These high-energy charged particles move out from their point of creation in curved paths because of the magnetic field.

*continued…*

 

Their path is visualized as they condense droplets from the saturated vapor.

Anderson observed an event in which a pair of charged particles was created at the same point, moving out in opposite directions and with equal but opposite curvature.

This corresponded to the earlier prediction by Dirac of a positively charged electron (a positron) based upon an extra solution to the equations of a relativistic invariant form of Schrodinger's Equation.

The electron and positron pair (pair production) was produced in a collision of a cosmic ray photon with a heavy nucleus in the vapor.

**7. C is correct.**

$E = h\nu$

$E = (6.626 \times 10^{-34}\ \text{J·s})·(6.43 \times 10^{14}\ \text{Hz}) / (1.60 \times 10^{-19}\ \text{J/eV})$

$E = 2.66\ \text{eV}$

**8. B is correct.**

The kinetic energy of the emitted electrons (KE) is the energy of the incident photon minus, at least, the work function of the photocathode surface.

Therefore:

$KE = 3.4\ \text{eV} - 2.4\ \text{eV}$

$KE = (1.0\ \text{eV})·(1.60 \times 10^{-19}\ \text{J / eV})$

$KE = 1.60 \times 10^{-19}\ \text{J}$

**9. D is correct.**

The energy of each incident photon is transferred to an electron, which must then overcome the material's work function to be ejected.

Therefore, to eject an electron:

$h\nu = hc / \lambda >$ Work function

$\lambda < hc$ / (Work function)

$\lambda < (6.626 \times 10^{-34}\ \text{J·s})·(3.00 \times 10^{8}\ \text{m/s}) / (2.9\ \text{eV} \cdot 1.60 \times 10^{-19}\ \text{J/eV})$

$\lambda < 428 \times 10^{-9}\ \text{m}$

The illumination range of 400 nm–700 nm that does not satisfy this requirement is:

$\lambda > 428\ \text{nm}$

 

**10. A is correct.**

Photons move at the speed of light because they are light.

Since the momentum of photon A is twice as great as the momentum of photon B, its energy is twice as great, and therefore its wavelength is half as great.

**11. A is correct.**

The Balmer formula for Hydrogen is:

$$1 / \lambda = (1 / 91.2 \text{ nm})\cdot(1 / m^2 - 1 / n^2)$$

The energy difference of the n = 20 and n = 7 state corresponds to a photon of wavelength:

$$1 / \lambda = (1 / 91.2 \text{ nm})\cdot(1 / 7^2 - 1 / 20^2)$$

$$\lambda = (91.2 \text{ nm}) / (1 / 7^2 - 1 / 20^2)$$

$$\lambda = 5092 \text{ nm}$$

The energy of a photon with wavelength λ is:

$$E = hc / \lambda$$

$$E = (6.626 \times 10^{-34} \text{ J}\cdot\text{s})\cdot(3.00 \times 10^8 \text{ m/s}) / (5092 \times 10^{-9} \text{ m})$$

$$E = (3.93 \times 10^{-20} \text{ J}) / (1.60 \times 10^{-19} \text{ J/eV})$$

$$E = 0.244 \text{ eV}$$

**12. D is correct.**

The *de Broglie wavelength* is given by:

$$\lambda = h / p$$

When the energy ($E = p^2 / 2m$ for a non-relativistic proton) is doubled, the momentum is increased by $\sqrt{2}$.

Therefore, its de Broglie wavelength decreases by $\sqrt{2}$.

**13. D is correct.**

Trivially, it is known that the incoming photon must lose energy to the electron in the scattering, and therefore its wavelength must increase.

There is only one answer with a longer wavelength.

*continued…*

 

Using the Compton equation at 120°:

$\Delta\lambda = h / m_0c\ (1 - \cos\theta)$

$\Delta\lambda = \lambda_{Compton}\ (1 - \cos\theta)$

where $\lambda_{Compton} = 2.43 \times 10^{-12}$ m

$\Delta\lambda = 0.00243$ nm (1.5)

$\Delta\lambda = 0.00365$ nm

$\lambda = 0.591$ nm + 0.00365 nm

$\lambda = 0.595$ nm

**14. C is correct.**

The *brightness* of a beam of light is linearly proportional to the energy of the photons in the beam and the number of photons in the beam.

If the *color* of the light beam, which is dependent on its frequency distribution or energy distribution, is unchanged, then the frequency and energy distribution of the light beam must be unchanged.

**15. B is correct.**

The Balmer formula for Hydrogen is:

$1 / \lambda = (1 / 91.2\ \text{nm})(1 / \text{m}^2 - 1 / \text{n}^2)$

$1 / \lambda = (1 / 91.2\ \text{nm})(1 / 4^2 - 1 / 9^2)$

$1 / \lambda = (1 / 91.2\ \text{nm})(1 / \text{m}^2 - 1 / \text{n}^2)$

$1 / \lambda = (1 / 1818\ \text{nm})$

The frequency of light is given by:

$\nu = c / \lambda$

$\nu = (3.00 \times 10^8\ \text{m/s}) / (1818 \times 10^{-9}\ \text{m})$

$\nu = 1.65 \times 10^{14}\ /\text{s} = 1.65 \times 10^{14}$ Hz

**16. D is correct.**

The *Balmer formula for Hydrogen* for the emission of radiation from the $n^{th}$ level down to the $m^{th}$ (i.e., n > m):

$$1 / \lambda = (1 / 91.2 \text{ nm})\cdot(1 / m^2 - 1 / n^2)$$

Therefore:

$$(1 / m^2 - 1 / n^2) = 91.2 \text{ nm} / 377 \text{ nm}$$

$$(1 / m^2 - 1 / n^2) = 0.2419$$

$$1 / n^2 = 1 / m^2 - 0.2419$$

For n to be real, it must be m = 1 or m = 2.

If m = 1:

$$1 / n^2 = 1 - 0.2419 = 0.7581$$

n is not an integer (i.e., the photon has less energy than the emission photon expected in a transition from n = 2 to m = 1).

Therefore, m must be 2 (i.e., the emission is from the n level down to the m = 2 level) and:

$$1 / n^2 = 1 / 4 - 0.2419 = 0.0081$$

$$n = 11$$

**17. C is correct.**

*Wein's displacement law* describes the wavelength of maximum emission of radiation of a black body at temperature T:

$$\lambda_{max} \cdot T = \text{constant} = 0.00290 \text{ m}\cdot\text{K}$$

At T = 5000K:

$$\lambda_{max} = 0.00290 \text{ m}\cdot\text{K} / 5000 \text{ K}$$

$$\lambda_{max} = 580 \text{ nm}$$

**18. D is correct.**

Dirac factored the Schrödinger equation into a simpler equation with two solutions; one solution described the electron, and the other described an identical particle with an opposite electric charge.

Four years later, such a "positively charged" electron was found in cloud chamber pictures of cosmic rays.

**19. B is correct.**

The electron absorbs the total energy of the photon and loses the work function energy as it escapes from the material.

Therefore, it has a maximum kinetic energy of:

Max KE = E – Work function

Max KE = 3.4 eV – 2.4 eV = 1.0 eV · ($1.60 \times 10^{-19}$ J/eV)

Max KE = $1.60 \times 10^{-19}$ J

**20. D is correct.**

Consider the problem from the center of the mass frame, a frame moving in the same direction as the incident photon. In this frame, the electron starts moving with the speed of the center of the mass frame and in the opposite direction of the center of the mass frame.

In the center of the mass frame, the electron is scattered with the same speed into one direction, and the photon is scattered in the opposite direction, the scattering angle.

Transforming back into the laboratory frame, the electron has a final velocity equal to the vector sum of its scattered velocity in the center of the mass frame and the velocity of the center of mass.

This sum is greatest when the scattering angle of the electron is in the same direction as the velocity of the center of mass frame, the direction of the incident photon. The photon is then scattered in the opposite direction of the center of the mass frame at a scattering angle of 180°.

If the velocity of the electron is maximal at that scattering angle, then the energy of the photon is minimal at that scattering angle.

Since wavelength is inversely proportional to energy, the maximal change in wavelength occurs when the photon is scattered at 180°.

===========================================================================

**Practice Set 2: Questions 21–40**

===========================================================================

**21. C is correct.**

This is a trick question that has nothing to do with the photocathode or the work function.

If the radiation has energy 3.5 eV, then its wavelength is given by:

$E = h\nu = hc / \lambda = 3.5 \text{ eV}$

$\lambda = hc / (3.5 \text{ eV})$

$\lambda = (6.626 \times 10^{-34} \text{ J·s})·(3.00 \times 10^{8} \text{ m/s}) / (3.5 \text{ eV})$

$\lambda = (6.626 \times 10^{-34} \text{ J·s})·(3.00 \times 10^{8} \text{ m/s}) / (3.5 \text{ eV})$

$\lambda = (5.679 \times 10^{-26} \text{ J·m /eV}) / (1.6 \times 10^{-19} \text{ J/eV})$

$\lambda = 355 \text{ nm}$

**22. A is correct.**

The energy of a photon is given by:

$E = h\nu$

$E = hc / \lambda$

Therefore, if the wavelength is doubled, the energy is halved.

**23. D is correct.**

The energy of each incident photon is transferred to an electron, which must then overcome the material's work function to be ejected.

Therefore:

$h\nu >$ Work function

$\nu > (2.8 \text{ eV} \cdot 1.60 \times 10^{-19} \text{ J/eV}) / (6.626 \times 10^{-34} \text{ J·s})$

$\nu > 6.8 \times 10^{-14} \text{/s}$

**24. B is correct.**

The de Broglie wavelength of a matter wave is:

$\lambda = h / p$

$\lambda = h / mv$

$v = h / (m\lambda)$

$v = (6.626 \times 10^{-34}\ \text{J·s}) / [(9.11 \times 10^{-31}\ \text{kg})·(380 \times 10^{-9}\ \text{m})]$

$v = 1.91 \times 10^{3}\ \text{m/s}$

**25. B is correct.**

Using the Compton Equation:

$\Delta\lambda = h / m_0c\ (1 - \cos\theta)$

$\Delta\lambda = \lambda_{\text{Compton}}\ (1 - \cos\theta)$

where $\lambda_{\text{Compton}} = 2.43 \times 10^{-12}\ \text{m}$

For $\theta = 90°$:

$\Delta\lambda = 2.43 \times 10^{-12}\ \text{m}$

and

$\lambda_{\text{scattered}} = \lambda_{\text{incident}} + \Delta\lambda$

$\lambda_{\text{scattered}} = (1.50 \times 10^{-10}\ \text{m}) + (2.43 \times 10^{-12}\ \text{m})$

$\lambda_{\text{scattered}} = 1.5243 \times 10^{-10}\ \text{m}$

**26. B is correct.**

As the *intensity of light* increases, the photon flux increases but not the energy of the photons.

The *kinetic energy* of the ejected electrons depends solely on the energy of the incident photons and the work function of the metal.

Since the energy of the incident photons is unchanged, the kinetic energy of the ejected electrons does not change.

However, the *probability of an electron being ejected*, and therefore the number of electrons ejected per second, depends on the flux of incident photons. It increases as the intensity of the light increases.

*continued…*

The electron is ejected at the same instance as the light is absorbed, independent of the intensity of the light.

Therefore, the time lag does not change.

*Note:* the time lag between the illumination of the surface (not the absorption of light) and the ejection of the first electron depends on the probability of ejection, which depends on the intensity of the incident light.

**27. D is correct.**

The *de Broglie wavelength* of a matter wave is:

$$\lambda = h / p$$

$$\lambda = h / (mv)$$

The *energy of a photon* with this wavelength is:

$$E = hv$$

$$E = hc / \lambda$$

Substituting in for λ:

$$E = hc / [h / (mv)]$$

$$E = mcv$$

$$E = (1.67 \times 10^{-27} \text{ kg})\cdot(3.00 \times 10^{8} \text{ m/s})\cdot(7.2 \times 10^{4} \text{ m/s})$$

$$E = 36.1 \times 10^{-15} \text{ J}$$

$$E = (36.1 \times 10^{-15} \text{ J}) / (1.60 \times 10^{-19} \text{ J/eV})$$

$$E = 225 \times 10^{3} \text{ eV}$$

**28. B is correct.**

The Compton effect measures the change in energy as an x-ray scatters off an electron.

The total momentum and total energy of the x-ray and the electron, initially at rest, must be conserved.

As the scattering angle of the x-ray increases monotonically, its change in momentum increases, and therefore the momentum imparted to the electron must increase.

If the momentum of the electron increases, then its energy must increase, and the energy of the X-ray decreases. If the energy of the X-ray decreases, then the frequency, which is proportional to its energy, must decrease.

**29. A is correct.**

The Rydberg formula for Hydrogen for the emission of radiation from the n[th] level down to m[th] (i.e., n > m) is:

$$1 / \lambda = (1 / 91.2 \text{ nm})\cdot(1 / m^2 - 1 / n^2)$$

If n = 16 (since the first spectral line is from n = 2) and m = 1, then:

$$1 / \lambda = (1 / 91.2 \text{ nm})\cdot(1 / 1^2 - 1 / 16^2)$$

$$\lambda = 91.2 \text{ nm} / (1 - 0.004)$$

$$\lambda = 91.6 \text{ nm}$$

**30. D is correct.**

If the frequency of light in a laser beam is doubled, the energy of each photon in that laser beam doubles since each photon has energy $E = h\nu$.

The wavelength of each photon is divided by 2, since $\lambda = c / \nu$.

The intensity of the laser beam is the power per unit area.

The power is the energy delivered per unit time.

The energy delivered is proportional to the energy of each photon times the number of photons. If the energy of each photon is doubled and the number of photons remains unchanged (assuming the area of the laser beam is unchanged), then the intensity doubles.

Therefore, I and III are correct.

**31. B is correct.**

Order diffraction occurs when the incident and scattered beams hit the crystal at the same angle (i.e., in this case, the crystal planes are oriented at 58° / 2 to the normal).

The neutron matter waves add coherently (i.e., constructive or destructive interference) if the extra distance that the neutrons must travel as they scatter off the next plane of the crystal is equal to an integer number of de Broglie wavelengths.

At 58°, the extra distance is:

$$2 \cdot 159.0 \text{ pm} \cdot \cos(58° / 2) = 278 \text{ pm}$$

The de Broglie wavelength of a matter wave is:

$$\lambda = h / p$$

*continued…*

 

Therefore:

$$p = h / \lambda$$

and

$$E = p^2 / 2m = (6.626 \times 10^{-34}\ \text{J·s} / 278 \times 10^{-12}\ \text{m})^2 / (2 \cdot 1.67 \times 10^{-27}\ \text{kg})$$

$$E = (1.70 \times 10^{-21}\ \text{J}) / (1.6 \times 10^{-19}\ \text{J/eV})$$

$$E = 0.0106\ \text{eV}$$

**32. A is correct.**

The energy of each incident photon is transferred to an electron, and the electron's energy goes into overcoming the photocathode's 2.5 eV work function.

The remaining energy is then stopped by the stopping potential.

The greatest amount of energy an electron can have comes from a photon with a wavelength of 360 nm.

The remaining energy is then:

$$\text{remaining energy} = hv - 2.5\ \text{eV}$$

$$\text{remaining energy} = hc / \lambda - 2.5\ \text{eV}$$

$$\text{remaining energy} = (6.626 \times 10^{-34}\ \text{J·s})\cdot(3.00 \times 10^{8}\ \text{m/s}) / (360 \times 10^{-9}\ \text{m}) - 2.5\text{eV}$$

$$\text{remaining energy} = [(5.52 \times 10^{-19}\ \text{J}) / (1.6 \times 10^{-19}\ \text{J/eV})] - 2.5\ \text{eV}$$

$$\text{remaining energy} = 3.45\ \text{eV} - 2.5\ \text{eV}$$

$$\text{remaining energy} = 0.95\ \text{eV}$$

The electron has a charge of –1 e.

Therefore, this energy can be stopped by a voltage of 0.95 V.

**33. D is correct.**

The de Broglie wavelength is given by:

$$\lambda = h / p$$

$$\lambda = (6.626 \times 10^{-34}\ \text{J·s}) / (1.95 \times 10^{-27}\ \text{kg·m/s})$$

$$\lambda = 340\ \text{nm}$$

 

**34. C is correct.**

The Balmer formula for Hydrogen for the emission of radiation from the $n^{th}$ level down to the $m^{th}$ (i.e., n > m):

$$1 / \lambda = (1 / 91.2 \text{ nm})\cdot(1 / m^2 - 1 / n^2)$$

If n = 9 and m = 6, then:

$$1 / \lambda = (1 / 91.2 \text{ nm})\cdot(1 / 6^2 - 1 / 9^2)$$

$$1 / \lambda = (1 / 91.2 \text{ nm})\cdot(0.01543)$$

Thus:

$$\nu = c / \lambda$$

$$\nu = (3.00 \times 10^8 \text{ m/s})\cdot(1 / 91.2 \text{ nm})\cdot(0.01543)$$

$$\nu = 5.08 \times 10^{13} \text{ /s}$$

$$\nu = 5.08 \times 10^{13} \text{ Hz}$$

**35. A is correct.**

$$E = h\nu$$

$$E = (6.626 \times 10^{-34} \text{ J}\cdot\text{s})\cdot(110 \text{ GHz})$$

$$E = (6.626 \times 10^{-34} \text{ J}\cdot\text{s})\cdot(110 \times 10^9 / \text{s})$$

$$E = 7.29 \times 10^{-23} \text{ J}$$

**36. C is correct.**

The visible spectrum ranges from 400 nm to 700 nm.

The Balmer formula for Hydrogen for the emission of radiation from the $n^{th}$ level down to the $m^{th}$ (i.e., n > m):

$$1 / \lambda = (1 / 91.2 \text{ nm})\cdot(1 / m^2 - 1 / n^2)$$

Consider emission from any level down to the m = 1 level.

The lowest energy is from the n = 2 level, and its wavelength is:

91.2 nm · (4 / 3) = 121 nm, which is not visible.

Therefore, no visible lines are radiating down to the m = 1 level.

Consider emission from any level down to the m = 3 level.

*continued…*

The highest energy comes from n = ∞ , and its wavelength is:

91.2 nm · 9 = 820 nm, which is not visible.

Therefore, no visible lines are radiating down to the m = 3 level, nor are any visible lines radiating down to any m level higher than 3.

Consider radiation from various levels down to the m = 2 level.

From n = 3, the wavelength is:

91.2 nm / (1 / 4 – 1 / 9) = 91.2 nm · 7.2 = 656.6 nm

From n = 4, the wavelength is:

91.2 nm / (1 / 4 – 1 / 16) = 91.2 nm · 5.333 = 486.4 nm

From n = 5, the wavelength is:

91.2 nm / (1 / 4 – 1 / 25) = 91.2 nm · 4.762 = 434.3 nm

From n = 6, the wavelength is:

91.2 nm / (1 / 4 – 1 / 36) = 91.2 nm · 4.5 = 410.4 nm

From n = 7, the wavelength is:

91.2 nm / (1 / 4 – 1 / 49) = 91.2 nm · 4.355 = 397.2 nm, which is not visible.

Therefore, there are 4 visible lines:

m = 2 to n = 3 656.6 nm

m = 2 to n = 4 486.4 nm

m = 2 to n = 5 434.3 nm

m = 2 to n = 6 410.4 nm

**37. B is correct.**

$E = h\nu$

$E = hc / \lambda$

$\lambda = hc / E$

$\lambda = (6.626 \times 10^{-34}\ \text{J·s})·(3.00 \times 10^{8}\ \text{m/s}) / (4.20\ \text{eV})$

$\lambda = (4.73 \times 10^{-26}\ \text{J·m/eV}) / (1.60 \times 10^{-19}\ \text{J/eV})$

$\lambda = 2.96 \times 10^{-7}\ \text{m}$

$\lambda = 296\ \text{nm}$

 

**38. C is correct.**

The energy of each incident photon is transferred to an electron, which must then overcome the material's work function to be ejected.

Therefore, if a wavelength of light is just able to eject an electron, then:

$hv = hc / \lambda$ = Work function

$(6.626 \times 10^{-34}\ \text{J·s})\cdot(3.00 \times 10^{8}\ \text{m/s}) / (500 \times 10^{-9}\ \text{m})$ = Work function

$(3.98 \times 10^{-19}\ \text{J}) / (1.6 \times 10^{-19}\ \text{J/eV})$ = Work function

2.48 eV = Work function

**39. C is correct.**

In the Bohr theory, there is a fixed number of de Broglie wavelengths of the electron in orbit.

This number is the principal quantum number:

$\text{n} = 2\pi r / \lambda$

The de Broglie wavelength is given by:

$\lambda = h / p$

Therefore:

$\text{n} = (2\pi / h)rp$

or:

$\text{n}^2 = (2\pi / h)^2 r^2 p^2$

The attractive force of the proton keeps the electron in a circular orbit.

The attractive force is proportional to $1 / r^2$, and if that force keeps the electron in orbit, $1 / r^2$ must be proportional to $v^2 / r$.

Therefore,

$v^2$ (or $p^2$) is proportional to $1 / r$

Since $r^2p^2$ is proportional to $\text{n}^2$, and $p^2$ is proportional to $1 / r$, then $r$ (i.e., $r^2 \cdot 1 / r$) is proportional to $\text{n}^2$.

**40. B is correct.**

Heisenberg's Uncertainty Principle states that:

$$\Delta p \Delta x \geq h / 2\pi$$

Therefore:

$$m \Delta v \Delta x \geq h / 2\pi$$

$$\Delta v \geq (6.626 \times 10^{-34}\ \text{J·s}) / (2\pi \cdot 0.053\ \text{nm} \times 1.67 \times 10^{-27}\ \text{kg})$$

$$\Delta v \geq 1.19 \times 10^{3}\ \text{m/s}$$

---

**Practice Set 3: Questions 41–60**

---

**41. C is correct.**

The Balmer formula for Hydrogen for the emission of radiation from the $n^{th}$ level down to the $m^{th}$ (i.e., $n > m$):

$$1 / \lambda = (1 / 91.2 \text{ nm})\cdot(1 / m^2 - 1 / n^2)$$

$$1 / \lambda = (1 / 91.2 \text{ nm})\cdot(1 / 9^2 - 1 / 11^2)$$

$$\lambda = (91.2 \text{ nm}) / (0.00408)$$

$$\lambda = 22{,}300 \text{ nm}$$

**42. D is correct.**

Heisenberg's Uncertainty Principle states that:

$$\Delta E \Delta t \geq h/2\pi$$

Therefore:

$$\Delta t \geq (h / 2\pi) / \Delta E$$

$$\Delta t \geq (6.626 \times 10^{-34} \text{ J}\cdot\text{s} / 2\pi) / (10^{-18} \text{ J})$$

$$\Delta t \geq 1.05 \times 10^{-16} \text{ s}$$

**43. C is correct.**

The energy of each incident photon is transferred to an electron, which must then overcome the material's work function to be ejected.

Therefore, if a wavelength of light can eject an electron with energy 2.58 eV, then:

$$hv = hc / \lambda = \text{Work function} + 2.58 \text{ eV}$$

$$(6.626 \times 10^{-34} \text{ J}\cdot\text{s})\cdot(3.00 \times 10^8 \text{ m/s}) / (240 \times 10^{-9} \text{ m}) = \text{Work function} + 2.58 \text{ eV}$$

$$\text{Work function} = [(8.28 \times 10^{-19} \text{ J}) / (1.6 \times 10^{-19} \text{ J/eV})] - 2.58\text{eV}$$

$$\text{Work function} = 2.60 \text{ eV}$$

**44. C is correct.**

The de Broglie wavelength of a matter wave is:

$\lambda = h / p$

$\lambda = h / (mv)$

$\lambda = (6.626 \times 10^{-34} \text{ J·s}) / [(1.30 \text{ kg})·(28.10 \text{ m/s})]$

$\lambda = (6.626 \times 10^{-34} \text{ J·s}) / [(1.30 \text{ kg})·(28.10 \text{ m/s})]$

$\lambda = 1.81 \times 10^{-35} \text{ m}$

**45. D is correct.**

By doubling the frequency of the light, the energy of each photon doubles, but not the number of photons (hence B is wrong).

Although doubling the energy of each photon would cause more electrons to be ejected, it does not necessarily, double that number since the number is not linearly proportional to the incident energy (hence A is not always true).

The kinetic energy of the ejected electrons would more than double since the initial kinetic energy is less than the energy of the incident photons - the work function of the surface reduced it (hence C is wrong).

The kinetic energy would increase by at least 2 but not necessarily by 4 (hence D is wrong).

**46. D is correct.**

The Balmer formula for Hydrogen is:

$1 / \lambda = (1 / 91.2 \text{ nm})·(1 / m^2 - 1 / n^2)$

From the n = 3 level, the transition to the n = 2 level has the lowest energy and, therefore the longest wavelength.

The Balmer formula for n = 3, m = 2 is:

$1 / \lambda = (1 / 91.2 \text{ nm})·(1 / 2^2 - 1 / 3^2)$

$1 / \lambda = (1 / 91.2 \text{ nm})·(5 / 36)$

$\lambda = 656 \text{ nm}$

 

**47. C is correct.**

Photons are light, and the speed of light is invariant. Therefore, I is wrong.

The energy of the photons in a beam of light determines the color of the light, and if the color is unchanged, then the average energy of the photons is unchanged, making II wrong.

Changing the brightness of a beam of light increases the number of photons in the beam of light.

**48. B is correct.**

Using the Compton Equation:

$$\Delta\lambda = h / m_0c\ (1 - \cos\theta)$$

$$\Delta\lambda = \lambda_{Compton}\ (1 - \cos\theta)$$

where $\lambda_{Compton} = 2.43 \times 10^{-12}$ m

At $\theta = 180°$, $\Delta\lambda = \lambda$.

Thus:

$$\lambda = 2\lambda_{Compton}$$

$$\lambda = 4.86 \times 10^{-12}\ \text{m}$$

**49. D is correct.**

The energy of the photons in the beam depends linearly on the frequency of the beam.

Since the frequency of beam B is twice the frequency of beam A, the photons in beam B have twice the energy as the photons in beam A.

The intensity depends on the energy of the photons and the number of photons per second carried by the beam.

Nothing is known about the intensity or the number of photons per second.

**50. A is correct.**

The Balmer formula for Hydrogen is:

$$1 / \lambda = (1 / 91.2\ \text{nm})\cdot(1 / m^2 - 1 / n^2)$$

From the n = 3 level, the transition to the n = 1 level has the greatest energy; therefore, the shortest wavelength.

*continued…*

The Balmer formula for n = 3, m = 1 is:

$$1 / \lambda = (1 / 91.2 \text{ nm})\cdot(1 / 1^2 - 1 / 3^2)$$

$$1 / \lambda = (1 / 91.2 \text{ nm})\cdot(8 / 9)$$

$$\lambda = 102.6 \text{ nm}$$

**51. D is correct.**

The uncertainty principle states that the uncertainty in position and momentum must be at least:

$$\Delta x \Delta p \approx h / 2\pi$$

$$\Delta p \approx h / (2\pi\Delta x)$$

If:

$$\Delta x = 5.0 \times 10^{-15} \text{ m}$$

Then:

$$\Delta p \approx 6.626 \times 10^{-34} \text{ J}\cdot\text{s} / (2\pi \cdot 5.0 \times 10^{-15} \text{ m})$$

$$\Delta p \approx 2.11 \times 10^{-20} \text{ kg}\cdot\text{m/s}$$

The uncertainty in the (non-relativistic) kinetic energy of the proton is then:

$$\Delta\text{KE} \approx (\Delta p)^2 / (2m)$$

$$\Delta\text{KE} \approx [h / (2\pi\Delta x)]^2 / (2m)$$

$$\Delta\text{KE} \approx [(6.626 \times 10^{-34} \text{ J}\cdot\text{s}) / (2\pi \cdot 5.0 \times 10^{-15} \text{ m})]^2 / (2 \cdot 1.67 \times 10^{-27} \text{ kg})$$

$$\Delta\text{KE} \approx (1.33 \times 10^{-13} \text{ J})$$

Converting to MeV:

$$\Delta\text{KE} \approx (1.33 \times 10^{-13} \text{ J}) / (1.6 \times 10^{-13} \text{ J/MeV})$$

$$\Delta\text{KE} \approx (0.83 \text{ MeV})$$

**52. B is correct.**

The electron absorbs the full energy of the photon, losing 2.4 eV as it escapes from the photocathode and an additional 1.1 eV of kinetic energy after it escapes from the photocathode.

This allows it to be stopped only with a potential exceeding 1.1 volts.

It therefore absorbs 2.4 eV + 1.1 eV = 3.5 eV from the incident photon.

*continued…*

A photon with energy 3.5 eV has a wavelength given by:

$$hv = hc / \lambda = E$$

$$\lambda = hc / E = (6.626 \times 10^{-34}\ J \cdot s) \cdot (3.00 \times 10^{8}\ m/s) / (3.5\ eV \cdot 1.60 \times 10^{-19}\ J/eV)$$

$$\lambda = 355\ nm$$

**53. B is correct.**

The *Compton effect* measures the reduction in wavelength of an x-ray beam as it scatters off a sample.

The reduction in wavelength can be explained if the incident x-ray beam is considered as being made up of individual particles, each with momentum and energy, and electrons scatter those particles in the sample. Hence, the Compton effect demonstrates the particle nature of electromagnetic radiation.

While the energy content and momenta of the individual x-rays are included in the scattering calculation, they are not directly demonstrated in the Compton effect.

**54. B is correct.**

The de Broglie wavelength of a matter wave is:

$$\lambda = h / p$$

$$\lambda = h / (m\gamma v)$$

(Since $v / c < 0.1$, this will equal about 1.005 and can be ignored)

$$\lambda = (6.626 \times 10^{-34}\ J \cdot s) / [(9.11 \times 10^{-31}\ kg) \cdot (2.5 \times 10^{7}\ m/s)]$$

$$\lambda = 29 \times 10^{-12}\ m$$

$$\lambda = 29\ pm$$

**55. D is correct.**

The energy uncertainty is given by the uncertainty principle:

$$\Delta E \Delta t \geq h / 2\pi$$

$$\Delta E \geq (6.626 \times 10^{-34}\ J \cdot s / 2\pi) / (30 \times 10^{-12}\ s)$$

$$\Delta E \geq (3.5 \times 10^{-24}\ J)$$

Converting to eV ($1.6 \times 10^{-19}$ J /eV):

$$\Delta E \geq 2.2 \times 10^{-5}\ eV$$

**56. A is correct.**

$E = hf$

If a photon has energy greater than 5.37 eV, this additional amount imparts KE to the electron, which is not necessary to ionize it.

$f = E_{transition} / h$

$f = (5.37\ \text{eV}) / (4.14 \times 10^{-15}\ \text{eV·s})$

$f = 1.3 \times 10^{15}\ \text{Hz}$

**57. C is correct.**

The energy of transition is the same as the energy of the photon.

$f = c / \lambda$

$E_{ph} = hf$

$E_{ph} = h(c / \lambda)$

$E_{ph} = (6.63 \times 10^{-34}\ \text{J·s})·[(3 \times 10^8\ \text{m/s}) / (1.25 \times 10^{-7}\ \text{m})]$

$E_{ph} = 1.6 \times 10^{-18}\ \text{J}$

The new energy level differs from the 0 J ground state by $1.6 \times 10^{-18}$ J.

The ground state represents the lowest energy; the excited state must be positive.

**58. A is correct.**

Find energy needed to ionize the electron:

$E_n = -13.6\ \text{eV}\ (1 / n^2 - 1 / \infty)$

$E_n = -13.6\ \text{eV} / n^2$

$E_2 = (-13.6\ \text{eV}) / 2^2$

$E_2 = -3.4\ \text{eV}$

The electron must absorb a photon of 3.4 eV to ionize it:

$E = hc / \lambda$

$\lambda = hc / E$

$\lambda = (4.135 \times 10^{-15}\ \text{eV·s})·(3 \times 10^8\ \text{m/s}) / (3.4\ \text{eV})$

$\lambda = 365\ \text{nm}$

**59. B is correct.**

By the *Born Rule*, the probability of obtaining any possible measurement outcome is equal to the square of the wave function.

**60. B is correct.**

Speed is constant for electromagnetic waves and only changes due to the transmission medium, not the frequency or wavelength of the wave.

Blue photons have higher energy and thus higher frequencies than red light due to:

$$E = hf$$

However, the frequency is inversely proportional to wavelength:

$$\lambda = c / f$$

Thus, blue photons have shorter wavelengths than red photons due to their higher frequencies.

*Notes for active learning*

# APPENDIX

# Periodic Table of the Elements

Key:

| atomic number<br>**Symbol**<br>name<br>abridged standard atomic weight |
|---|

| 1 | 2 | 3 | 4 | 5 | 6 | 7 | 8 | 9 | 10 | 11 | 12 | 13 | 14 | 15 | 16 | 17 | 18 |
|---|---|---|---|---|---|---|---|---|---|---|---|---|---|---|---|---|---|
| 1<br>**H**<br>hydrogen<br>1.0080<br>± 0.0002 | | | | | | | | | | | | | | | | | 2<br>**He**<br>helium<br>4.0026<br>± 0.0001 |
| 3<br>**Li**<br>lithium<br>6.94<br>± 0.06 | 4<br>**Be**<br>beryllium<br>9.0122<br>± 0.0001 | | | | | | | | | | | 5<br>**B**<br>boron<br>10.81<br>± 0.02 | 6<br>**C**<br>carbon<br>12.011<br>± 0.002 | 7<br>**N**<br>nitrogen<br>14.007<br>± 0.001 | 8<br>**O**<br>oxygen<br>15.999<br>± 0.001 | 9<br>**F**<br>fluorine<br>18.998<br>± 0.001 | 10<br>**Ne**<br>neon<br>20.180<br>± 0.001 |
| 11<br>**Na**<br>sodium<br>22.990<br>± 0.001 | 12<br>**Mg**<br>magnesium<br>24.305<br>± 0.002 | | | | | | | | | | | 13<br>**Al**<br>aluminium<br>26.982<br>± 0.001 | 14<br>**Si**<br>silicon<br>28.085<br>± 0.001 | 15<br>**P**<br>phosphorus<br>30.974<br>± 0.001 | 16<br>**S**<br>sulfur<br>32.06<br>± 0.02 | 17<br>**Cl**<br>chlorine<br>35.45<br>± 0.01 | 18<br>**Ar**<br>argon<br>39.95<br>± 0.16 |
| 19<br>**K**<br>potassium<br>39.098<br>± 0.001 | 20<br>**Ca**<br>calcium<br>40.078<br>± 0.004 | 21<br>**Sc**<br>scandium<br>44.956<br>± 0.001 | 22<br>**Ti**<br>titanium<br>47.867<br>± 0.001 | 23<br>**V**<br>vanadium<br>50.942<br>± 0.001 | 24<br>**Cr**<br>chromium<br>51.996<br>± 0.001 | 25<br>**Mn**<br>manganese<br>54.938<br>± 0.001 | 26<br>**Fe**<br>iron<br>55.845<br>± 0.002 | 27<br>**Co**<br>cobalt<br>58.933<br>± 0.001 | 28<br>**Ni**<br>nickel<br>58.693<br>± 0.001 | 29<br>**Cu**<br>copper<br>63.546<br>± 0.003 | 30<br>**Zn**<br>zinc<br>65.38<br>± 0.02 | 31<br>**Ga**<br>gallium<br>69.723<br>± 0.001 | 32<br>**Ge**<br>germanium<br>72.630<br>± 0.008 | 33<br>**As**<br>arsenic<br>74.922<br>± 0.001 | 34<br>**Se**<br>selenium<br>78.971<br>± 0.008 | 35<br>**Br**<br>bromine<br>79.904<br>± 0.003 | 36<br>**Kr**<br>krypton<br>83.798<br>± 0.002 |
| 37<br>**Rb**<br>rubidium<br>85.468<br>± 0.001 | 38<br>**Sr**<br>strontium<br>87.62<br>± 0.01 | 39<br>**Y**<br>yttrium<br>88.906<br>± 0.001 | 40<br>**Zr**<br>zirconium<br>91.224<br>± 0.002 | 41<br>**Nb**<br>niobium<br>92.906<br>± 0.001 | 42<br>**Mo**<br>molybdenum<br>95.95<br>± 0.01 | 43<br>**Tc**<br>technetium<br>[97] | 44<br>**Ru**<br>ruthenium<br>101.07<br>± 0.02 | 45<br>**Rh**<br>rhodium<br>102.91<br>± 0.01 | 46<br>**Pd**<br>palladium<br>106.42<br>± 0.01 | 47<br>**Ag**<br>silver<br>107.87<br>± 0.01 | 48<br>**Cd**<br>cadmium<br>112.41<br>± 0.01 | 49<br>**In**<br>indium<br>114.82<br>± 0.01 | 50<br>**Sn**<br>tin<br>118.71<br>± 0.01 | 51<br>**Sb**<br>antimony<br>121.76<br>± 0.01 | 52<br>**Te**<br>tellurium<br>127.60<br>± 0.03 | 53<br>**I**<br>iodine<br>126.90<br>± 0.01 | 54<br>**Xe**<br>xenon<br>131.29<br>± 0.01 |
| 55<br>**Cs**<br>caesium<br>132.91<br>± 0.01 | 56<br>**Ba**<br>barium<br>137.33<br>± 0.01 | 57-71<br>lanthanoids | 72<br>**Hf**<br>hafnium<br>178.49<br>± 0.01 | 73<br>**Ta**<br>tantalum<br>180.95<br>± 0.01 | 74<br>**W**<br>tungsten<br>183.84<br>± 0.01 | 75<br>**Re**<br>rhenium<br>186.21<br>± 0.01 | 76<br>**Os**<br>osmium<br>190.23<br>± 0.03 | 77<br>**Ir**<br>iridium<br>192.22<br>± 0.01 | 78<br>**Pt**<br>platinum<br>195.08<br>± 0.02 | 79<br>**Au**<br>gold<br>196.97<br>± 0.01 | 80<br>**Hg**<br>mercury<br>200.59<br>± 0.01 | 81<br>**Tl**<br>thallium<br>204.38<br>± 0.01 | 82<br>**Pb**<br>lead<br>207.2<br>± 1.1 | 83<br>**Bi**<br>bismuth<br>208.98<br>± 0.01 | 84<br>**Po**<br>polonium<br>[209] | 85<br>**At**<br>astatine<br>[210] | 86<br>**Rn**<br>radon<br>[222] |
| 87<br>**Fr**<br>francium<br>[223] | 88<br>**Ra**<br>radium<br>[226] | 89-103<br>actinoids | 104<br>**Rf**<br>rutherfordium<br>[267] | 105<br>**Db**<br>dubnium<br>[268] | 106<br>**Sg**<br>seaborgium<br>[269] | 107<br>**Bh**<br>bohrium<br>[270] | 108<br>**Hs**<br>hassium<br>[269] | 109<br>**Mt**<br>meitnerium<br>[277] | 110<br>**Ds**<br>darmstadtium<br>[281] | 111<br>**Rg**<br>roentgenium<br>[282] | 112<br>**Cn**<br>copernicium<br>[285] | 113<br>**Nh**<br>nihonium<br>[286] | 114<br>**Fl**<br>flerovium<br>[290] | 115<br>**Mc**<br>moscovium<br>[290] | 116<br>**Lv**<br>livermorium<br>[293] | 117<br>**Ts**<br>tennessine<br>[294] | 118<br>**Og**<br>oganesson<br>[294] |

| | | | | | | | | | | | | | | |
|---|---|---|---|---|---|---|---|---|---|---|---|---|---|---|
| 57<br>**La**<br>lanthanum<br>138.91<br>± 0.01 | 58<br>**Ce**<br>cerium<br>140.12<br>± 0.01 | 59<br>**Pr**<br>praseodymium<br>140.91<br>± 0.01 | 60<br>**Nd**<br>neodymium<br>144.24<br>± 0.01 | 61<br>**Pm**<br>promethium<br>[145] | 62<br>**Sm**<br>samarium<br>150.36<br>± 0.02 | 63<br>**Eu**<br>europium<br>151.96<br>± 0.01 | 64<br>**Gd**<br>gadolinium<br>157.25<br>± 0.08 | 65<br>**Tb**<br>terbium<br>158.93<br>± 0.01 | 66<br>**Dy**<br>dysprosium<br>162.50<br>± 0.01 | 67<br>**Ho**<br>holmium<br>164.93<br>± 0.01 | 68<br>**Er**<br>erbium<br>167.26<br>± 0.01 | 69<br>**Tm**<br>thulium<br>168.93<br>± 0.01 | 70<br>**Yb**<br>ytterbium<br>173.05<br>± 0.02 | 71<br>**Lu**<br>lutetium<br>174.97<br>± 0.01 |
| 89<br>**Ac**<br>actinium<br>[227] | 90<br>**Th**<br>thorium<br>232.04<br>± 0.01 | 91<br>**Pa**<br>protactinium<br>231.04<br>± 0.01 | 92<br>**U**<br>uranium<br>238.03<br>± 0.01 | 93<br>**Np**<br>neptunium<br>[237] | 94<br>**Pu**<br>plutonium<br>[244] | 95<br>**Am**<br>americium<br>[243] | 96<br>**Cm**<br>curium<br>[247] | 97<br>**Bk**<br>berkelium<br>[247] | 98<br>**Cf**<br>californium<br>[251] | 99<br>**Es**<br>einsteinium<br>[252] | 100<br>**Fm**<br>fermium<br>[257] | 101<br>**Md**<br>mendelevium<br>[258] | 102<br>**No**<br>nobelium<br>[259] | 103<br>**Lr**<br>lawrencium<br>[262] |

International Union of Pure and Applied Chemistry (IUPAC), 4 May 2022

## Common Equations

Throughout the test the following symbols have the definitions specified unless otherwise noted.

| | | | |
|---|---|---|---|
| L, mL | = liter(s), milliliter(s) | mm Hg | = millimeters of mercury |
| g | = gram(s) | J, kJ | = joule(s), kilojoule(s) |
| nm | = nanometer(s) | V | = volt(s) |
| atm | = atmosphere(s) | mol | = mole(s) |

**ATOMIC STRUCTURE**

$E = h\nu$

$c = \lambda\nu$

$E$ = energy
$\nu$ = frequency
$\lambda$ = wavelength

Planck's constant, $h = 6.626 \times 10^{-34}$ J s

Speed of light, $c = 2.998 \times 10^{8}$ m s$^{-1}$

Avogadro's number $= 6.022 \times 10^{23}$ mol$^{-1}$

Electron charge, $e = -1.602 \times 10^{-19}$ coulomb

**EQUILIBRIUM**

$K_c = \dfrac{[C]^c[D]^d}{[A]^a[B]^b}$, where $a\,A + b\,B \rightleftarrows c\,C + d\,D$

$K_p = \dfrac{(P_C)^c(P_D)^d}{(P_A)^a(P_B)^b}$

$K_a = \dfrac{[H^+][A^-]}{[HA]}$

$K_b = \dfrac{[OH^-][HB^+]}{[B]}$

$K_w = [H^+][OH^-] = 1.0 \times 10^{-14}$ at 25°C

$= K_a \times K_b$

$pH = -\log[H^+]$, $pOH = -\log[OH^-]$

$14 = pH + pOH$

$pH = pK_a + \log\dfrac{[A^-]}{[HA]}$

$pK_a = -\log K_a$, $pK_b = -\log K_b$

Equilibrium Constants

$K_c$ (molar concentrations)
$K_p$ (gas pressures)
$K_a$ (weak acid)
$K_b$ (weak base)
$K_w$ (water)

**KINETICS**

$\ln[A]_t - \ln[A]_0 = -kt$

$\dfrac{1}{[A]_t} - \dfrac{1}{[A]_0} = kt$

$t_{1/2} = \dfrac{0.693}{k}$

$k$ = rate constant
$t$ = time
$t_{1/2}$ = half-life

**GASES, LIQUIDS, AND SOLUTIONS**

$$PV = nRT$$

$$P_A = P_{total} \times X_A\text{, where } X_A = \frac{\text{moles A}}{\text{total moles}}$$

$$P_{total} = P_A + P_B + P_C + \ldots$$

$$n = \frac{m}{M}$$

$$K = °C + 273$$

$$D = \frac{m}{V}$$

$$KE \text{ per molecule} = \frac{1}{2}mv^2$$

Molarity, $M$ = moles of solute per liter of solution

$$A = abc$$

$P$ = pressure
$V$ = volume
$T$ = temperature
$n$ = number of moles
$m$ = mass
$M$ = molar mass
$D$ = density
$KE$ = kinetic energy
$v$ = velocity
$A$ = absorbance
$a$ = molar absorptivity
$b$ = path length
$c$ = concentration

Gas constant, $R = 8.314 \text{ J mol}^{-1} \text{K}^{-1}$
$= 0.08206 \text{ L atm mol}^{-1} \text{K}^{-1}$
$= 62.36 \text{ L torr mol}^{-1} \text{K}^{-1}$
1 atm = 760 mm Hg
= 760 torr
STP = 0.00°C and $10^5$ Pa

**THERMOCHEMISTRY/ ELECTROCHEMISTRY**

$$q = mc\Delta T$$

$$\Delta S° = \sum S° \text{ products} - \sum S° \text{ reactants}$$

$$\Delta H° = \sum \Delta H_f° \text{ products} - \sum \Delta H_f° \text{ reactants}$$

$$\Delta G° = \sum \Delta G_f° \text{ products} - \sum \Delta G_f° \text{ reactants}$$

$$\Delta G° = \Delta H° - T\Delta S°$$
$$= -RT \ln K$$
$$= -nFE°$$

$$I = \frac{q}{t}$$

$q$ = heat
$m$ = mass
$c$ = specific heat capacity
$T$ = temperature
$S°$ = standard entropy
$H°$ = standard enthalpy
$G°$ = standard free energy
$n$ = number of moles
$E°$ = standard reduction potential
$I$ = current (amperes)
$q$ = charge (coulombs)
$t$ = time (seconds)

Faraday's constant, $F$ = 96,485 coulombs per mole of electrons

$$1 \text{ volt} = \frac{1 \text{ joule}}{1 \text{ coulomb}}$$

## FLUID MECHANICS AND THERMODYNAMICS

| | |
|---|---|
| Density | $\rho = \frac{m}{V}$ |
| Pressure | $P = \frac{F}{A}$ |
| Absolute Pressure | $P = P_0 + \rho g h$ |
| Buoyant Force | $F_b = \rho V g$ |
| Fluid Continuity Equation | $A_1 v_1 = A_2 v_2$ |
| Bernoulli's Equation | $P_1 + \rho g y_1 + \frac{1}{2}\rho v_1^2 = P_2 + \rho g y_2 + \frac{1}{2}\rho v_2^2$ |
| Heat Conduction | $\frac{Q}{\Delta t} = \frac{kA\Delta T}{d}$ |
| Thermal Radiation | $P = e\sigma A(T^4 - T_C^4)$ |
| Ideal Gas Law | $PV = nRT = Nk_BT$ |
| Average Energy | $K = \frac{3}{2}k_BT$ |
| Work | $W = -P\Delta V$ |
| Conservation of Energy | $\Delta E = Q + W$ |
| Linear Expansion | $\Delta l = \alpha l_o \Delta T$ |
| Heat Engine Efficiency | $n_c = \lvert W/Q_H \rvert$ |
| Carnot Heat Engine Efficiency | $n_c = \frac{T_H - T_C}{T_H}$ |
| Energy of Temperature Change | $Q = mc\Delta T$ |
| Energy of Phase Change | $Q = mL$ |

$A$ = area

$c$ = specific heat

$d$ = thickness

$e$ = emissivity

$F$ = force

$h$ = depth

$k$ = thermal conductivity

$K$ = kinetic energy

$l$ = length

$L$ = latent heat

$m$ = mass

$n$ = number of moles

$n_c$= efficiency

$N$ = number of molecules

$P$ = pressure or power

$Q$ = energy transferred to system by heating

$T$ = temperature

$t$ = time

$E$ = internal energy

$V$ = volume

$v$ = speed

$W$ = work done on a system

$y$ = height

$\sigma$ = Stefan constant

$\alpha$ = coefficient of linear expansion

$\rho$ = density

## QUANTUM MECHANICS

| | | |
|---|---|---|
| Photon Energy | $E = hf$ | $B$ = Balmer constant |
| | | $c$ = speed of light |
| Photoelectric Electron Energy | $K_{max} = hf - \phi$ | $E$ = energy |
| Electron Wavelength | $\lambda = \frac{h}{p}$ | $f$ = frequency |
| | | $K$ = kinetic energy |
| Energy Mass Relationship | $E = mc^2$ | $m$ = mass |
| | | $p$ = momentum |
| Rydberg Formula | $\frac{1}{\lambda} = R(\frac{1}{n_f^2} - \frac{1}{n_i^2})$ | $R$ = Rydberg constant |
| | | $v$ = velocity |
| Balmer Formula | $\lambda = B(\frac{n^2}{n^2 - 2^2})$ | $\lambda$ = wavelength |
| Lorentz Factor | $\gamma = \frac{1}{\sqrt{1 - \frac{v^2}{c^2}}}$ | $\phi$ = work function |
| | | $\gamma$ = Lorentz factor |

# ACS Physical Chemistry Preparation and Test-Taking Strategies

## Test preparation strategies

The best way to do well in ACS Physical Chemistry is to be good at chemistry. There is no way around knowing the subject; proper preparation is key to success. Prepare to answer as many questions as possible with confidence.

### *Study in advance*

Assuming you completed all necessary classwork and did well, devote 2 to 3 months to studying. The information is manageable by studying regularly before the test.

Cramming is not a successful tactic. However, do not study too far in advance. Studying more than four months ahead is not advised and may result in fatigue and poor knowledge retention.

### *Develop a realistic study and practice schedule*

Cramming eight hours a day is unfeasible and leads to burnout, which is detrimental to performance.

Commit to a realistic study and practice schedule.

### *Remove distractions*

During this preparation period, temporarily eliminate distractions.

However, balance is critical, and it is crucial not to neglect physical well-being and social or family life.

Prepare with full intensity, but do not jeopardize your health or emotional well-being.

### *Develop an understanding over memorization*

When studying, devote time to each topic.

After a study session, write a short concept outline to clarify relationships and increase knowledge retention.

### *Make flashcards*

Avoid commercial flashcards because making cards helps build and retain knowledge.

Consider using self-made flashcards to develop knowledge retention and quiz what you know.

***Find a study partner***

Occasionally, studying with a friend to prepare for the test can motivate and provide accountability.

Explaining concepts to another improves and fine-tunes your understanding, integrates knowledge, bolsters competence, and identifies deficiencies in comprehension.

***Take practice tests***

Do not take practice tests too early.

First, develop a comprehensive understanding of concepts.

In the last weeks, use practice tests to fine-tune your final preparation. If you do not score well on practice tests, you want time to improve without stress.

Alternate studying and practicing to increase knowledge retention and identify areas for study.

Taking practice tests accustoms you to the challenges of test-taking.

## Test day strategies

***Be well-rested and eat the right foods***

Get a full night's sleep before the test for proper mental and physical capacity. If you are up late the night before, you will have difficulty concentrating and focusing on the test.

Avoid foods and drinks that lead to drowsiness (carbohydrates and protein).

Avoid drinks high in sugar, causing glucose to spike and crash.

***Pack in advance***

Check what you are allowed to bring to the test. Pay attention to the required check-in items (e.g., confirmation and identification).

Pack the day before to avoid the stress of not frantically looking for things on test day.

***Arrive at the testing center early***

Starting right is an advantage. Allow time to check in and remain calm before the test begins.

Plan your route to the center correctly without additional challenges and unnecessary stress. Map and test the route to the center in advance and determine needed parking.

If unfamiliar with the test location, visit before the test day to practice and avoid travel errors.

 

***Maintain a positive attitude***

Avoid falling into a mental spiral of negative emotions. Too much worry leads to underperformance.

If you become anxious, chances are higher for lower performance in preparation and during the test.

Inner peace helps during preparation and the high-stakes test.

To do well on the test requires logical, systematic, and analytical thinking, so relax and remain calm.

Do not be concerned with other test-takers.

Someone proceeding rapidly through the exam may be rushing or guessing on questions.

***Take breaks***

Do not skip the available timed breaks. Refreshing breaks help you finish strong.

Eat a light snack to replenish your energy. Your mind and body will appreciate the available breaks.

The best approach to any test is *not* to keep your head down the whole session.

While there is no time to waste, breathe deeply for a few seconds between questions.

Momentarily clear your thoughts to relax your mind and muscles.

## Time management strategies

***Timing***

Besides good preparation, time management is a critical strategy for the exam.

Timed practice is not the objective at this stage.

While practicing, note how many questions you would have completed in the allotted time.

***Average time per question***

In advance, determine the average time allotted for each question.

Use two different approaches depending on which preparation phase you are working.

During the first preparation phase, acquire, fortify, and refine your knowledge.

During final practice, use the time needed to develop analytical thought processes related to specific questions.

***Work systematically***

Note your comprehension compared to the correct answers to learn and identify conceptual weaknesses.

Do not overlook explanations to questions as a source of content, analysis, and interdependent relationships.

During the second preparation phase, do not spend more than the average allotted time on each question when taking practice tests.

Pace your response time to develop a consistent pace and complete the test within the allotted time.

If you are time-constrained during the final practice phase, work more efficiently, or your score will suffer.

***Focus on the easy questions and skip the unfamiliar***

Easy or difficult questions are worth the same points. Score more points for three quickly answered questions than one hard-earned victory.

Answer familiar and easy questions to maximize points if time runs out.

***Identify strengths and weaknesses***

Skip unfamiliar questions in the first round, as challenging questions require more time.

In the second review, questions you cannot approach systematically or lack fundamental knowledge will not be answered through analysis.

Use the elimination and educated guessing strategy to select an answer and move on to another question.

***Do not overinvest in any question***

Some questions consume more time than average and make you consider investing more.

Stop thinking that investing more time in challenging questions is productive.

Do not get entangled with questions while losing track of time.

The test is timed, so do not spend too much time on any questions.

 

***Look at every question on the exam***

It is unfortunate not to earn points for a question you could have quickly answered because you did not see it.

If you are in the first half of the test and spending more than the average on a question, select the best option, note the question number, and move on.

Do not rush through the remaining questions, causing you to miss more answers.

If time allows, return to marked questions and take a fresh look.

However, do not change the original answer unless you have a reason to change it.

University studies on students show that hastily changing answers often replace correct and incorrect answers.

## Multiple-choice questions

***Answer all questions***

How many questions are correct, not how much work went into selecting the answers matters. An educated guess earns the same points as an answer known with confidence.

On the test, you need to think and analyze information quickly.

The skill of analyzing information quickly cannot be gained from a college course, prep course, or textbook.

Working efficiently and effectively is a skill developed through focused effort and applied practice.

***Strategic approach***

Strategies, approaches, and perspectives for answering multiple-choice questions help maximize points.

Many strategies seem like common sense. However, these helpful approaches might be overlooked under the pressure of a timed test.

While no strategy replaces comprehensive preparation, apply probability for success on unfamiliar questions.

***Understand the question***

Know what the question is asking before selecting an answer.

It is surprising when students do not read (and reread) carefully and rush to select the wrong answer.

The test-makers anticipate hasty mistakes, and many enticing answers include specious choices.

A successful student reads and understands the question precisely before looking at the answers.

***Focus on the answer***

Separate the vital information from distracters and understand the design and thrust of the question.

Answer the question and not merely pick a factually accurate statement or answer a misconstrued question.

Rephrasing the question helps articulate precisely what the correct response requires.

When rephrasing, do not change the question's meaning; assume it is direct and to the point as written.

After selecting the answer, review the question and verify that the choice selected answers the question.

***Answer the question before looking at choices***

This valuable strategy is applicable if the question asks for generalized factual details.

Form a thought response first, then look for the choice that matches your preordained answer.

Select the predetermined statement as it is likely correct.

***Factually correct, but wrong***

Questions often have incorrect choices that are factually correct but do not answer the question.

Predetermine the answer and do not choose merely a factually correct statement.

Verify that the choice answers the question.

***Do not fall for the familiar***

When in doubt, it is comforting to choose what is familiar. If you recognize a term or concept, you may be tempted to pick that choice impetuously.

However, do not go with familiar answers merely because they are familiar.

Before selecting the answer and how it relates to the question, think about it.

***Experiments questions***

Determine the purpose, methods, variables, and controls of the experiment.

Understanding the information presented helps answer the question.

With multiple experiments, understand variations of the same experiment by focusing on the differences.

For example, focus on changes between the first and second experiments, second and third, and first and third.

Direct comparison between experiments helps organize and apply the information to the answer.

***Words of caution***

The words *"all," "none,"* and *"except"* require attention. Be alert with questions containing these words, as the answer may not be apparent on the first read of the question.

***Double-check the question***

After selecting an answer, return to the question to ensure the selected choice answers the question as asked.

***Fill the answers carefully***

Many mistakes happen when filling in answers.

Filling answers correctly is simple but crucial.

Note the question number and enter answers accordingly. If you skip a question, skip it on the answer sheet.

## Quantitative multiple-choice questions

***Know the equations***

Many exam questions require scientific equations, so understand when to use each.

As you work with this book, apply formulas and equations and use them in many questions.

***Manipulate the formulas***

Know how to rearrange the formulas.

Many questions require manipulating equations to calculate the correct answer.

Familiarity includes manipulating the terms, understanding relationships, and isolating variables.

### *Estimating*

Estimating helps choose the answer quickly for quantitative questions with a sense of the order of magnitude.

Often, estimation enables the correct answer to be identified quickly compared to the time needed for calculations.

Estimating is especially applicable to questions where the answer choices have different orders of magnitude.

It saves time when estimating instead of computing the solution.

### *Evaluate the units*

For quantitative problems, analyze the units to build relationships between the question and the correct answer.

Understand what value is sought and eliminate wrong choices with improper units.

### *Make visual notes*

Write, draw, or graph the question to determine the information provided, the objective, and the concept tested.

A chart or table often makes the solution apparent, even if a question does not require a graphic answer.

## Elimination strategies

If the correct answer is not immediately apparent, use the process of elimination.

Use the strategy of educated guessing by eliminating one or two answers.

Usually, at least one answer choice is easily identified as wrong. Eliminating one choice increases the odds of selecting the correct one.

### *Process of elimination*

Eliminate choices:

- Use proportionality for quantitative questions to eliminate choices that are too high or too low.
- Eliminate answers that are *almost right* or *half right.*

  Consider *half right* as *wrong* since distractor choices are purposely included.

- If two answers are direct opposites, the correct answer is likely one of them.

  Note if they are direct opposites or if another reason indicates they are correct. Therefore, eliminate the other choices and narrow the search for the correct one.

- For numerical questions, eliminate the smallest and largest numbers (unless for a reason).

***Roman numeral questions***

Roman numeral questions present several statements and ask which is/are correct.

These questions are tricky for most test-takers because they have more than one potentially correct statement.

Roman numeral questions are often included in combinations with more than one answer.

Eliminating a wrong Roman numeral statement eliminates choices that include it.

## Educated guessing

***Correct ways to guess***

Do not assume you must get every question right; this adds unnecessary stress during the exam.

You will (likely) need to guess for some questions.

Answer as many questions correctly as possible without wasting time.

Random guessing does not help with challenging questions. Use educated guessing after elimination.

***Playing the odds***

Guessing is a form of "partial credit" because while you might not be sure of the correct answer, you have the relevant knowledge to identify some wrong choices.

There is a 25% chance of correctly guessing random responses since questions have four choices. Therefore, the odds are guessing 1 question correctly to 3 incorrectly.

### *Guessing after elimination of answers*

After eliminating one answer as wrong, you have a 33% chance of a lucky guess. Therefore, your odds move from 1 question right to 2 questions wrong.

While this may not seem like a dramatic increase, it can make an appreciable difference in your score.

Confidently eliminating two wrong choices increases the chances of guessing correctly to 50%!

When using elimination:

- Do not rely on gut feelings alone to answer questions quickly.

  Understand and recognize the difference between *knowing* and *having gut feelings* about the answer.

  Gut feelings should sparingly be used after the process of elimination.

- Do not fall for answers that sound "clever," and choose "bizarre" answers.

  Choose them only with a reason to believe they may be correct.

### *Eliminating Roman numeral choices*

A workable strategy for Roman numeral questions is to guess the wrong statement.

For example: A. I only

B. III only

C. I and II only

D. I and III only

Notice that statement II does not have an answer dedicated to it. This indicates that statement II is likely wrong and eliminates choice C, narrowing your search to three choices.

However, if you are confident that statement II is the answer, do not apply this strategy.

*ACS Physical Chemistry Review* provides comprehensive and targeted coverage of physical chemistry topics tested on the ACS exam. The content covers foundational principles and theories necessary to understand the material and answer test questions.

*ACS Organic Chemistry Practice Questions* provides high-yield practice questions covering topics tested on the ACS Organic Chemistry exam. Develop the ability to apply your knowledge and quickly choose the correct answer to increase your test score.

**Visit our Amazon**

Made in the USA
Las Vegas, NV
09 May 2025

21882565R00391